AF572533

Macromolecular Symposia 207

Fundamentals and Applications of Polymer Gels

Oaxaca City, Mexico
December 5–7, 2001

Symposium Editor:
R.A. Siegel, Minnesota, USA

pp. 1–268 · February 2004
ISBN 3-527-31040-1

Macromolecular Symposia publishes lectures given at international symposia and is issued irregularly, with normally 14 volumes published per year. For each symposium volume, an Editor is appointed. The articles are peer-reviewed. The journal is produced by photo-offset lithography directly from the authors' typescripts.
Further information for authors can be found at http://www.ms-journal.de.
Suggestions or proposals for conferences or symposia to be covered in this series should also be sent to the Editorial office (E-mail: macro-symp@wiley-vch.de).

Macromolecular Symposia:
Annual subscription rates 2004
Macromolecular Full Package: including Macromolecular Chemistry & Physics (18 issues), Macromolecular Rapid Communications (24), Macromolecular Bioscience (12), Macromolecular Theory & Simulations (9), Macromolecular Materials and Engineering (12), Macromolecular Symposia (14):

Europe	Euro	6.424 / 7.067
Switzerland	Sfr	11.534 / 12.688
All other areas	US$	7.948 / 8.743

print only **or** electronic only / print **and** electronic

Postage and handling charges included. All Wiley-VCH prices are exclusive of VAT.
Prices are subject to change.

Single issues and back copies are available. Please ask for details at: service@wiley-vch.de.

Orders may be placed through your bookseller or directly at the publishers:
WILEY-VCH Verlag GmbH & Co. KGaA, P. O. Box 10 11 61, 69451 Weinheim, Germany, Tel. +49 (0) 62 01/6 06-400, Fax +49 (0) 62 01/60 61 84. E-mail: service@wiley-vch.de

For USA and Canada: Macromolecular Symposia (ISSN 1022-1360) is published with 14 volumes per year by WILEY-VCH Verlag GmbH Co. KGaA, Boschstr. 12, 69451 Weinheim, Germany. Air freight and mailing in the USA by Publications Expediting Inc., 200 Meacham Ave., Elmont, NY 11003, USA. Application to mail at Periodicals Postage rate is pending at Jamaica, NY 11431, USA. POSTMASTER please send address changes to: Macromolecular Symposia, c/o Wiley-VCH, III River Street, Hoboken, NJ 07030, USA.

Printing: Strauss Offsetdruck, Mörlenbach. Binding: J. Schäffer, Grünstadt

Macromolecular Symposia

Articles published on the web will appear several weeks before the print edition. They are available through:

www.ms-journal.de **www.interscience.wiley.com**

Fundamentals and Applications of Polymer Gels
PPF-7 Microsymposium, Oaxaca City (Mexico), 2001

Author Index

TOYOICHI TANAKA—THREADS OF RESEARCH

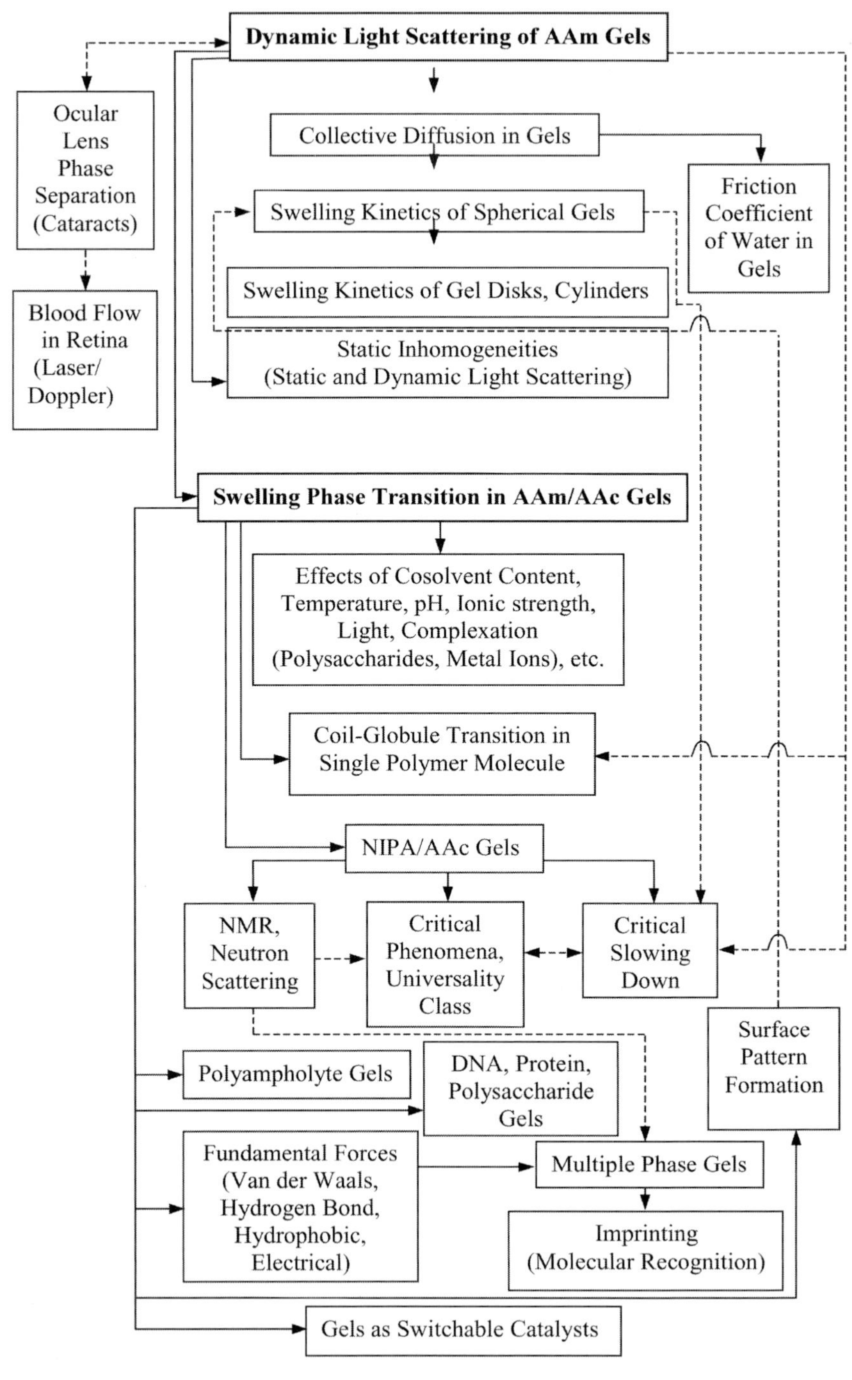

Preface

Early in the year 2000, plans were initiated to include a microsymposium on polymer gels as a feature of the 7th Pacific Polymer Conference in Oaxaca, Mexico. It was agreed among the organizers that Professor Toyoichi Tanaka of the Massachusetts Institute of Technology (MIT) should be featured as keynote speaker. Professor Tanaka was regarded internationally as the father of modern research in polymer gels. However, before Professor Tanaka could be contacted, the shocking news was received that he had passed away on May 20, 2000, while playing tennis. He was 54 years old. To mourn this tremendous loss to the gel community and celebrate Professor Tanaka's outstanding career, it was decided to dedicate the Oaxaca microsymposium to his memory.

Many leaders in the gel field from around the world whom we were able to contact agreed to attend this microsymposium, which was held December 5 – 7, 2001. Support for speaker travel from outside North America was provided by the American Chemical Society and by the Gel Group of the Society of Polymer Science of Japan, while sponsorship of publicity and a social event, which was attended by members of Professor Tanaka's family, was provided by MIT. The Mexican Polymer Society waived registration fees for many of the invited speakers.

The original scientific program was packed, and it was expected that a very tight schedule would have to be enforced. Then world events intervened. In the aftermath of the September 11 attacks, a number of speakers from outside North America were unable to attend, either because of official visa and travel restrictions that were suddenly in place, or due to discouragement of travel to North America by the speakers' home countries. Fear that the microsymposium's quality would be compromised was unfounded however. The gaps in the schedule due to cancellations allowed for extended discussions after presentations, and the schedule became much more flexible. The feel was more of a Gordon Conference than a tightly run national or international meeting of a scientific society.

Prior to Professor Tanaka's entry into the gel field, most understanding of hydrated, crosslinked polymer networks was phenomenological, and experiments and theories concerning gel behavior were confined almost exclusively to macroscopic phenomena, such as equilibrium swelling, viscoelastic properties, and molecular transport properties. This focus was consistent with the utilization of gels as ion exchangers, supports for gel permeation chromatography, ultrafiltration membranes, soft contact lenses, and vehicles for controlled drug delivery, and also with the recognition that hydrated biological connective tissues such as cornea, cartilage, and intercellular matrix, as well as many foods, share features with synthetic gel systems. The ability of hydrogels to perform chemomechanical energy conversions (e. g. artificial muscles) was studied by W. Kuhn, A. Katchalsky and others, but again the approach was macroscopic and phenomenological.

In 1974 Professor Tanaka, collaborating with L. Hocker and G. B. Benedek, published a seminal paper investigating the microscopic dynamics of polymer chains within gels, using dynamic light scattering. This work demonstrated that thermally driven motions of these chains could be understood as a collective diffusion that results from a balance of elastic restoring forces due to the polymer network and viscous forces as a result of the relative motion of the polymer against the solvent that swells the network. While the scattering measurements were performed with the gel in a static environment, Professor Tanaka showed in 1979 that collective diffusion coefficients determined by dynamic light scattering could be used to predict the kinetics of gel swelling. Thus, dynamics probed at the microscale were linked, for the first time, with macroscopic dynamical phenomena in gels.

Professor Tanaka's other early major discovery was the existence of first-order volume phase transitions in gels. While the possibility of such transitions had been suggested by K. Dusek and D. Patterson in 1968 based on an analysis of Flory-Rehner theory, the first experimental evidence was found, serendipidously, when Professor Tanaka noticed the divergence of scattered light intensity of poly-(acrylamide) gels in supercooled water. To prove such divergence was not due to ice formation, light scattering in gels in a series of water/acetone cosolvents was studied at room temperature. Surprisingly, it was found that when a critical amount of acetone was added, the gels collapsed. It was later shown that the presence of a small amount of ionized acrylic acid was crucial for this first order collapse phase transition. Subsequent work showed that first order swelling/collapse phase transitions could also be brought about by changes in other environmental parameters such as temperature, pH, and ionic strength.

These two early lines of work led to broad and deep investigations of the properties of gels. The accompanying "flowchart" provides a nonexhaustive overview of Professor Tanaka's *oeuvre*, with some attempt to link various topics. Solid arrows indicate an ordered relation, while dashed arrows indicate a "lateral" association or application of a technique to study a topic. Readers may prefer other relational connections. Abbreviations are: AAm (acrylamide), AAc (acrylic acid), and NIPA (*N*-isopropyl acrylamide). The latter monomer proved to be an extremely useful tool, since gels containing it exhibit first order transitions in pure water with changes in temperature (they are of the lower critical transition type).

After Professor Tanaka's death, his colleagues at MIT collaborated to publish his final work, which included several articles dealing with pattern formation, molecular recognition, and switchable catalysis by gels. The first contribution in this volume summarizes much of the work on molecular recognition, which was perhaps the topic that captured Professor Tanaka's interest most during the final period of his life. Other topics in this volume deal with both macroscopically and microscopically observable properties of gels, as well as potential applications in technology. While Professor Tanaka's research was focused on the basic physics of gels, he was also very interested in applications. All contributors were influenced, one way or another, by Professor Tanaka, and this volume is dedicated to his memory.

The editors acknowledge the contributions of Professor Yoshihito Osada, who was instrumental in the initiation and organization of the Oaxaca microsymposium, but was unable to attend.

R. A. Siegel, A. Licea-Claveríe, A. Grosberg

Multiple-Contact Adsorption of Target Molecules by Heteropolymer Gels†

Kenji Ito,[*1] *Jeffrey Chuang,*[2] *Carmen Alvarez-Lorenzo,*[3] *Tsuyoshi Watanabe,*[4] *Nozomi Ando,*[5] *Alexander Yu. Grosberg*[6]

[1]National Institute of Advanced Industrial Science and Technology, Tsukuba, Ibaraki 305–8565, Japan; E-mail: k-ito@aist.go.jp

[2]Department of Biochemistry and Biophysics, University of California, San Francisco, California 94143, USA

[3]Departamento de Farmacia y Tecnología Farmacéutica, Facultad de Farmacia, Universidad de Santiago de Compostela, 15706-Santiago de Compostela, Spain

[4]Fundamental Research Laboratory, SUMITOMO BAKELITE Co., Ltd., Totsuka-ku, Yokohama-shi, Kanagawa 245–0052, Japan

[5]Department of Physics, Cornell University, Ithaca, New York 14853, USA

[6]Department of Physics, University of Minnesota, Minneapolis, Minnesota 55455, USA

Summary: We examined adsorption of target molecules through a multiple-contact interaction in a thermo-sensitive heteropolymer gel which can undergo a volume transition at 34 °C in water. Multi-valent anionic target molecules were adsorbed electrostatically by monovalent cationic adsorbing sites in the gel. The overall affinity (*SK*) between the gel and the target molecule was calculated from the initial slope of the Langmuir adsorption isotherm: $[T_{ads}] = KS[T_{sol}]/(1 + K[T_{sol}])$, where S, K, $[T_{sol}]$ and $[T_{ads}]$ represent the number of adsorption sites per unit volume of the gel, the effective binding constant, the equilibrium target concentration in the external solution, and the target concentration adsorbed in the gel, respectively. The affinity for the collapsed gel at 60 °C was studied in terms of the concentrations of three factors: the adsorber ([Ad]), the coexistent salt (replacement ion, [Re]), and the cross-linker ([Xl]). We found that the relationship between the affinity and these factors can be summarized by the following formula first suggested by T. Tanaka:

$$\text{Affinity, } SK \propto \left(\frac{[\text{Ad}]}{[\text{Re}]}\right)^{p} \exp\left[-c(p-1)\frac{[\text{Xl}]}{[\text{Ad}]^{2/3}}\right].$$

Here c is a constant and p is the number of the contact points between the adsorber units and the target molecule.

†This work was largely carried out in the Department of Physics and Center for Materials Science and Engineering, Massachusetts Institute of Technology, Cambridge, Massachusetts 02139.

 DOI: 10.1002/masy.200450301

Introduction

Polymer gels belong to the privileged class of the most actively studied polymer systems. The main immediate motivation for most gel studies has been various technological applications. At the same time, T. Tanaka (1947–2000), as one of the leaders of the science of gels, wanted to look further. His ambitious goal was to mimic some properties of proteins using gels as a model system.

In some instances, gels do exhibit features of protein-like behavior. For example, in one experiment, a gel was created with a catalytic activity that could be switched on and off by an infinitesimal change in temperature or solvent composition.[1,2] The on-off behavior apparently resulted from the fact that several functional monomers were able to cluster into a catalytic center in the collapsed state of the gel, while they were unable to do so in the swollen state. In general, a key to the design of protein-like gels is to understand how functional monomers in the gel cluster and interact with substrate molecules. This realization led Tanaka to formulate the problem of multiple-point adsorption by a heteropolymer gel.

For the last few years, we performed an extensive study of the adsorption of such substrates, or 'target molecules,' into heteropolymer gels. In these systems, a main gel component monomer (e.g. *N*-isopropylacrylamide, NIPA) was polymerized with a small amount of monovalent functional monomer (e.g. methacrylamidopropyl trimethylammonium cation, $MAPTA^+$) and cross-linker (e.g. *N,N*-methylene-bis(acrylamide), BIS). Multivalent target molecules (e.g. 1,3,6,8-pyrene tetrasulfonate anion, Py^{4-}) with the opposite charge sign were then adsorbed into the gel. The experiments characterized whether adsorption of the targets was through multiple adsorber contacts, or by single adsorbers, and under what conditions the different types of adsorption would occur.

Recent investigations have confirmed that several adsorbers take part in the adsorption of one target molecule in the collapsed state of gels, while only a single monomer does in the swollen state.[3] Such conclusions can be drawn from the dependence of the affinity on the adsorber concentration in the gel. The affinity for the collapsed gel is much stronger than that for the swollen gel above certain adsorber concentrations. The results suggest that the binding sites in the gel behave like active sites on a protein, in that they can catch and release target molecules. Tests of the affinity dependence on the coexistent salt concentration in the outer solution supported such a mechanism.[4]

The cross-linker dependence of the affinity can be used to characterize the binding sites in the gel as well. If a target molecule is adsorbed by several adsorber monomers, then the polymer chains in the gel will be constrained, and the system will suffer a loss of entropy. This entropy loss can be observed as a decrease of the affinity with an increase of the cross-linker concentration in the gel. The quantitative relationship between the affinity and cross-linker concentration can be used to relate the number of contact points to the microscopic properties of the gel.

In this paper, we establish a simple model, proposed by Prof. Tanaka, describing the adsorption of target molecules through a multiple-contact interaction. This model provides an expression for the affinity, which successfully explains the experimentally determined dependence of the affinity on several factors: the concentration of adsorber monomers in the gel, the concentration of replacement ions (the coexistent salt) in solution, and the concentration of cross-linker in the gel. These experiments are detailed in the sections that follow.

Tanaka equation

Here we provide a summary for what has been called the Tanaka equation, which relates the target molecule affinity of a random gel to several key experimental parameters. The Tanaka expression for the affinity of a gel to adsorb target molecules from a solution is

$$\text{Affinity} = \frac{[\text{Ad}]^p}{p[\text{Re}]^p}\exp(-\beta p \varepsilon)\exp\left[-c(p-1)\frac{[\text{Xl}]}{[\text{Ad}]^{2/3}}\right]. \qquad (1)$$

The concentration of adsorbing monomers in the gel is denoted by [Ad], and [Re] is the concentration of replacement molecules, i.e. salt molecules that each bind with a single site on the target molecule when it is not bound to adsorbing monomers. p is the number of bonds between separate adsorbing monomers and the target molecule. β is the Boltzmann factor $1/k_BT$, and ε is the difference in binding energy of a target molecule with a) an adsorbing monomer or b) a replacement molecule. [Xl] is the concentration of cross-linker (e.g. BIS). c is a constant which can be estimated from the persistence length and concentration of the main component of the gel chains (e.g. NIPA).

The Tanaka equation has done a good job of explaining how adsorber monomer concentration, salt concentration, and cross-linker concentration control the affinity. The main assumption in eq.(1) is that adsorption of target molecules is dominated by one value of p at a time. In general, the validity of this assumption depends on the concentrations of the various components of the gel and the target-monomer binding energy.

Langmuir adsorption expressed by

$$[\mathrm{T}_{\mathrm{ads}}] = S\frac{K[\mathrm{T}_{\mathrm{sol}}]}{K[\mathrm{T}_{\mathrm{sol}}]+1} \tag{2}$$

has been used to interpret the binding affinity of a gel for a target molecule in the experiments described below. Here $[\mathrm{T}_{\mathrm{ads}}]$ is the concentration of target molecules adsorbed into the gel, $[\mathrm{T}_{\mathrm{sol}}]$ is the concentration of target molecules in solution, and S is the concentration of binding sites. K is the binding constant — it can be thought of as the affinity per binding site. SK is the overall affinity of all binding sites in the gel for the target molecule and is equivalent to the affinity mentioned in eq.(1).

The overall binding affinity SK can be determined from the partition function that sums over the different possible states of the target molecule — 0 adsorbers bound, 1 adsorber bound, ... p_{max} adsorbers bound. The partition function will be of the form $Z = Z_0 + Z_1 + Z_2 + ... + Z_{p_{\mathrm{max}}}$, with Z_p indicating the term of the partition function in which a target molecule is bound by p adsorbing monomers. Z_0 corresponds to the case of the target molecule being completely unbound. From the equation of the Langmuir isotherm (eq.(2)), we can see that $SK[\mathrm{T}_{\mathrm{sol}}]$ is the concentration of bound target molecules, and SK should be the ratio of bound to unbound molecules. Therefore $SK = (Z_1 + Z_2 + ... + Z_{p_{\mathrm{max}}})/Z_0$.

We now give a qualitative explanation of the various terms in the Tanaka equation. We first consider the power-law dependence of the affinity SK on [Ad]. For a target molecule to be adsorbed, there must be p adsorbing monomers clustered together to simultaneously bind it. The probability of such a cluster existing at a given point in a random gel is a product of the probabilities for each of the adsorbing monomers. Therefore the dependence goes as $[\mathrm{Ad}]^p$. Each of these clusters requires p adsorbing monomers, hence SK is proportional to $1/p$.

The affinity is proportional to $[\mathrm{Re}]^{-p}$ because these replacement molecules (typically salt ions) act as competitors to the adsorbing monomers. In solution, a target molecule may either be adsorbed into the gel or bound by p replacement molecules. Binding to the replacement molecules prevents adsorption by the gel. In order for a target to be bound to p replacement molecules, the replacements must cluster around the target. This creates a power-law dependence similar to that for target molecule adsorption by adsorbing monomers, but with an opposite sign exponent. Note that we have assumed that each replacement molecule binds to one site on the target molecule, as do the adsorbing monomers. If the adsorber monomers and replacement molecules have different valences, these exponents should be modified.

The energetic attraction of a target molecule to p adsorbing monomers is encapsulated in the term $\exp(-\beta p \varepsilon)$. This is a Boltzmann probability based on a binding energy ε per adsorbing monomer.

The dependence of the affinity, *SK*, on the cross-links can be explained as follows. The adsorber units in the gel can move rather freely within a certain volume determined by the cross-link density. It has been established that, below a certain length scale associated with the cross-link density, the gel behaves like a liquid. Beyond that length scale, however, the gel behaves as an elastic solid body. The adsorber units cannot diffuse further than that length scale.

We therefore envision that each adsorber is at one end of a fictitious Gaussian chain with a length half the average polymer length between the nearest cross-links (Xl), $l = nb = ([\mathrm{NIPA}]/[\mathrm{Xl}]/2)a$. Here n is the number of monomer segments of persistent length b contained in the chain. If there are [NIPA]/2[Xl] monomers between the cross-link and an adsorbing monomer group, then $n = [\mathrm{NIPA}]/(2m[\mathrm{Xl}])$ and $b = ma$, where m is the number of monomers involved in the persistent length and a is the length of each monomer. At a concentration of [Ad] of adsorbing monomers, the average spatial distance between adsorbing monomers is $R = [\mathrm{Ad}]^{-1/3}$. For a molar concentration C_{ads} this corresponds to $R = 1\ \mathrm{cm}/(C_{\mathrm{ads}}N_{\mathrm{A}})^{1/3}$, where N_{A} is the Avogadro number. The fictitious Gaussian chain represents the restricted ability of the adsorber groups to diffuse within a certain volume in the gel.

We expect that the probability for two adsorber monomers to meet should be proportional to the Boltzmann factor of the entropy loss associated with the formation of one pair of adsorbers, $P = P_0\exp(-R^2/nb^2) = P_0\exp(-c[\text{Xl}]/[\text{Ad}]^{2/3})$, where the quantity c is determined by the persistence length, the number of monomers in a persistence length, and the concentration of the main component of the chains through the relation $c = 2m/([\text{NIPA}]b^2)$. Since the adsorption of a divalent target by two adsorbers brings together each end from two fictitious Gaussian polymers, the affinity should be proportional to this probability. If more than two contact points are expected, the equation can be generalized as

$$SK \propto \exp\left[-c(p-1)\frac{[\text{Xl}]}{[\text{Ad}]^{2/3}}\right], \quad (3)$$

where p is the number of contact points. In other words, the cross-links will frustrate the formation of a target-binding site requiring multiple ($p > 1$) adsorber monomers, and the frustration will increase with p. On the other hand, if the target molecule adsorbs by a single point only ($p = 1$), SK should be independent of the cross-linker concentration.

Together these contributions make up the Tanaka equation eq.(1) for binding of a target molecule by p adsorber monomers. It can be shown by consideration of the relative weights of the terms Z_p, that for a given set of experimental parameters, the affinity SK will almost always be dominated by one value of p, either $p = 1$ or $p = p_{max}$, depending on the parameters chosen. Therefore, the Tanaka equation is quite general, and one can ignore the contributions of different types of adsorption. At the gel volume phase transition, the concentrations of adsorber, cross-linker, and salt, as well as the temperature of the system may be altered. To make a gel in which p changes across the volume phase transition, one should choose components such that the difference in $Z_{p_{max}}$ and Z_1 changes sign across the transition.

To conclude the theoretical section, we should discuss the important condition that salt concentration should be large enough. That is necessary for the applicability of the above theoretical considerations, because we have assumed everywhere that Coulomb interactions were effectively short ranged, which is only true if they are sufficiently Debye-Hückel screened. This requirement is safely met under the conditions of the experiments described below. However, as a matter of principle, it should be understood that at smaller salt concentration a completely new set of phenomena may arise, such as Coulomb correlation effects between target molecules (see the review article[5] and references therein). These effects comprise an interesting subject of future studies, but we leave them outside the scope of the present paper following the idea that mainly the single particle behavior with respect to target molecules is of interest in the context of the attempts to mimic protein-style behavior.

Experimental verification of the Tanaka equation

Methods

Gel preparation The gels were prepared by free radical polymerization using 6 M *N*-isopropyl-acrylamide (NIPA, which was kindly supplied by Kohjin Co., Japan), 0–120 mM methacrylamidopropyl trimethylammonium chloride (MAPTAC, Mitsubishi Rayon Co. Ltd., Japan) and 5–200 mM cross-linker *N,N*-methylene-bis(acrylamide) (BIS, Polysciences Inc., PA). After the monomers were dissolved in dimethyl-sulfoxide (DMSO), 10 mM 2,2-azobisisobutyronitrile (AIBN, initiator) was added, and the solutions were immediately transferred to test tubes in which micropipettes of inner diameter of approximately 0.5 mm were placed. The solutions filled the micropipettes, and were then degassed under vacuum for a few seconds. The polymerization was carried out at 60 °C for 24 hours. After gelation was completed, the micropipettes were crushed. To remove unreacted chemicals, the gels were washed with large amounts of 100 mM hydrochloric acid (HCl) and sodium chloride (NaOH) aqueous solutions and

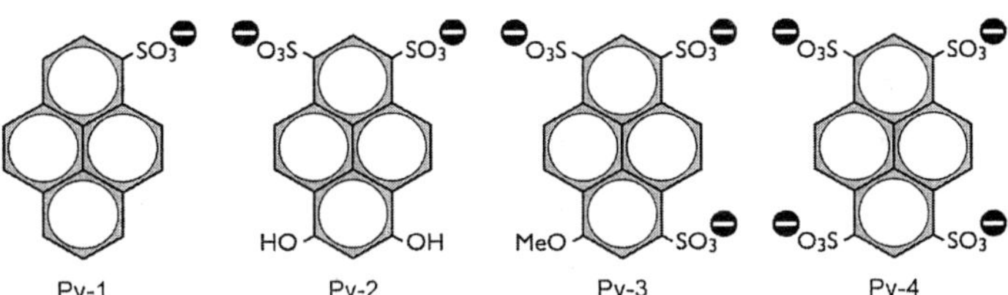

Figure 1: Chemical structure of targets.

then with deionized distilled water for ten days. Finally, the gels were removed from the solution and dried under vacuum for one week. The obtained gel can undergo a volume phase transition at 34 °C in water, and the gel is in the swollen state below that temperature, while is in the collapsed state above that.[6] Then we confirmed that a swelling ratio for all collapsed gels at 60 °C was 1, meaning the same as that at preparation, irrespective of the experimental conditions such as adsorber concentration

(0–120 mM), salt concentration (27–200 mM), and cross-linker concentration (5–200 mM).

Adsorption studies As target molecules, several different types of pyrene sulfonate derivatives were used: 1-pyrene sulfonic acid sodium salt (Py-1), 6,8-dihydroxy-pyrene-1,3-disulfonic acid disodium salt (Py-2), 8-methoxy pyrene-1,3,6-trisulfonic acid trisodium salt (Py-3), and 1,3,6,8-pyrene tetrasulfonic acid tetrasodium salt (Py-4), portraying in Figure 1. These chemicals present 1 (Py-1), 2 (Py-2), 3 (Py-3) or 4 (Py-4) anionic charges, which can interact electrostatically with a cationic charged site such as on $MAPTA^+$.

Pieces of cylindrical gel (5–20 mg dry weight) were placed in 1.5–4 mL target aqueous solution with the concentration of the targets ranged from 2 μM to 0.5 mM. The solutions also contained NaCl of a prescribed concentration to provide monovalent

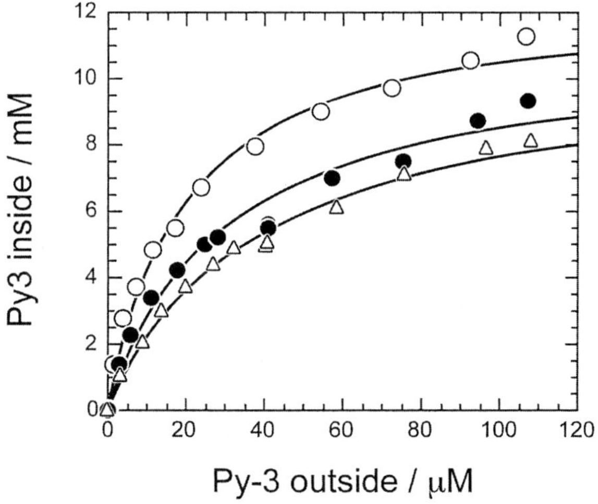

Figure 2: Typical adsorption isotherms of the collapsed gel at 60 °C for the adopted binding systems. The plot shows the isotherms for the gels with the following experimental conditions as: The target molecule is Py-3. The concentrations of the main monomer (NIPA), the adsorber (MAPTAC) and the replacement ion (NaCl) are 6 M, 60 mM and 80 mM, respectively. The BIS concentrations are 5 mM (open circle), 100 mM (closed circle) and 200 mM (triangle). All isotherms observed in the present work were well fit by the Langmuir equation eq.(2).

chloride ions to replace the target molecules. The samples were kept swollen (20 °C) or shrunken (60 °C) for 48 hours. Equilibrium concentration of the target molecules in the medium was measured using UV spectrophotometer. The amount of the adsorbed target molecules was then evaluated as the difference between the initial and the final quantities in the outer solution. These amounts were then converted to units of moles of target molecule per volume of gel at the time of synthesis. The adsorption isotherms were analyzed in terms of the Langmuir equation, $[T_{ads}] = KS[T_{sol}]/(1 + K[T_{sol}])$. Once again, in this equation, $[T_{ads}]$ is the amount of target adsorbed per unit volume of gel in the shrunken state, $[T_{sol}]$ is the final equilibrium concentration in the solvent, S is the number of adsorbing sites per unit volume of gel, and K is the affinity of one adsorption site for a target molecule. From the slope and the intercept at zero $[T_{sol}]$, both S and K, and the overall affinity SK were deduced. Typical adsorption results are shown in Figure 2, which displays the adsorber dependence of the adsorption of Py-3 for a collapsed gel. All the adsorption isotherms were well fit by the Langmuir equation.

Adsorber dependence

The dependences of the affinity for Py-3 and Py-4 as a function of the MAPTAC concentration are shown in Figure 3. At MAPTAC concentrations below 10 mM, the absorption for the collapsed gels becomes independent of MAPTAC concentration, indicating that the gel major component NIPA contributes more to the absorption of

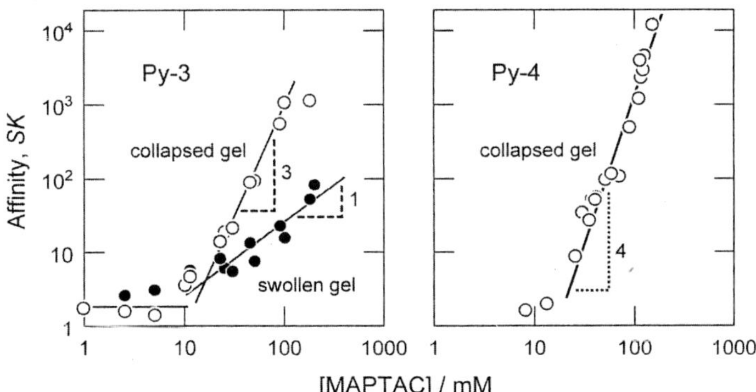

Figure 3: Adsorber dependence of affinity for Py-3 and -4 adsorptions. The data were quoted from Ref. [3].

pyranine (due to a hydrophobic interaction) than do the $MAPTA^+$ groups in that regime. Above these MAPTAC concentrations, both log-log plots show a straight line with slope three for Py-3 and with slope four for Py-4. In this experiment, the coexistent salt concentration of the system was kept at a constant (100 mM) so that the Donnan potential, which might cause a nonspecific adsorption of the target molecule, was swept out. Thus, it is concluded that the power-law relationships observed above the MAPTAC concentration (~10 mM) are due to three- and four-point absorptions, respectively. Absorption sites are formed when three (or four) equivalent adsorbing molecules ($MAPTA^+$) gather to capture one Py-3 (or Py-4) molecule. The obtained power laws are explained well by the Tanaka equation.

On the other hand, the log-log slope for the swollen gel is one, indicating that $MAPTA^+$ absorbs the target molecule with a single contact. Single-point absorption is favored because the $MAPTA^+$ monomers are well separated from one another. The slope returns to 3 or 4 upon shrinking, showing recovery of the multipoint binding sites.

These results suggest the following picture. In the gel's collapsed state, the adsorber monomer units can be clustered to form sites in which a target molecule can bind with several adsorber monomer units simultaneously. If the gel is made to swell, the adsorber groups separate, and it becomes entropically unfavorable for the multipoint adsorption complex to assemble.

Replacement ion dependence

Figure 4 shows the affinity SK of the collapsed gel for the various target molecules as a function of NaCl concentration. In the adopted system, concentrations of the adsorber ([MAPTAC]) and the cross-linker ([BIS]) were fixed at 40 mM and 10 mM, respectively. The affinities correlated very strongly with the NaCl concentration. For instance, in the case of Py-3, when the NaCl concentration was increased by a factor of three, the affinity decreased by more than two orders of magnitude. More importantly, the plots clearly show the linear relationships between ln([NaCl]) and ln(SK) for the different target molecules. In each case, the log-log plot has a slope of $-p_{max}$, where p_{max} is the number of charged groups on the target molecule Py-p_{max}. The evidence obtained here follows well our prediction described in the theory section, which means that the chloride ions of NaCl act as replacement ions for the charged groups on a target molecule and they substitute for the target molecules in binding to the p_{max} adsorbers, preventing adsorption of some of the targets.

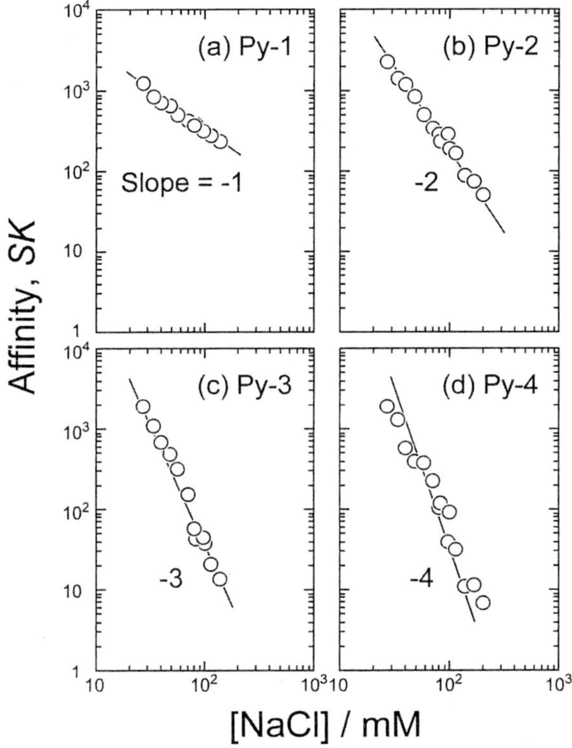

Figure 4: Replacement ion dependence of affinity for Py-1 to -4 adsorptions. The data were quoted from Ref. [4].

We also considered whether it is solely the ratio of adsorber to replacement concentration that determines the affinity, or whether the absolute value of the salt concentration plays some role. For example, it is conceivable that there is some sort of coulombic screening that does not follow the mass action law. To fix these expectations, we examined the dependence of the affinity for Py-3 and Py-4 on the ratio of the adsorber concentration [Ad] to the external salt concentration [Re] as shown Figure 5.

The two variables, [Ad] and [Re], were tested in separate experiments. In one, the replacement ion concentration [Re] = [NaCl] = 80 mM was fixed while [Ad] was varied. In the other, [Ad] = 40 mM was fixed while [Re] was varied. In both cases, [BIS] was fixed to 10 mM. The affinity curves from the two experiments are practically coincident.

This experimental evidence suggests that the affinity is universally determined by the ratio of adsorber to replacement concentration, that is, affinity $\propto$ ([Ad]/[Re])^p. Here, it is confirmed that the power-law relationship followed the proposed model.

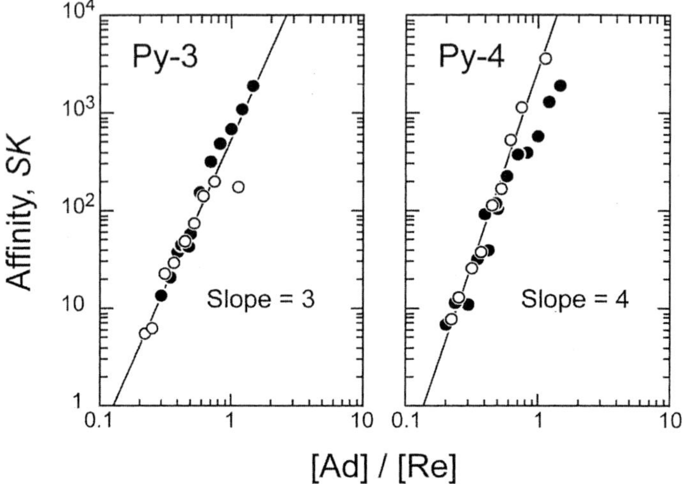

Figure 5: Dependence of the affinity on the ratio of the adsorber to the replacement ion for Py-3 and -4 adsorptions. Open and closed circles represent the variables of the adsorber and replacement ion, respectively.

Cross-linker dependence

Figure 6 shows the dependence of the affinity *SK* on the degree of cross-linking for several target molecules that can establish different numbers of contact points (p). For each target molecule, the affinities exponentially decrease as the cross-linker, BIS, is increased. This effect is especially significant for the cases of contact numbers above two. As can be seen, the slope for the target with p = 1 is almost negligible, while for increasing p the slope becomes larger. The initial slopes were well fit by an exponential function. This exponential nature can be understood by the discussion described above. Adsorption of the target molecules by multiple contacts is affected by the required entropy loss of the polymer chains.

Let us consider the situation more quantitatively from a microscopic point of view. Figure 7 shows a plot between the number of contact points and the exponential decay rate defined as

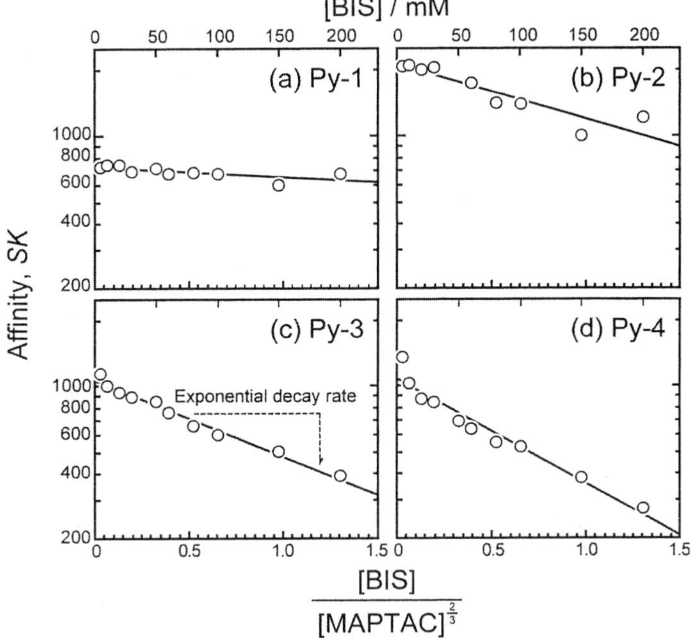

Figure 6: Series of semi-log plots for the affinity vs the BIS concentration for a collapsed gel at 60 °C. The concentrations of MAPTAC and NaCl are 60 mM and 80 mM, respectively. The lower horizontal label shows the ratio of BIS concentration to that of MAPTAC to the two-thirds power, while the upper label shows the corresponding BIS concentration.

$$\text{slope} = -\frac{\partial(\ln SK)}{\partial\left(\frac{[\text{BIS}]}{[\text{MAPTAC}]^{2/3}}\right)}, \tag{4}$$

which is obtained from a fit of each plot in Figure 6 to eq.(3). The plot gives a linear relationship in which the slope is 0.32. Following the proposed model, the obtained slope should associate with the parameter c in eq.(1), though there is a correction factor because here the concentrations are reported in moles per liter, rather than particles per liter. This parameter c can be used to calculate the persistence length $b = ma$ for the polymer chains of the presented system. $b \sim 2.9$ nm is obtained with $c \sim 0.32$, which is within a reasonable range.[7]

To test the effect of the adsorber concentration on the frustration, we examined the dependence of the affinity on the degree of cross-linking at various adsorber

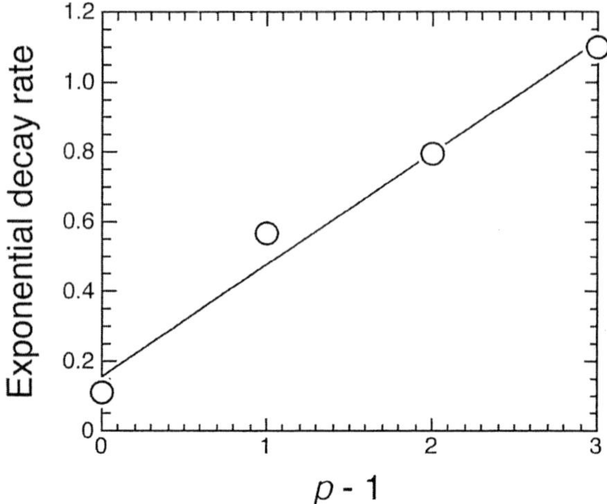

Figure 7: Plot of observed exponential decay rate vs $p - 1$, where p is the number of contact points on the target molecule. The decay rates were obtained through a curve fitting using eq.(3) on the data of Figure 6. The plot shows a good linearity, with a slope of 0.32.

concentrations ([MAPTAC] = 30–60 mM) and a NaCl concentration of 30 mM, for a target molecule (Py-4) that can establish four contact points (p = 4). The gels were prepared with 6 M NIPA. The results are shown in Figure 8. As was mentioned in a

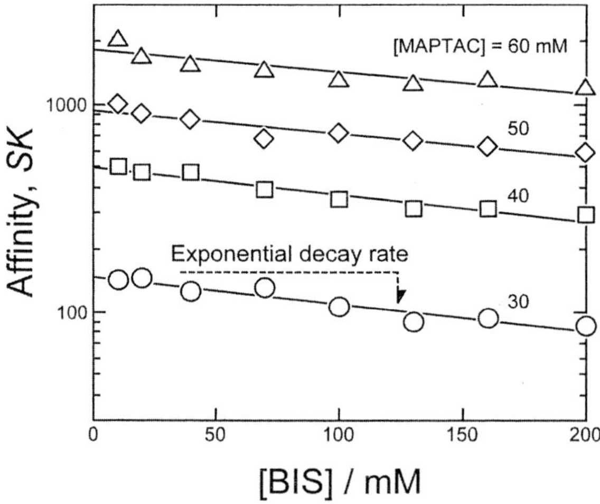

Figure 8: Dependence of the affinity on the degree of cross-linking at various adsorber concentrations, at a NaCl concentration of 30 mM for Py-4 adsorption by collapsed gels at 60 °C. The affinity decreases exponentially with cross-linker concentration, in accordance with eq.(1).

previous section, increasing the amount of adsorber groups increases *SK*, since the saturation capacity of the gel and the probability to accomplish the acquired number, i.e. p of the adsorbers per adsorbing site becomes larger. The slope of the cross-linker-affinity log-log plot should be affected by the adsorber concentration according to eq.(1), in which $[\mathrm{Ad}]^p$ should lead stronger contribution to the affinity than the term, $\exp(-[\mathrm{MAPTAC}]^{-2/3})$ in the given conditions. As the result, we obtained four apparently parallel curves showing the exponential decrease of *SK* with the cross-linking density for gels prepared at four different MAPTAC concentrations. However, $\partial(\ln SK)/\partial([\mathrm{BIS}])$ decreases with ln([MAPTAC]) as shown in Figure 9. According to eq.(3), the quantity,

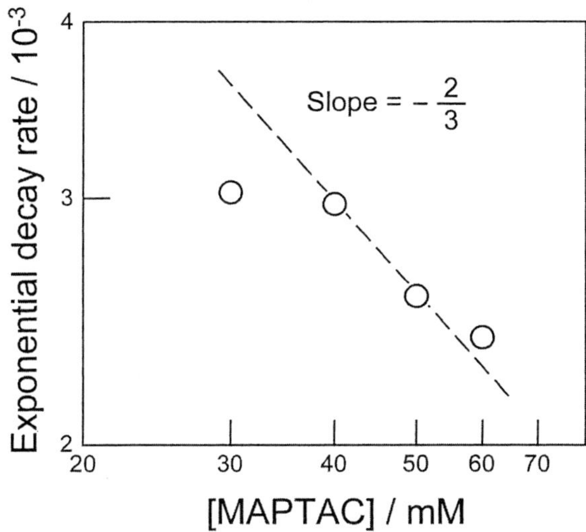

Figure 9: Double log plot of exponential decay rate versus the adsorber concentration.

$\partial(\ln SK)/\partial([\mathrm{BIS}])$, should be proportional to a value $-3c[\mathrm{MAPTAC}]^{-2/3}$ ($p = 4$). There is reasonable agreement with the predicted slope of –2/3, particularly above [MAPTAC] ≥ 40 mM. The negative deviation at [MAPTAC] = 30 mM may be caused by a non-specific interaction with the NIPA main-chain, which is more relevant at smaller values of *SK*.

Conclusion

In this work, we have formulated a theory for the adsorption of target molecules into a random heteropolymer gel and have verified the theory by a large series of experiments.

The gel has been modeled as a set of adsorbing monomers connected by Gaussian chains to fixed cross-linking points. The theory can be summarized by the Tanaka equation, which predicts power-law dependence of the gel affinity with regard to the adsorbing monomer concentration, salt concentration, and cross-linker concentration. Based on this model, we have established that for gels of the appropriate composition, adsorption occurs through multiple contacts in the collapsed state, but through single contacts in the swollen state. Thus the model allows us to quantify the affinity with respect to parameters such as composition of the gel, the number of contact points, and the target molecule chosen. Furthermore, the model provides the basis for understanding imprinted gels,[8] which, through control of the monomer sequence, can be designed to have even higher affinity and specificity for target molecules.[9] Such imprinted gels offer the promise of specific molecular recognition, and will hopefully lead to the development of true protein-like functionality in a synthetic polymer.

[1] T. Tanaka, T. Enoki, A. Y. Grosberg, S. Masamune, T. Oya, Y. Takeoka, K. Tanaka, C. Wang, and G. Wang, *Ber. Bunsenges. Phys. Chem.* **1998**, *102*, 1529.
[2] G. Wang, K. Kuroda, T. Enoki, A. Yu Grosberg, S. Masamune, T. Oya, Y. Takeoka, and T. Tanaka, *Proc. Natl. Acad. Sci.* **2000**, *97*, 9861.
[3] T. Oya, T. Enoki, A. Y. Grosberg, S. Masamune, T. Sakiyama, Y. Takeoka, K. Tanaka, G. Wang, Y. Yilmaz, M. S. Feld, R. Dasari, and T. Tanaka, *Science* **1999**, *286*, 1543.
[4] T. Watanabe, K. Ito, C. Alvarez-Lorenzo, A. Yu. Grosberg, and T. Tanaka, *J. Chem. Phys.* **2001**, *115*, 1596.
[5] A. Grosberg, T. Nguyen, and B. Shklovskii, *Rev. Mod. Phys.*, **2002**, *74*, 329.
[6] H. G. Schild, *Prog. Polym. Sci.* **1992**, *17*, 163.
[7] A. Yu. Grosberg and A. R. Khokhlov, "*Statistical Physics of Macromolecules*," AIP, New York 1994.
[8] T. Enoki, K. Tanaka, T. Watanabe, T. Oya, T. Sakiyama, Y. Takeoka, K. Ito, G. Wang, M. Annaka, K. Hara, R. Du, J. Chuang, K. Wasserman, A. Y. Grosberg, S. Masamune, and T. Tanaka, *Phys. Rev. Lett.* **2000**, *85*, 5000; C. Alvarez-Lorenzo, O. Guney, T. Oya, Y. Sakai, M. Kobayashi, T. Enoki, Y. Takeoka, T. Ishibashi, K. Kuroda, K. Tanaka, G. Wang, A. Y. Grosberg, S. Masamune, and T. Tanaka, *Macromolecules* **2000**, *33*, 8693; C. Alvarez-Lorenzo, O. Guney, T. Oya, Y. Sakai, M. Kobayashi, T. Enoki, Y. Takeoka, T. Ishibashi, K. Kuroda, K. Tanaka, G. Wang, A. Y. Grosberg, S. Masamune, and T. Tanaka, *J. Chem. Phys.* **2001**, *114*, 2812.
[9] The review including the presented work can be found in: K. Ito, J. Chuang, C. Alvarez-Lorenzo, T. Watanabe, N. Ando, and A. Yu. Grosberg, *Prog. Polym. Sci.* **2003**, *28*, 1489.

Microrheology of Chemically Crosslinked Polymer Gels by Diffusing-Wave Spectroscopy

Tetsuharu Narita,[1] *Alexandra Knaebel,*[1] *Jean-Pierre Munch,*[1]
Miklós Zrínyi,[2] *Sauveur Jean Candau**[1]

[1] Laboratoire de Dynamique des Fluides Complexes, 4, rue Blaise Pascal, 67070 Strasbourg Cedex, France
[2] Department of Physical Chemistry, Technical University of Budapest, H-1521 Budapest, Hungary

Summary: The Brownian motion of probe particles in aqueous solutions of poly(vinyl alcohol)(PVA) and in chemically crosslinked PVA gels has been studied by diffusing-wave spectroscopy (DWS). At long time scales the measurements allow us to determine the effect of the crosslinking ratio on the macroscopic viscosity of sols and the shear modulus of gels. The local shear modulus of gels as obtained from the characteristic length of the Brownian cage was found to agree with that measured by classical rheometry and dynamic light scattering (DLS). These microrheological techniques were applied to two polymer gel systems. Substrate induced gradient structure of hydrogels was studied from a microrheological point of view using DLS. It is clearly seen that hydrophobic substrate induces weakly crosslinked network formation at the interface region up to a few millimeters as expected from other experimental facts. Magnetic particle motion in gels under external magnetic field was investigated by DWS. The translational motion of the magnetic particles in gels due to the alternating magnetic force can be detected and found to be superimposed on the relaxation due to the thermal motion.

I. Introduction

Microrheological techniques have been applied to the study of soft-matter systems since the late 1920s. Recently, several advanced techniques such as diffusing-wave spectroscopy (DWS), particle-tracking microrheology or atomic force microscopy measurements have been developed.[1] The main advantage of these microrheological techniques over the macroscopic rheometry is the possibility to study the "in situ" viscoelastic behavior of materials in a large range of frequencies extending up to 1 kHz. DWS measurements are sensitive to fluctuations in the multiple-scattering media at length scales as small as 1 nanometer, which is smaller than the wavelength of the light, because the decorrelation of speckle patterns of multiply scattered light intensity requires only very small displacement of the scatterers. By single and multiple dynamic light scattering measurements, it is possible to determine the mean-square displacement $<\Delta r^2(t)>$ of probe particles from the autocorrelation function of the scattered light

 DOI: 10.1002/masy.200450302

intensity, over a wide range of time (from 10^{-6} to 10^{3} s) and length (larger than 100 nm for single, from 0.3 to 300 nm for multiple, approximately) scales.[2]

Experimental studies using single and multiple dynamic light scattering of the viscoelasticity of synthetic and biological polymer solutions, emulsions, colloidal suspensions, and gels have been reported.[3-8]

In this paper we report DWS measurements in chemically crosslinked gels under three different experimental conditions. We first compare the values of the shear modulus determined by DWS and classical rheometry in the bulk of the sample. We then present two specific applications of the microrheological techniques to gel systems: micro-rheological investigation of gradient gel structure induced by substrates, and microscopic motion of magnetic particles in gels under an external magnetic field.

II. DWS Measurements of Probe Particle Motion in Chemically Cross-linked Gels: Elasticity of the Gels

We have performed DWS measurements on poly(vinyl alcohol) (PVA) solutions and chemically crosslinked gels containing latex probe particles. Chemically crosslinked PVA (molecular weight: 95.000, degree of deacetylation: 95 mol%, from Acros Organics, USA) gels were prepared from PVA solutions at the desired concentration (2 – 10 %) by adding glutaraldehyde (GA, 8 % aqueous solution, from Sigma, USA), at acidic condition (pH = 2, adjusted by HCl). The crosslinking ratio, R_c, was defined as the molar ratio of GA to repeating units of PVA. DWS measurements were carried out on PVA samples containing polystyrene microspheres (diameter: 107 or 535 nm, from PolySciences). The particles were dispersed in PVA solutions at a concentration of 1 % prior to crosslinking. PVA gels were prepared in light scattering cells (21 mm in diameter, 2 mm in thickness).

The laser beam at λ = 488 nm from an argon ion laser (SpectraPhysics 2020) was expanded to about 1 cm diameter at the sample. The diffused light was collected by a single-mode optic fiber placed in the axis of the beam for transmission geometry, or placed in the backscattering cone for backscattering geometry. The intensity autocorrelation function of the diffused light was calculated using a BI9000 correlator. In order to obtain the ensemble average of the scattered intensity, the optical fiber detecting the light was oscillated at 10 mHz, so that the scattering signal from different positions was recorded. A detailed theoretical presentation of DWS in non-ergodic systems was reported previously.[8,9]

The experimentally determined quantity is the intensity autocorrelation function of the scattered intensity $I(t)$,

$$g_T^{(2)}(\tau) = \frac{< I(t+\tau) \cdot I(t) >}{< I(t) >^2}. \qquad (1)$$

The dynamic structure factor obtained in the transmission geometry from the Siegert relation, $g_T^{(2)}(t) = 1 + \beta \left| g_T^{(1)}(t) \right|^2$, where β is an experimental factor (≈ 1), is given by:

$$g_T^{(1)}(t) = \frac{\left(\frac{L/l^* + 4/3}{z_0/l^* + 2/3}\right)\left[\sinh\left(\frac{z_0}{l^*}\sqrt{k_0^2 \Delta r^2(t)}\right) + \frac{2}{3}\sqrt{k_0^2 \Delta r^2(t)}\cosh\left(\frac{z_0}{l^*}\sqrt{k_0^2 \Delta r^2(t)}\right)\right]}{\left(1 + \frac{4k_0^2 \Delta r^2(t)}{9}\right)\sinh\left(\frac{L}{l^*}\sqrt{k_0^2 \Delta r^2(t)}\right) + \frac{4}{3}\sqrt{k_0^2 \Delta r^2(t)}\cosh\left(\frac{L}{l^*}\sqrt{k_0^2 \Delta r^2(t)}\right)} \qquad (2)$$

where L is the thickness of the sample, l^* is the scattering mean free path (determined experimentally), l_T is the characteristic length scale probed by DWS in the transmission geometry, which is expressed as $l_T = l^*/k_0 L$, z_0 is the penetration depth, and $k_0 = 2\pi/\lambda$.

Figure 1(a) shows the dynamic structure factors obtained in transmission geometry for dispersions of latex particles in water, a PVA solution and gels. For water and the polymer solution the structure factor goes to zero at long time, which means that the relaxation of the concentration fluctuations is total in the experimental time scale so that the system can be considered as ergodic. For the polymer solution, the representative curve is shifted towards the right with respect to that of water, indicating a slower diffusion of the particles through the medium. For the gels however, the dynamic structure factor does not go to zero at infinite time, which indicates the existence of frozen-in concentration fluctuations. In that case the systems are non-ergodic, the time average of the correlation function of the field scattered by a given scattering volume differing from the ensemble average obtained by scanning the sample. It is the latter quantity which has a physical meaning for the analysis of the gel dynamics. The plateau value increases upon increasing the crosslinking ratio, indicating that the particle motion is more restricted due to an increased modulus of the gel.

The dynamic structure factor was numerically inverted, using eq.(2) into the mean-square displacement of the probe particles, $<\Delta r^2(t)>$.[10] Figure 1(b) shows $<\Delta r^2(t)>$ as a function of time for the latex particles in PVA systems.

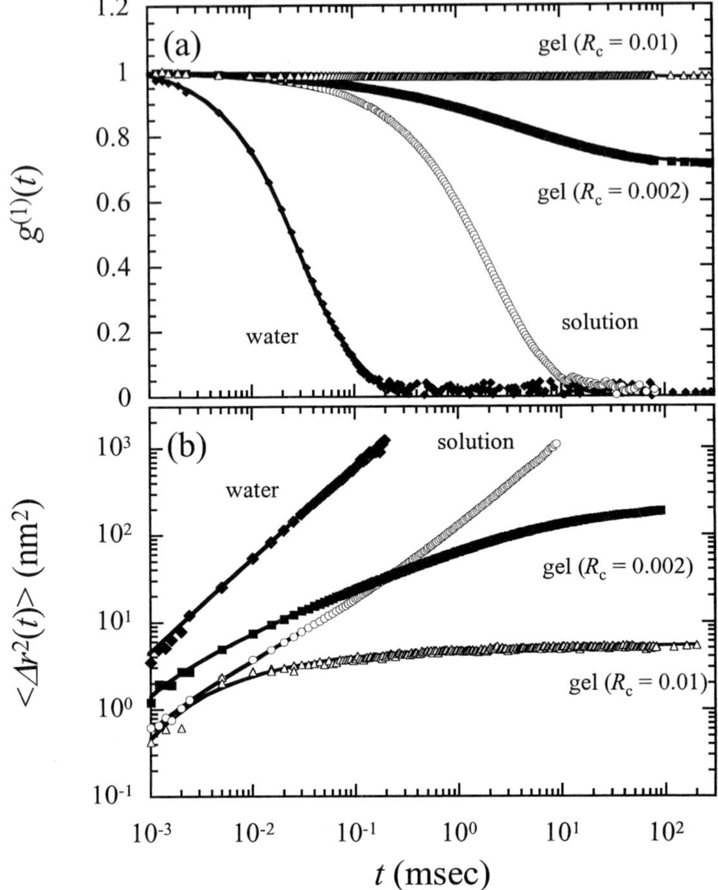

Figure 1. (a) Normalized dynamic structure factors for a PVA solution at $C_p = 5$ % and gels at the same concentration containing latex particles (radius: 535 nm) measured by DWS. (b) Mean-square displacement of latex particles in PVA solutions and gels.

For solutions, $<\Delta r^2(t)>$ is proportional to the time at long times, indicating that the particles undergo Brownian diffusion with $<\Delta r^2(t)> = 6Dt$ where D is the translational diffusion constant of the latex in the solution.[3] One can estimate the macroscopic viscosity of the system by using the Stokes-Einstein equation. The results thus obtained agree with those measured by macroscopic rheometry (data not shown). At short times, the shape of the $<\Delta r^2(t)>$ curves reflects the viscoelasticity of the sample.[3,11] For concentrations larger than 5 %, which correspond to the regime where the polymer chains are entangled, one observes an inflexion point in the curve reflecting a confinement of the latex particles in the entanglement network.

For gels, at long time scales, $<\Delta r^2(t)>$ levels off at a certain value that decreases upon increasing crosslinking ratio. The mean square displacement of the particles in gels is limited due to caging, and the maximum displacement is determined by the competition between the thermal energy of the particles and the elasticity of the network that is considered as a continuous medium since the particle size is much larger than the mesh size. $<\Delta r^2(t)>$ can be expressed by this equation,[12]

$$< \Delta r^2(t) >= 6\delta^2 \left[1 - \exp\left[-(t/\tau_c)^\gamma\right]\right], \quad (3)$$

where δ and τ_c are respectively the characteristic size and time of the Brownian cage, which are the characteristic spatial length and time required for the particles to explore the cage. The short time behavior of the particles is subdiffusive, and expected to reflect the local dynamics of the gels.

The local shear modulus μ of the medium can be estimated from the plateau value of the mean-square displacement δ^2, arguing that the work necessary for a particle of radius R to move over δ is equal to the thermal energy k_BT of the particle, so that[3]

$$\mu = \frac{k_B T}{6\pi R \delta^2}. \quad (4)$$

As discussed by Schnurr et al,[13] the above expression is strictly valid only at time scales that are short enough so that the solvent and the network move as one at scales large when compared with the mesh size of the network. In this limit the gel behaves as an incompressible fluid.

In order to evaluate the validity of the DWS measurements of the shear modulus, a comparison of the values of the local storage shear modulus obtained by DWS with those measured by macroscopic rheometry was made.

Figure 2 shows a comparison between the modulus μ and the macroscopic value G' measured by classical rheometry for PVA gels at two concentrations and various crosslinking ratios. One observes a good agreement between the two sets of data for gels with $C_p = 8$ %. For these gels the increase of the modulus with crosslinking ratio is slightly less than the linear dependence expected from rubber elasticity theory.

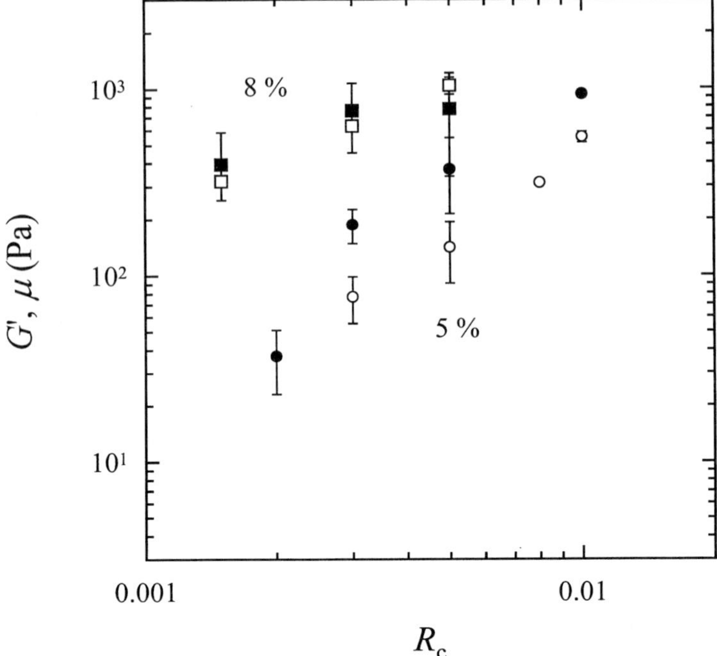

Figure 2. Comparison of the microscopic shear modulus μ with the macroscopic storage modulus G' measured at 92 Hz of PVA gels as a function of the crosslinking ratio R_c. Open circles: $C_p = 5$ %, macroscopic (without latex); filled circles: $C_p = 5$ %, microscopic; open squares: $C_p = 8$ %, macroscopic (without latex); filled squares: $C_p = 8$ %, microscopic.

For the gels with $C_p = 5$ % the agreement between G' and μ is not as satisfactory. This might be due to the increased difficulty in performing accurate measurements by mechanical rheometry when the gels are soft, that is when the polymer concentration and/or the crosslinking ratio are low. The large variation of G' and μ with the crosslinking ratio, already reported for other systems[14] is generally explained by the formation of dangling chains at low crosslinking ratio and trapped entanglements at high crosslinking ratio.

We also performed a comparative study using DWS and DLS techniques. It is possible to investigate the cage dynamics by DLS when the characteristic length of the cage is large enough, that is for gels with low crosslinking ratio and/or low polymer concentration provided that the size of the probe particle is small enough. Usually, the decay function of DLS for a gel containing probes is bimodal, as the fluctuations of polymer concentration and the motion of confined particles are observed simultaneously. The characteristic time of the cage is independent of q, while the fluctuations of the gel are diffusive as inferred from the q^2 dependence of the relaxation time. The fluctuating and frozen-in components of light scattered by both the latex and gel were determined by fitting a two component exponential function. The fraction of the frozen-in intensity scattered by the latex A_G can be related to δ, according to

$$A_G = \exp(-q^2\delta^2)\,. \tag{5}$$

From the value of δ, the shear modulus μ was determined by DLS and compared with that obtained by DWS. In Table 1, the shear moduli μ of a PVA gel (C_p = 2 %, R_c = 0.01), measured by DWS and DLS using various sizes of probes, are shown. The comparison between the values of μ for a PVA gel with C_p = 2 % and R_c = 0.01 measured by DLS and those determined by DWS, indicates that the two methods are equivalent and that the addition of concentrated latex to the gel does not affect significantly the modulus.

method	latex conc. (%)	latex diameter (nm)	μ (Pa)
DLS	0.01	107	36
DWS	1		43
		535	31

Table 1. Local shear modulus μ of PVA gel (2 %, R_c = 0.001) determined by DWS and DLS using latexes with different size.

III. Applications

In this section we present two examples of microrheological measurements of gels that emphasize the specific possibilities of the light scattered based microrheological techniques .

III-1. Gradient Gel Structure Induced by Substrate Effect

The interfacial properties of hydrogels made by radical polymerization of water soluble vinyl monomers depend on whether the substrate used for the preparation of the gel is hydrophobic (tetrafluoroethylene, polystyrene, hydrophobically modified glass, etc) or hydrophilic (glass, mica, etc).[15-17] In particular, gels synthesized on hydrophobic substrates exhibit a reduced friction coefficient against glass.[15] This behavior was interpreted by assuming that the surface of the gel made on a hydrophobic substrate consists of a loosely crosslinked network structure with dangling chains.

In situ monitoring of the refractive index change of solutions under polymerization reaction by electronic speckle pattern interferometry showed the presence of a polymer concentration gradient over a spatial range extending up to 1 mm from the interface between the gel and the substrate.[18] However, the refractive index profile does not give any information on the actual structure of the gel.

Viscoelastic measurements in gels allow us to probe directly the network structure since the presence of dangling chains should contribute to increase the local viscosity and decrease the modulus. In this section, we report on microrheological measurements by dynamic light scattering (DLS) of confined probe particles along the concentration gradient within the gel, taking advantage of the spatial resolution of the microrheological technique.

Measurements of the single scattering from a small amount of trapped particles dispersed within polyacrylamide (PAAm) gels before crosslinking reaction were performed. PAAm gel was synthesized by redox radical polymerization of acrylamide (2.5 %) in the presence of methylene bis(acrylamide) as a crosslinker (1 mol% of acrylamide), ammonium persulfate as an initiator (1 mM) and tetramethylethylene diamine as a redox initiator (0.6 v%) under nitrogen atmosphere at room temperature. All the chemicals were used as received. Before the redox initiator was added, oxygen in the solution was removed by purging nitrogen for 5 min. All the solutions were filtered through 0.45 μm Millipore filter. Polystyrene microspheres (Polysciences, Inc., USA) with radius 535 nm was used as probe particles (volume fraction $\phi = 10^{-4}$). For DLS measurement, the gel was prepared in a glass tube (12 mm in diameter). As a hydrophilic substrate, the bottom of the glass tube was used, and as a hydrophobic substrate, a piece of tetrafluoroethylene (teflon) plate was set at the bottom. The glass tube and the teflon plate were sonicated in ethanol and then in detergent aqueous solution, followed by rinsing in distilled water. All the DLS measurements were made at 25 °C, at scattering angle 60°, triplicated and averaged. The scattering cell was vertically moved with a resolution of 10 μm from the surface of the substrate up to 6 mm within the gel. An arrangement using two pinholes (diameter 200 μm) and a lens in front of the photomultiplier, provides a detection volume of (200 μm)^3. This determines the spatial resolution of the depth profiling. The experimental setup for the DLS measurements is shown in Figure 3(a).

Height profiles of mean-square characteristic length δ^2, (which is inversely proportional to the local shear modulus) and the characteristic time τ_c (proportional to the local friction coefficient of the network which the particles experience) were determined and plotted in Figure 4. Both the characteristic length and time in the gel made on glass are

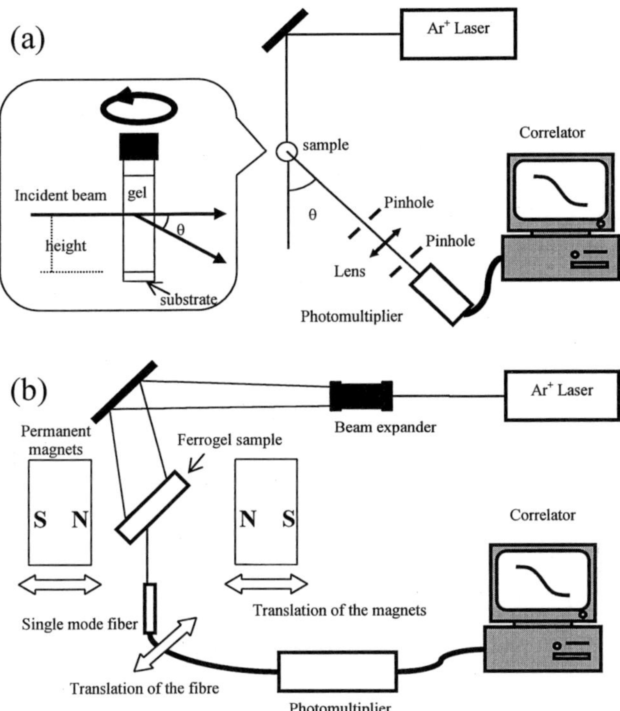

Figure 3. (a) Schematic illustrations of the experimental setup for DLS measurements of substrate induced gradient structure of PAAm gels. A gel is synthesized in a glass tube having a teflon substrate at the bottom (or the bottom was used as a glass substrate). In order to obtain ensemble average over different positions of the gel at the same height, the tube was continuously rotated. (b) DWS measurements under external oscillating magnetic fields. Ferrogel samples are fixed between two permanent magnets mounted on a piezoelectric actuator. Then the magnets were sinusoidally oscillated to create an alternating magnetic field. In order to obtain an ensemble average, a single-mode optical fibre was periodically translated to scan the sample.

independent of the distance to the substrate. On the other hand, those of the gel made on teflon increase upon approaching the substrate. This suggests that the crosslinking density of the network decreases in the vicinity of the interface since δ^2 depends on the crosslinking density. The characteristic length increases over a 4 mm depth, while the characteristic time increases in a 1 mm thick layer close to the substrate. The local viscosity is likely to be less sensitive to the variation of crosslinking density except in the very close vicinity of the interface where the presence of many dangling chains might increase the friction exerted on the probe particles.

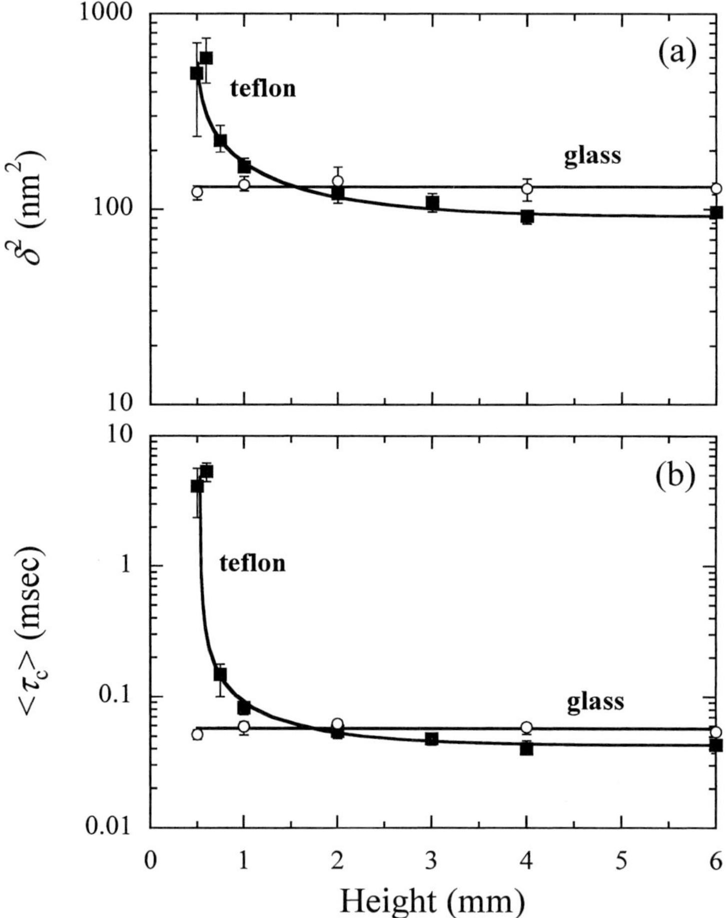

Figure 4. Height dependences of (a) the local mean square characteristic length and (b) the local characteristic time of PAAm gels (1 mol%) made on glass (open symbols) and teflon (filled symbols).

This study is a striking illustration of the potential of the microrheological techniques to investigate the local viscoelasticity of materials. Also the results reported here confirm the intriguing observation that the hydrophobicity of the substrate perturbs the gelation process over a surprisingly large spatial range, producing a gradient of the crosslink density.

III-2. Magnetically Sensitive Gels

In DWS measurements, one observes usually the thermal fluctuations of the confined probes. Motion of the probes induced by macroscopic external forces such as an oscillating shear flow have been reported.[6] However, DWS experiments using direct microscopic manipulation of the probe by external forces have not been reported. Here we present a study of the particle displacement induced by an external magnetic field. This work is the microscopic analogue of the rheological studies of magnetic field-sensitive gels containing magnetic particles dispersed homogeneously in a polymeric network. The particle motion in response to a non-uniform magnetic field induces elongation, contraction or bending of the gels with short response time.[19,20]

Chemically crosslinked PVA gels containing magnetic particles were studied. Magnetic particles were prepared by emulsion polymerization of styrene monomers in the presence of magnetite particles. Magnetite (Fe_3O_4) sol was prepared from $FeCl_2$ and $FeCl_3$ in aqueous solution using sodium hydroxide as a precipitating agent. Obtained stabilized magnetite sol has a concentration of 15.2 m% and an average particle diameter of 10 nm. 60 cm^3 of styrene monomers was emulsified in 120 cm^3 of distilled water, in the presence of 1.89 g of dihexyl sodium sulfosuccinate, 0.25 g of potassium hydrogen carbonate and 5 g of the magnetite sol under gentle stirring for 2 hours at constant temperature of 60 °C. The solution was then heated to 80 °C and an initiator solution (1.6 g of potassium peroxidisulfate dissolved in 20 cm^3 of distilled water) was added. The polymerization reaction was carried out for 20 h at 80 °C. After polymerization the magnetic particles were purified by evaporating the unreacted monomers and by dialyzing against water. The obtained particles have average radius of 1.3 μm. The particles were homogeneously dispersed in PVA solutions prior to the crosslinking reaction.

We have analyzed the particle motion induced by the inhomogeneous magnetic field, with a set-up in which the gel was fixed between two permanent magnets translated by means of a piezoelectric actuator, at various frequencies (Figure 3(b)). To perform an ensemble averaging of the correlation function of the multiple scattered intensity, the sample was scanned periodically at a frequency of 10^{-4} Hz over a displacement of 1 mm by means of an optical fiber. Dynamical structure factors obtained both in the absence of magnetic field and with a magnetic field of intensity B = 25 mT at various frequencies are presented in Figure 5.

In the absence of magnetic field the dynamical structure factor exhibits only the classical relaxation mode corresponding to the Brownian diffusion of the magnetic

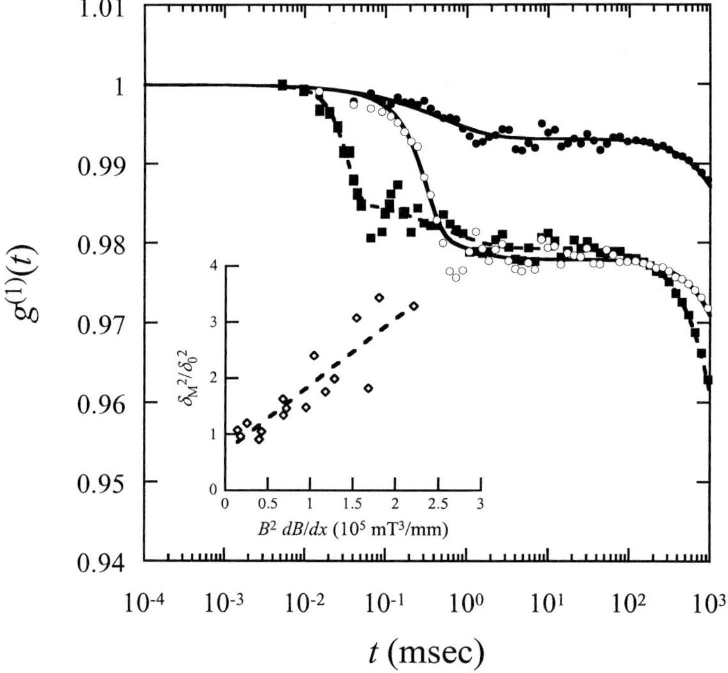

Figure 5. Normalized field correlation function of a ferrogel under oscillatory magnetic field of various frequencies. The oscillatory magnetic field is made by translating a pair of permanent magnets (in a sinusoidal wave with amplitude 500 μm). The frequencies are: 3 Hz (open squares); 1 Hz (closed squares); 0.1 Hz (open circles). Results without magnetic field: closed circles. The translation of the detection fiber that allows to measure ensemble averages, leads to a cut off of the correlation function at about 10^3 msec. Inset: Normalized mean square characteristic spatial length δ_M^2 as a function of B^2dB/dx.

particles. This mode can be fitted by a stretched exponential with an exponent $\beta = 0.7$ and an average relaxation time $\tau_0 \cong 1$ msec. At low frequencies of the magnetic fields (≤ 0.1 Hz) we still see a single decay mode but it becomes sharper upon increasing the frequency. At frequencies larger than 0.1 Hz one observes ,in addition to the Brownian motion, a second faster decay, sharper than a single relaxation process.

The sharpness of this mode increases upon increasing the frequency of the magnetic field whereas the characteristic time τ_M (taken as the time at half decay) is inversely proportional to the frequency. This mode is likely associated with the uni-dimensional motion of the particles along the field gradient.

From the curves of Fig. 5, one can also determine the intermediate value g_{int} of the plateau following the faster decay. By analogy with what was previously discussed for particles undergoing Brownian motion, g_{int} can be described as a function of the characteristic spatial length δ_M explored by the particles:

$$g_{\text{int}} = \exp\left(-\sqrt{\frac{\delta_M{}^2 L^2}{6l^{*2}}\frac{2\pi}{\lambda}}\right), \tag{6}$$

where L is the thickness of the gel, l^* is the scattering mean free path, that is determined as explained in the literature,[21] and λ is the wavelength of the laser, respectively.

From the dynamical structure factors given in Fig. 5, we have calculated the values of $\delta_M{}^2$ as a function of the magnetic force F, which at low field intensity B can be written as:

$$F \propto B^2 \, \mathrm{d}B/\mathrm{d}x. \tag{7}$$

The inset of Figure 5 shows normalized mean square characteristic spatial length, $\delta_M{}^2$ as a function of B^2 dB/dx. It is found that within the experimental accuracy the mean square length explored by the particles varies linearly with the magnetic force. Also $\delta_M{}^2$ was found to be independent to the first order of the frequency of the magnetic field.

IV. Conclusions

Single and multiple dynamic light scattering can be a powerful tool for a microrheological investigation of polymer gels. We studied cage dynamics of probe particles confined in chemically crosslinked gels. We found that the value of the local shear modulus of the chemically crosslinked PVA gels determined by diffusing-wave spectroscopy is equivalent to that obtained by classical mechanical rheometry and single light scattering.

These techniques were used to investigate the gradient structure of polymer gels induced by the substrates. We determined the height profiles of the mean-square characteristic length and the characteristic time of the probe particles motion in polyacrylamide gels made on hydrophobic and hydrophilic substrates. It is shown that the gel surface region made on hydrophobic substrates has lower modulus and higher viscosity indicating that the gel surface region has a loosely crosslinked network structure with dangling chains.

We also outlined the possibilities offered by the DWS technique to investigate the motion of magnetic particles confined in a gel, under an inhomogeneous magnetic field. It was shown that the application of a periodic magnetic field to a ferrogel induces a sharp decay in the autocorrelation function with a characteristic time that is a decreasing function of the frequency of the magnetic field. This decay can be assessed to a uni-dimensional motion of the magnetic particles along the gradient field.

[1] F.C.MacKintosh, C.F. Schmidt, *Curr. Opin. Colloid Interface Sci.*, **1999**, *4*, 300

[2] J.L. Harden, V. Viasnoff, *Curr. Opin. Colloid Interface Sci.*, **2001**, *6*, 438

[3] T.G. Mason, D.A. Weitz, *Phys. Rev. Lett.*, **1995**, *74*, 1250

[4] A. Palmer, T.G. Mason, J. Xu, S.C. Kuo, D. Wirtz, *Biophys. J.*, **1999**, *76*, 1063

[5] T. Gisler, D.A. Weitz, *Phys. Rev. Lett.*, **1999**, *82*, 1606

[6] P. Hébraud, F. Lequeux, J.P. Munch, D.J. Pine, *Phys. Rev. Lett.*, **1997**, *78*, 4657

[7] J.G.H. Joosten, E.T.F. Gelladé, P.N. Pusey, *Phys. Rev. A*, **1990**, *42*, 2161.

[8] G. Nisato, P. Hébraud, J.-P. Munch, S.J. Candau, *Phys. Rev. E*, **2000**, *61*, 2879.

[9] T. Narita, A. Knaebel, J.-P. Munch, S.J. Candau, *Macromolecules*, **2001**, 34, 8224.

[10] D.A. Weitz, D.J. Pine, In *Dynamic Light Scattering: The Method and Some Applications*, W. Brown, Ed., Clarendon, Oxford, **1993**, pp652.

[11] A. Palmer, T.G. Mason, J. XU, S.C. Kuo, D. Wirtz, *Biophys. J.*, **1999**, 76, 1603

[12] P.N. Pusey, W. Van Megen, *Physica A*, **1989**, 157, 705

[13] B. Schnurr, F. Gittes, F.C. MacKintosh, C.F. Schmidt, *Macromolecules*, **1997**, 30, 7781

[14] A. Silberberg, In *Polyelectrolyte Gels*, R.S. Harland, R.K. Prod'homme, Eds., American Chemical Society, Washington DC, **1992**, pp146 and 150.

[15] J.P. Gong, T. Narita, T. Kurokawa, G. Kagata, Y. Osada, G. Nishimura, M. Kinjo, *J. Am. Chem. Soc.* **2001**, *123*, 5582.

[16] T. Narita, A. Hirai, J. Xu, J.P. Gong, Y. Osada, *Biomacromolecules* **2000**, *1*, 162.

[17] X.M. Zhang, J. Xu, K. Okawa, Y. Katsuyama, J.P. Gong, Y. Osada, K.S. Chen, *J. Phys. Chem. B* **1999**, *103*, 2888.

[18] A. Kii, J. Xu, J.P. Gong, Y. Osada, X.M. Zhang, *J. Phys. Chem. B 2001*, *105*, 4565.

[19] M. Zrínyi, D. Szabó, and H.-G. Kilian, *Polymer Gels Networks* **1998**, *6*, 441.

[20] M. Zrínyi, L. Barsi, and A. Büki, *Polymer Gels Networks* **1997**, *5*, 415.

[21] E.M. Furst and A.P. Gast, *Phys. Rev. E*, **1998**, *58*, 3372.

Pressure Effects on the Inhomogeneities of Environmentally Sensitive Polymer Gels

Mitsuhiro Shibayama,[*1] *Shinichi Takata,*[1] *Chikanobu Nakamoto*[2]

[1] Neutron Scattering Laboratory, Institute for Solid State Physics, University of Tokyo, Tokai, Ibaraki 319-1106, Japan

[2] Industrial Technology Center of Wakayama Prefecture, Ogura, Wakayama, 649-6261, Japan

Summary: The effects of pressure at gel preparation on the inhomogeneities were investigated by small-angle neutron scattering. Poly(N-isopropylacrylamide) (PNIPA) gels, known as thermosensitive polymer gels undergoing volume transition, were prepared at various pressures ranging from 0.1 to 300 MPa. The scattering intensity increased with increasing P_{prep}, up to 200 MPa, then decreased by further increase in P_{prep}. The degree of inhomogeneities evaluated by the ratio of the static and dynamic correlators exhibited a similar behavior to that of the ensemble average light scattered intensity, $\langle I \rangle_E$. The physical meaning of the gel inhomogeneities is discussed from the viewpoint of swelling thermodynamics. It will be shown that the increase in the swelling degree is due not to an increase in the miscibility of PNIPA with water but to the increase in inhomogeneities.

Introduction

Poly (*N*-isopropylacrylamide) (PNIPA) hydrogels are known to be temperature-sensitive gels.[1,2] Upon a slight increase in temperature above the transition temperature, T_C ($\approx$ 34 °C),[3,4] PNIPA gels undergo a shrinking transition. The environment sensitivity is mainly ascribed to the nature of strong hydrophobicity of the *N*-isopropyl group located on the side chain. Hydrophobicity is related to ordering of water molecules surrounding hydrophobic groups. That is, water molecules around an *N*-isopropyl group are ordered and form an iceberg structure[5] in order to minimize the number of contacts with the hydrophobic group. However, this ordered structure is broken by thermal motion of water molecules when temperature is increased above the so-called volume phase transition temperature T_C. This leads to demixing between PNIPA and water. This volume transition is induced by pressure as well.

Since the iceberg formation involves positive change in volume, pressurizing is expected to decrease the LCST (the lower critical solution temperature), resulting in a shrinking transition. However, contrary to this speculation, a swelling transition is commonly observed.[6-8] According to Kato, the Flory interaction parameter is a

 DOI: 10.1002/masy.200450303

quadratic function of pressure, which results in a reentrant swelling-shrinking transition as pressure is increased.[8] Zhong et al. reported that the pressure sensitivity of PNIPA gels is a consequence of their temperature sensitivity and is a result of the increase in its LCST with temperature.[7] We reported that the structure and dynamics of PNIPA gels are more or less frozen during the preparation process of the gel.[9] For example, when a PNIPA gel is prepared at a high pressure, P_{prep}, it keeps a memory of the circumstance of gel preparation even after releasing the pressure.[10] In this paper, we discuss the P_{prep} dependence on the microscopic structure of PNIPA gels on the basis of small-angle neutron scattering (SANS) data. It will be shown that there is a critical pressure at which inhomogeneities diverge.

Experimental Section

A series of poly(*N*-isopropylacrylamide) (PNIPA) gels were prepared by redox polymerization. Prescribed amounts of NIPA monomer (purchased from ACROS, U.S.A.) and *N,N'*-methylenebisacrylamide (BIS; cross-linker, Wako Chemicals) were dissolved in distilled water (total volume 50 ml), whose concentrations were 700 and 8.63 mM, respectively. By adding 1.7 mM of ammonium persulfate (initiator; Wako Chemicals) and 3.0 mM of *N,N,N',N'*-tetramethylenediamine (accelerator; Wako Chemicals) to the solution, polymerization was initiated. The polymerization was carried out in a 10 mm test tube at room temperature (20 ±1°C) under several pressures which were applied by a wet-bag type cold isostatic pressing equipment (Dr. CIP®; KOBE STEEL LTD) for 24 h. The preparation pressure, P_{prep}, was varied from 0.1 to 300 MPa.

The SANS experiments were carried out on the research reactor, SANS, at The Institute of Solid State Physics, The University of Tokyo. A flux of cold neutrons with a wavelength of 7.0 Å irradiated the sample, and the scattering intensity profile was collected with an area detector of 128 x 128 pixels (0.5 mm/pixel). The sample-to-detector distance was set to be 4 m, which covered the accessible q range being 0.01 to 0.078 Å^{-1}. Here, q is the scattering vector. The sample was placed in a brass chamber with quartz windows and the chamber was thermo-regulated within an error of ± 0.1 °C at the sample position with a NESLAB 110 water-circulating bath. The sample thickness was 4 mm. The scattering intensities were circularly averaged by taking account of the detector inhomogeneities, corrected for cell scattering, fast neutrons, transmission, and incoherent scattering, and then calibrated to the absolute intensities with a polyethylene standard sample (Lupolen).

Results and Discussion

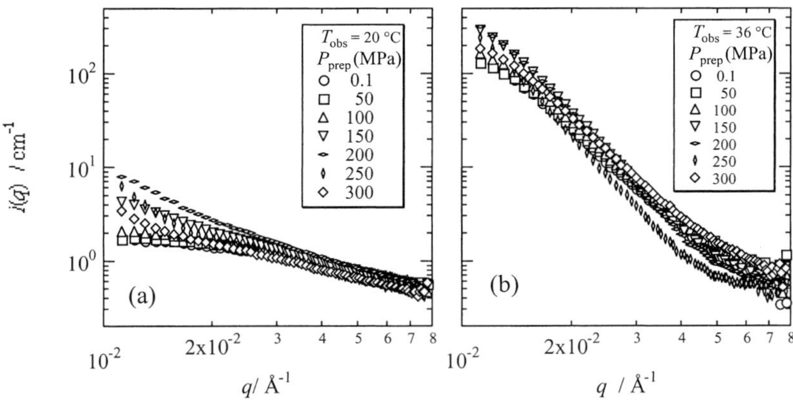

Figure 1. SANS scattering function, $I(q)$, of PNIPA gels prepared at various pressures, P_{prep}. The observation temperatures were (a) T_{obs} = 20 and (b) 36 °C.

Figure 1 shows the SANS intensity profiles, $I(q)$, of PNIPA gels prepared at various pressures, P_{prep}'s, where q is the momentum transfer. The observation temperatures, T_{obs}, were (a) 20 and (b) 36 °C. Fig. 1a shows that $I(q)$ depends on P_{prep}, at low q regions, while no P_{prep} dependence is resolved at high q regions ($q \geq 0.04$ Å^{-1}). When the observation temperature was increased to T_{obs} = 36 °C, $I(q)$ increased markedly at all q regions and the P_{prep} dependence becomes less significant (Fig. 1b). This figure indicates that the temperature effect is much larger than the pressure effect. Note that $I(q)$ at T_{obs} = 20 °C became strongest at $P_{prep} \approx 200$ MPa.

Figure 2 shows the P_{prep} dependence of $I(q = 0.011$ Å^{-1}) observed at various T_{obs}'s. This shows two important features of PNIPA gels. One is that $I(q)$ increases with increasing T_{obs}, and the other is that $I(q)$ has an maximum (or $I(q)$ is divergent) at $P_{prep} \approx 200$ MPa, irrespective of T_{obs}. The PNIPA gels underwent phase separation at $P_{prep} \geq 200$ MPa.

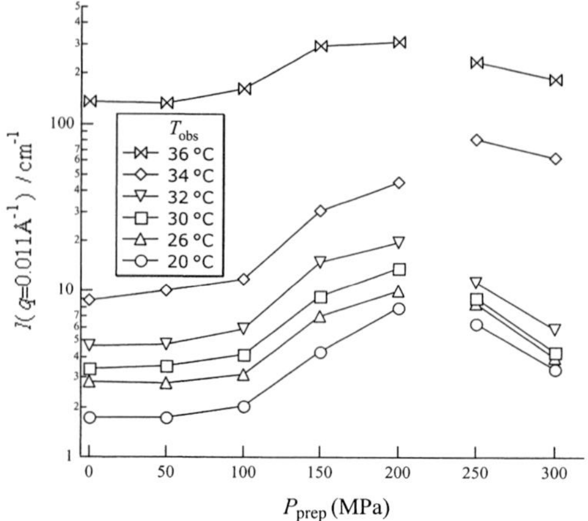

Figure 2. P_{prep} dependence of $I(q = 0.011\ Å^{-1})$ of PNIPA gels observed at various observation temperatures, T_{obs}.

Figure 3a shows the P_{prep} dependence of ensemble-average light scattering intensity, $\langle I \rangle_E$.[10] $\langle I \rangle_E$ was obtained by measuring scattering intensity at 100 different sampling points in a gel. Here, similarly to $I(q)$ observed by SANS, a divergence in $\langle I \rangle_E$ is observed. Shrinking kinetics experiments on this series of gels indicate that the PNIPA gels underwent phase separation for $P_{prep} > 170$ MPa and the shrinking rate is increased by a factor of 10^3.[10] The SANS results obtained here are in good accordance with these results. The inset of Fig. 3a shows a $1/\langle I \rangle_E$ vs. P_{prep} plot, in which $\langle I \rangle_E$ diverges at $P_{prep} = 168$ MPa.

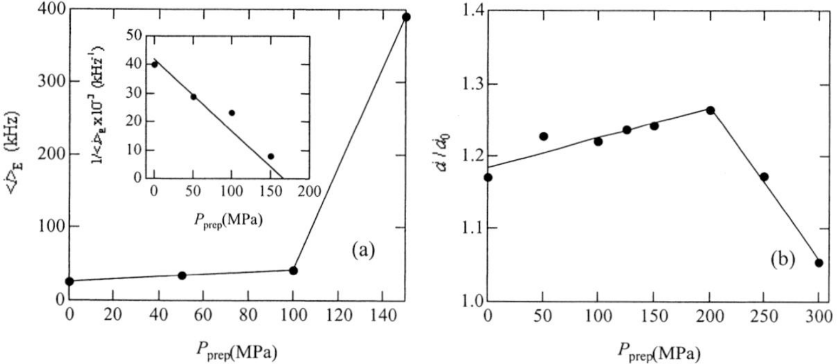

Figure 3. P_{prep} dependence of (a) ensemble average scattering intensity, $\langle I \rangle_E$, and (b) equilibrium swelling ratio, d/d_0.

Figure 3b shows the equilibrium swelling ratio, d/d_0, of the PNIPA gels prepared at various P_{prep}'s, where d and d_0 are the diameters of the gel at observation and preparation, respectively. Interestingly, d/d_0 increases with P_{prep} and has a maximum around $P_{prep} \approx 170$ MPa. The increase in d/d_0 means that the gel becomes more inhomogeneous with increasing P_{prep} and the effective number of cross-links is greatly reduced by formation of polymer-rich domains. The subsequent decrease in d/d_0 results from phase separation of the gel, resulting in a lack of domain-connecting chains responsible for swelling. This speculation is supported by the experimental evidence for the preparation temperature dependence of the equilibrium swelling.[11]

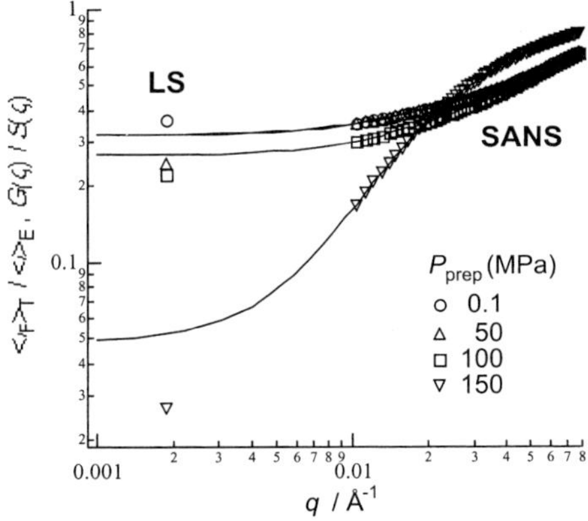

Figure 4. q dependence of the degree of inhomogeneities, $G(q)/S(q)$ and $<I_F>_T/<I>_E$ evaluated by SANS and LS, respectively. The solid lines are the theoretical functions calculated with the Panyukov-Rabin theory.

Figure 4 shows the q dependence of the static inhomogeneities. The q range is converted from the SANS to LS regimes. The solid lines are the fit with Panyukov-Rabin theory[12] on the structure factor of polymer gels, which theoretically takes account of static inhomogeneities. Hence, the total scattering intensity is written by,

$$I(q) = KS(q) = K[G(q) + C(q)] \qquad (1)$$

where K, $S(q)$, $G(q)$, and $C(q)$, are the proportionality constant, the structure factor, the thermal correlator, and the static correlator, respectively. The ratio, $G(q)/S(q)$, is a measure of the degree of inhomogeneities, taking smaller values with larger inhomogeneities. Though the details of the theory are described elsewhere,[12,13] the extraporation of $G(q)/S(q)$ fits well to $<I_F>_T/<I>_E$ for $P_{prep} \leq 150$ MPa.

The divergence of light scattering intensity of PNIPA in water was first reported by Otake et al. for PNIPA aqueous solutions.[14] Our LS results show that $<I>_E$ increases steeply for $P_{prep} > 100$ MPa. As shown in the inset of Fig. 3a, the plot of $1/<I>_E$ vs P_{prep} indicates a linear relationship between $1/<I>_E$ vs P_{prep} and $<I>_E$ diverges at $P_{prep} = P_{prep,C} = 168$ MPa. Although, the reason for the presence of such a critical pressure has not been elucidated at all, it may be related to the unique properties of water molecule itself, as pointed out by Tanaka,[15] who proposed a two state model for liquid water, i.e., the presence of two competing orderings in water. One is density ordering and the other is bond ordering. This model features a crossover pressure ($\approx$200 MPa). It is not clear that this crossover pressure is related to the critical pressure disclosed in this work. A further investigation to clarify this issue is now in progress.

Conclusion

The microscopic structure of PNIPA gels have been investigated with small-angle neutron scattering as a function of gel preparation pressure, P_{prep}. The scattering intensity, $I(q)$, was found to be P_{prep} dependent, particularly in the low q region. $I(q)$ increased with increasing temperature. $I(q)$ exhibited a strong P_{prep} dependence and was largest for the gel prepared at $P_{prep} = 200$ MPa. The degree of imhomogeneities was found to be a smooth function of q in the scattering regions of SANS and LS, and the inhomogeneities increase with lowering q.

Acknowledgment

This work is partially supported by the Ministry of Education, Science, Sports and Culture, Japan (Grant-in-Aid, 12450388 and 13031019 to M.S.).

[1] T. Tanaka, Sci. Am. 1981, 244, 110.
[2] M. Shibayama, T. Tanaka, Adv. Polym. Sci. 1993, 109, 1.
[3] Y. Hirokawa, T. Tanaka, J. Chem. Phys. 1984, 81, 6379.
[4] T. Tanaka, D. J. Fillmore, S. T. Sun, I. Nishio, G. Swislow, A. Shah, Phys. Rev. Lett. 1980, 45, 1636.
[5] C. Tanford, Physical Chemistry of Macromolecules, Wiley, NY 1961.
[6] E. Kato, T. Kitada, C. Nakamoto, Macromolecules 1993, 26, 1758.
[7] X. Zhong, Y.-X. Wang, S.-C. Wang, Chem. Eng. Sci. 1996, 51, 3235.
[8] E. Kato, J. Chem. Phys. 1997, 106, 3792.
[9] M. Shibayama, S. Takata, T. Norisuye, Physica A 1998, 249, 245.
[10] C. Nakamoto, T. Motonaga, M. Shibayama, Macromolecules 2000, 34, 911.
[11] S. Takata, M. Shibayama, T. Norisuye, Polymer 2002, 43, 3101.
[12] S. Panyukov, Y. Rabin, Physics Report 1996, 269, 1.
[13] S. Takata, T. Norisuye, M. Shibayama, Macromolecules 2002, 35, 4779.
[14] K. Otake, R. Karaki, T. Ebina, C. Yokoyama, S. Takahashi, Macromolecules 1993, 26, 2194.
[15] H. Tanaka, J. Chem. Phys. 2000, 112, 799.

A Hybrid Polymer Gel and Its Static Nonergodicity

Yue Zhao,[1] *Chi Wu**[1,2]

[1] The Open Laboratory of Bond Selective Chemistry, Department of Chemical Physics, University of Science and Technology of China, Hefei, Anhui, China
[2] Department of Chemistry, The Chinese University of Hong Kong, Shatin, N.T., Hong Kong
* The Hong Kong address should be used for correspondence.

Summary: We used a thermally reversible hybrid gel made of billions of *physically* jam-packed, swollen, thermally sensitive poly(*N*-isopropyl-acrylamide) *chemical* microgels. Laser light scattering study on a series of such hybrid gels formed at different gelling rates and temperatures revealed that the position-dependence of the scattering speckle pattern (static nonergodicity) came from large voids formed during the sol-gel transition. With proper preparation, such nonergodicity is completely removed, indicating that the static nonergodicity generally observed in a gel is not intrinsic, but comes from clustering "island" structures formed during the gelation process.

Introduction

A gel is generally considered to be a three-dimensional polymer network swollen by a large amount of solvent. Polymer chains in chemical gels are crosslinked by covalent bonds. A chemical gel is normally formed by a copolymerization of monomers with a crosslinking agent.[1-4] In contrast, the formation of a physical gel usually starts with a polymer solution. In physical gels, polymer chains are interconnected via physical interactions (e.g., hydrogen bonding, electrostatic interaction, and coordination) and the gelation is often induced by an alternation of physical conditions, such as temperature and ionic strength.[5] It has been well known that in the formation of chemical gels, short linear chains are first formed, then large branched clusters, and finally a network in which the interconnected clusters are looked like "islands" in a three-dimensional solvent "sea."[6-10] Such an inhomogeneous clustering structure should be absent when long polymer chains uniformly distributed in a solution are randomly associated to form a physical gel.

Recently, it has been shown that there also exists a scattering speckle pattern when an aqueous mixture of poly(vinyl alcohol) and Congo red is gelled at 43 °C or below.[5] The question is whether this static nonergodicity is intrinsic for all thermally reversible physical gels or only for this particular one. In order to have a better understanding of

 DOI: 10.1002/masy.200450304

the gelation of physical gels, we have developed a novel hydrogel made of billions of thermally sensitive poly(*N*-isopropylacrylamide) (PNIPAM) microgels that jam-pack when they swell at ~25 °C or lower temperatures. Inside each microgel, the polymer chains are *chemically* crosslinked, but the interaction between the microgels is purely *physical*. This is why it is named "hybrid". It resembles a glass in which spherical swollen soft microgels replaces small atoms or molecules. It is actually formed via a volume concentration-induced gelation.

There are several advantages of using this hybrid gel as a model system to study the sol-gel transition and structures of physical gels. Advantages include: the structural inhomogeneity from chemical crosslinking is limited to length scales much smaller than the wavelength of our laser light, so that individual swollen microgels are transparent; the microgel building blocks are narrowly distributed and well characterized; and gelation is completely thermally reversible without any hysteresis. Using a combination of static and dynamic laser light scattering, we studied a series of such hybrid gels formed at different gelling rates and temperatures. It has been found that the observed scattering speckle pattern in thermally reversible physical gels is not intrinsic, but strongly depends on whether large voids were formed during gelation.

Experimental

Narrowly distributed spherical poly(*N*-isopropylacrylamide) (PNIPAM) microgels were prepared by emulsion polymerization of *N*-isopropylacrylamide in the presence of *N,N*'-methylenebis-(acrylamide) as crosslinker and anionic surfactant (sodium dodecyl sulfate, SDS) as dispersant. The unreacted monomers and SDS were removed by a three successive centrifugation-decantation-redispersion cycles in deionized water. The zeta potential of the resultant dispersion was close to zero. The synthesis has been detailed before [11,12]. A commercial LLS spectrometer (ALV/SP-125) equipped with an ALV-5000 multi-τ digital time correlator and a solid-state laser (DPSS, output power ≅ 400 mw at λ = 532 nm) was used in this study. The details of LLS instrumentation and theory can be found elsewhere.[13-15]

Results and Discussion

Figure 1 shows that in the gel state, the time average scattered intensity $\langle I \rangle_T$ randomly varies with the sample position. In contrast, the scattering speckle pattern disappears at

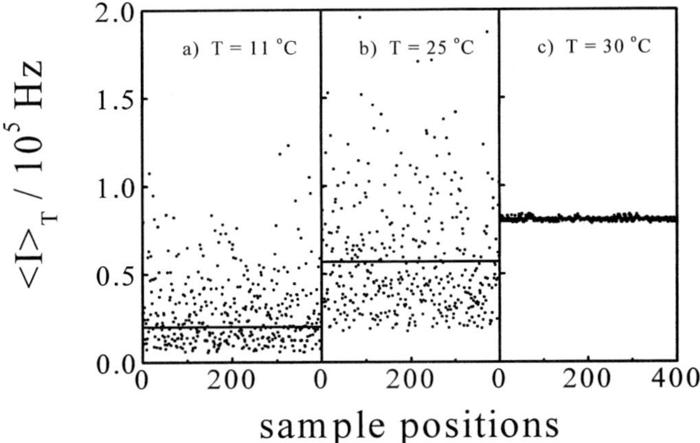

Figure 1. Sample position dependence of time-average scattered intensity $<I>_T$ of hybrid PNIPAM gel after the microgel dispersion was quickly cooled down from 40 °C to different setting temperatures, where each solid line represents an ensemble-average intensity $<I>_E$ defined as $(\sum_i <I>_{T,i})/N$ with N being the total number of the sample positions measured.

30 °C because the bulk hybrid gel melts into a microgel dispersion (i.e., the sol state) due to the shrinking of individual microgels. It is worth noting that in the temperature range 15-33 °C, the microgels in a dilute dispersion can reversibly swell and shrink with a size change between ~45 nm and ~106 nm, corresponding to a 13-fold variation in volume. This is why swelling could induce the sol-gel transition. Note that $<I>_E$ increases with the temperature, very differently from previous observations in which $<I>_E$ decreased when a physical gel melted into individual less scattered polymer chains,[5] but similarly to a concentration induced sol-gel transition. This difference can be attributed to the fact that in this study, individual collapsed microgels are stronger scattering objects than the swollen gel network, especially when the temperature approaches its LCST.

Besides $<I>_T$, for each chosen sample position, we also measure a normalized intensity-intensity time correlation function S(q, τ) defined as[16,17]

$$\mathbf{S(q,\tau)} = \frac{\langle \mathbf{I(q,0)I(q,\tau)} \rangle}{\langle \mathbf{I(q,0)} \rangle^2} - \mathbf{1} \qquad (1)$$

where q is the scattering vector and τ is the delay time. We found that in the gel state, the plot of "log[S(q, τ)] vs log(τ)" had a typical power-law behavior with a critical exponent close to ~0.8, higher than ~0.4 reported in [5], indicating that the hybrid gel better resembles a glass.[18] As shown in [16,17,19,20], for a nonergodic gel, $\langle I\rangle_T$ is the summation of the static component $\langle I_s\rangle_T$ and the dynamic component $\langle I_d\rangle_T$, so that

$$S(q,\tau) = \left(\frac{\langle I_d\rangle_T}{\langle I\rangle_T}\right)^2 \exp(-2Dq^2\tau) + 2\left(\frac{\langle I_d\rangle_T}{\langle I\rangle_T}\right)\left[1 - \left(\frac{\langle I_d\rangle_T}{\langle I\rangle_T}\right)\right]\exp(-Dq^2\tau) \qquad (2)$$

where D is the collective diffusion coefficient. Therefore, for a nonergodic system at a given q, one has four different scattered intensities, $\langle I\rangle_T$, $\langle I_s\rangle_T$, $\langle I_d\rangle_T$ and $\langle I\rangle_E$, but the former three are not independent. $\langle I\rangle_T$ and $\langle I_s\rangle_T$ varies with the sample position, but not $\langle I\rangle_E$ and $\langle I_d\rangle_T$ by their definitions. Only in an ergodic system is $\langle I_s\rangle_T = 0$, so that we have $\langle I\rangle_T = \langle I_d\rangle_T = \langle I\rangle_E$. Experimentally, the initial slope of "lnS(q, τ) vs τ" leads to an apparent diffusion coefficient D_A in the range of $D/2 < D_A < D$. It has been shown that D and $\langle I_d\rangle_T$ are related to D_A and $\langle I\rangle_T$ as[2,7]

$$\frac{\langle I\rangle_T}{D_A} = \frac{2\langle I\rangle_T}{D} - \frac{\langle I_d\rangle_T}{D} \qquad (3)$$

For each chosen sample position at a given q, we can measure one $\langle I\rangle_T$ from static LLS and calculate one D_A from S(q, τ) measured in dynamic LLS. Figure 2 shows that D_A decreases as $\langle I\rangle_T$ increases and approaches a constant when $\langle I\rangle_T$ is sufficiently high. The inset is a corresponding plot on the basis of eq 3. A least-squares fit of the data leads to D and $\langle I_d\rangle_T$, respectively, from its slope and intercept.

Figure 3 shows that when T < 25 °C, $\langle I\rangle_E$ increases with the temperature, but $\langle I_d\rangle_T$ remains nearly constant; while in the higher temperature range, $\langle I_d\rangle_T$ increases sharply and catches $\langle I\rangle_E$ at ~30 °C, indicating a progressive melting of the hybrid gel into individual collapsed microgels, i.e., from a nonergodic bulk gel to a completely ergodic microgel dispersion. The extrapolations of $\langle I_d\rangle_T$ in the two temperature regions leads to a temperature close to the gelation temperature (T_{gel}) observed in a flow test. Note that a high value of $\langle I\rangle_E$ was often used in the past as an indication of the inhomogeneity.

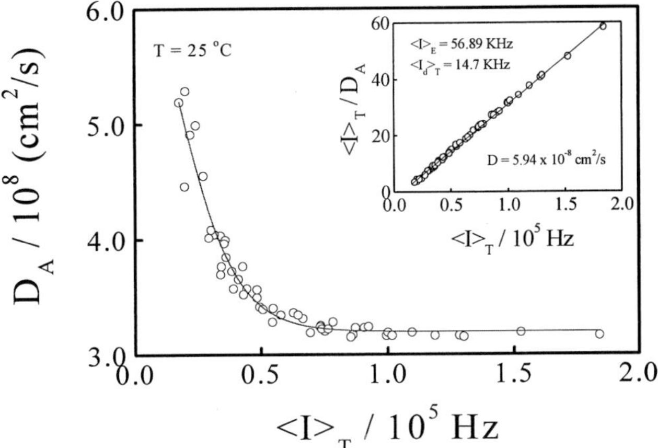

Figure 2. Time-average scattered intensity $<I_T>$ dependence of apparent diffusion coefficient D_A of hybrid PNIPAM gel at 25 °C, where D_A was obtained from the initial slope of the intensity-intensity time correlation function. The inset shows a decomposition plot on the basis of Eq 3.

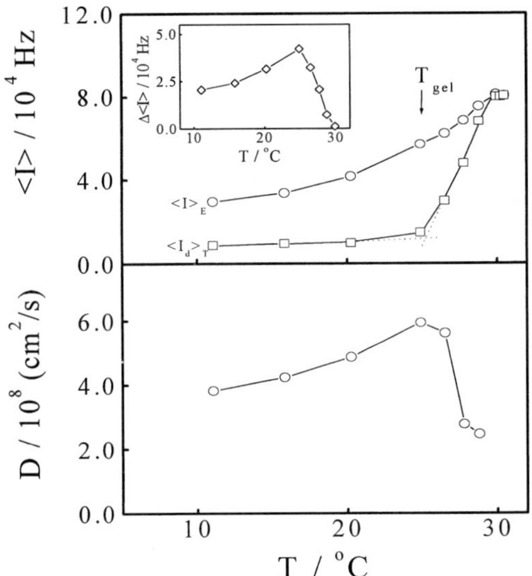

Figure 3. Temperature dependence of ensemble-average scattered intensity $<I>_E$, time average dynamic component of scattered intensity $<I_d>_T$ and collective diffusion coefficient D of hybrid PNIPAM gel. The arrow shows the gelation temperature T_{gel} observed by a flow test. The inset is the temperature dependence of the intensity difference (ΔI) between $<I>_E$ and $<I_d>_T$.

However, our result tells a different story because we know that the microgel dispersion at ~30 °C is homogeneous and ergodic even though $<I>_E$ is higher.

Careful examination of Figure 3 reveals that *it should be the difference between* $<I>_E$ *and* $<I_d>_T$ *that reflects the extent of the inhomogeneity of a give system, not* $<I>_E$ *alone.* The inset in Figure 3 shows that even before reaching the gelation threshold, the system already becomes inhomogeneous, i.e., $\Delta I>0$. As the temperature decreases, the intensity difference ΔI reaches a maximum at T_{gel} and then decreases in the lower temperature range. This is because large voids are formed at the gelation threshold and further swelling of individual microgels squeezes the voids between the jam-packed microgels, leading to a more uniform hybrid gel. The temperature dependence of D shows a similar tendency. The jump-wise decrease in D near T_{gel} was also observed for other physical gels,[5] but not for chemical gels,[2] probably because the formation of a chemical gel is irreversible and the size of the clusters always increases during gelation. In the present case, as the temperature increases, individual microgels shrink so that non-jammed microgels diffuse rapidly. Above the melting temperature, the gel network breaks into slowly diffusing large clusters so that D decreases. It is worth noting that the value of D is close to that of individual microgels in a dilute dispersion. Since individual swollen microgels are transparent to the light used, the inhomogeneities observed in Figure 1 must come from an imperfect packing of the microgels when they were suddenly jammed together. Therefore, if we slow down the cooling process, the swollen microgels would have time to arrange themselves into a more uniform structure. Ideally, if individual collapsed microgels could be closely stacked together at a high temperature before cooling them down, we might be able to obtain a uniform hybrid gel. Figure 4 shows a schematic of how the gel structure could be influenced by the gelation condition. The different inhomogeneities shown in Figure 4 have been confirmed in terms of the scattering speckle pattern and the difference between $<I>_E$ and $<I_d>_T$, as shown in Figure 5. It should be stated that the hybrid gel prepared under centrifugation showed a color if is was illuminated with white light even, though it could not completely disperse the white light as a prism or a crystal. This imperfect dispersion might be attributed to the polydispersity in the microgel size and a small refractive index contrast between water and smollen microgel networks, but it clearly indicates a certain degree of ordering of the microgels inside the hybrid bulk gel.

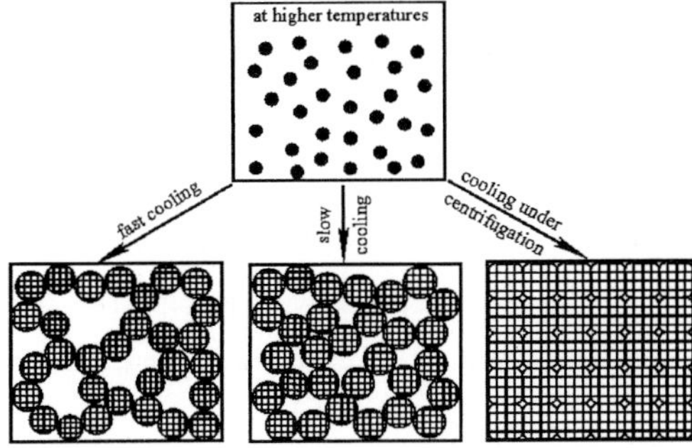

Figure 4. Schematic of the gelation condition dependence of the structure of the hybrid gel made of spherical thermally sensitive swollen PNIPAM microgels.

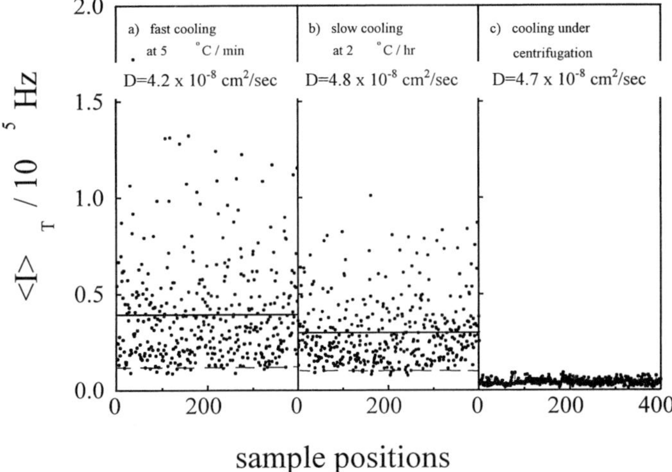

Figure 5. Sample position dependence of time-average scattered intensity $<I>_T$ of hybrid PNIPAM gel after the microgel dispersion was cooled down from 40 °C to 15 °C under different conditions, where the solid and dashed lines represent $<I>_E$ and $<I_d>_T$, respectively.

Note that Figures 1c and 5c are similar. Both of them display an ergodic behavior, i.e., $<I>_E = <I_d>_T$. However, it is worth noting that they were obtained from two completely different states; namely, Figure 1c shows a measurement of a uniform and ergodic microgel dispersion in which individual collapsed microgels are under random Brownian motions in water, while Figure 5c is from a uniform hybrid gel. To our knowledge, this is the first observation of such a uniform and ergodic hybrid gel.

Figure 5 clearly demonstrates that the static inhomogeneity of thermally reversible physical gels is not intrinsic, but is very strongly dependent of the gelation process. It is also interesting to note that D is nearly a constant, which indicates that the voids inside the gel network are much larger than the microgels and the gel network does not affect the relaxation of individual non-jammed microgels. As expected, for a given temperature, individual non-jammed microgels with the same size relax at the same rate. How the gel network is formed can only affects the static part, i.e., $\langle I_d \rangle$, but not D. This point has been overlooked in the past.

Conclusion

It is not difficult to imagine that below the gelation threshold, swollen clusters look like "islands" floating in a solvent "sea", but after gelation, the interconnected clusters become a "continent" (background) and the "sea" changes into many "lakes" (voids).[8,9,21,22] It is these large and non-uniform voids that lead to the scattering speckle pattern, and static nonergodicity. Our results suggest that this kind of static nonergodicity in both physical and chemical gels might have a similar origin. The only difference is that in chemical gels, the clusters are known to be branched polymer chains and microgels, but in physical gels, they could be formed via a non-uniform interchain association of polymer chains prior to the gelation threshold.

Acknowledgements. The financial support of the NNSF Fund (29974027), the CAS Bei Ren Project, and the Research Grants Council of the Hong Kong Special Administration Region Earmarked Grant (CUHK 4266/00P, 2160135) are gratefully acknowledged.

[1] E. Sato-Matsuo, M. Orkisz, S. T. Sun, Y. Li and T. Tanaka, *Macromolecules* **1994**, *27*, 6791 and references therein.
[2] M. Shibayama, T. Norisuye and S. Nomura, *Macromolecules* **1996**, *29*, 8746.
[3] M. Shibayama, S. Takata and T. Norisuye, *Physica A* **1998**, *249*, 245.
[4] M. Shibayama, Macromol. *Chem. Phys.* **1998**, *199*, 1.
[5] F. Ikkai and M. Shibayama, *Phys. Rev. Lett.* **1999**, *82*, 4946.
[6] M. Shibayama, M. Morimoto and S. Nomura, *Macromolecules* **1994**, *27*, 5060.
[7] M. Shibayama, Y. Fujikawa and S. Nomura, *Macromolecules* **1996**, *29*, 6535.
[8] C. Wu, J. Zuo and B. Chu, *Macromolecules* **1989**, *22*, 633.
[9] C. Wu, J. Zuo and B. Chu, *Macromolecules* **1989**, *22*, 838.
[10] C. Wu, B. Chu and G. Stell, *Makromol. Chem., Macromol. Symp.* **1989**, *45*, 75.
[11] S. Q. Zhou and B. Chu, *J. Phys. Chem. B* **1998**, *102*, 1364.

[12] C. Wu and S. Q. Zhou, *J. Polym. Sci., Polym. Phys. Ed.* **1996**, *34*, 1597.
[13] C. Wu and S. Q. Zhou, *Macromolecules* **1995**, *28*, 3375.
[14] B. Chu, *Laser Light Scattering* (Academic Press: New York, 1991), 2nd ed.
[15] B. Berne; R. Pecora, *Dynamic Light Scattering* (Plenum Press: New York, 1976).
[16] P. N. Pusey and W. van Mergen, *Physica A* **1989**, *157*, 705.
[17] J. G. H. Joosten, J. L. McCarthy and P. N. Pusey, *Macromolecules* **1991**, *24*, 6690.
[18] J. E. Martin, J. Wilcoxon and J. Odinek, *Phys. Rev. A* **1991**, *43*, 858.
[19] T. Tanaka, L. O. Hocker and G. B. Benedek, *J. Chem. Phys.* **1973**, *59,* 5151.
[20] F. Horkay. *Macromolecules* **1993**, *26*, 3375.
[21] T. Norisuye, M. Shibayama, S. Nomura, *Polymer* **1998**, *39,* 2769.
[22] T. Norisuye, M. Inoue, M. Shibayama, R. Tamaki, Y. Chujo, *Macromolecules* **2000**, *33*, 900.

Nanostructured Polymer Gels

Zhibing Hu

Departments of Physics and Chemistry, University of North Texas
P. O. Box 311427, Denton, TX 76203, USA

Summary: A brief review of our recent work on polymer gel opals is given in memory of late Professor Tanaka. The central idea is to first synthesize monodisperse polymer gel nanoparticles, then self-assemble them into a 3D network, and eventually covalently bond them. The covalent bonding contributes to the structural stability, while self-assembly provides them with crystal structures that diffract light, resulting in colors. N-isopropylacrylamide (NIPA) gel has been used as a model system for this study. The nanostructured NIPA gel, which contain up to 97 wt % water, displays a striking iridescence like precious opal but are soft and flexible like gelatin.

Introduction

With his unique story-telling style, late Professor Tanaka has shown us that polymer gels can swell or collapse reversibly in response to external stimuli.[1-3] His pioneering research not only opened a door for fundamental scientific study [4-7] and practical applications [8-14] but also inspired a new generation of polymer scientists. Here a brief review of our recent work [15-17] on nano-structured polymer gels is given in memory of late Professor Tanaka.

Polymer gels are a class of macromolecular networks that contain a large fraction of solvent such as water within their structure.[18] Polymer gels are usually formed by the free radical polymerization of monomers in the presence of a difunctional crosslinking agent and a solvent. They can be made either in bulk or in nano- or micro-particles. The bulk gels are easy to handle, but usually have very slow swelling rate and amorphous structures arising from randomly crosslinked polymer chains, while the gel nanoparticles react quickly to an external stimulus, but are too small for some practical applications. Recently, creating nanostructures in gels has attracted attentions. Such structures include gels with embedded self-assembled solid polymer spheres [19-20] and the complex formation of polyelectrolyte gels with oppositely charge surfactants. [21]

We have proposed to synthesize bulk gels by covalently bonding self-assembled gel nanoparticles (also called colloidal gel particles or microgels).[15-16] The covalent

 DOI: 10.1002/masy.200450305

bonding contributes to the structural stability, while self-assembly provides them with crystal structures that diffract light, resulting in colors. Such stable periodic 3D structures allow us to obtain useful functionality not only from the constituent gel building blocks but also from the long-range ordering that characterizes these structures.

A nanostructured polymer gel should have a two-level structural hierarchy as shown in Figure 1 (left panel): The primary network (I) consists of crosslinked polymer chains inside each nanoparticle, while the secondary network (II) is a crosslinked system of the nanoparticles. The mesh sizes of the primary and the secondary networks are typically around 1-10 nm and 10-500 nm respectively. The networks could have randomly packed structure (left panel, Fig. 1) or periodic structures (right panel, Fig. 1).

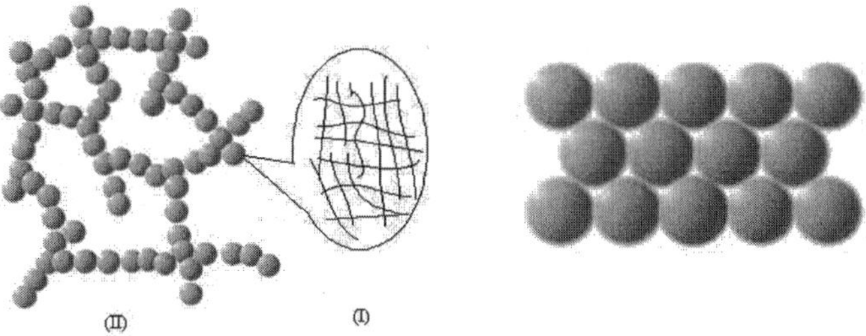

Figure. 1. Conceptual models. Left panel: The model of the two-level structural hierarchy for all gel nanoparticle networks: the primary network (I) consists of crosslinked polymer chains inside each nanoparticle, while the secondary network (II) is a system of the crosslinked nanoparticles. Right panel: The networks could have periodic structures.

We have chosen N-isopropylacrylamide (NIPA) gel as a model system for this study. Since Tanaka and co-workers' pioneering work,[3] the NIPA gel has been extensively studied in various structural forms including macrogel,[4-5, 11] porous gel,[12] and microgels.[22-25]

Experimental

N-isopropylacrylamide (NIPA) nanoparticles and two of its derivatives, with various particle sizes, were synthesized using an emulsion polymerization method:[22] One derivative was NIPA co-polymerized with acrylic acid (AA) and the other was NIPA co-polymerized with 2-hydroxyethyl acrylate (HEAc). The NIPA had thermally responsive properties,[3] while the AA and the HEAc provided carboxyl (-COOH) and hydroxyl (-OH) groups, respectively, for the crosslinking sites. The pure NIPA nanoparticles were used as a model system for the study of colloidal crystallization and the NIPA derivative nanoparticles were used as building blocks for synthesis of nanoparticle networks. The size distribution of nanoparticles was characterized using a light scattering spectrometer (ALV-5000). The hydrodynamic radius of the resultant nanoparticles in water was narrowly distributed with a size variance of about 1%.

Scheme 1

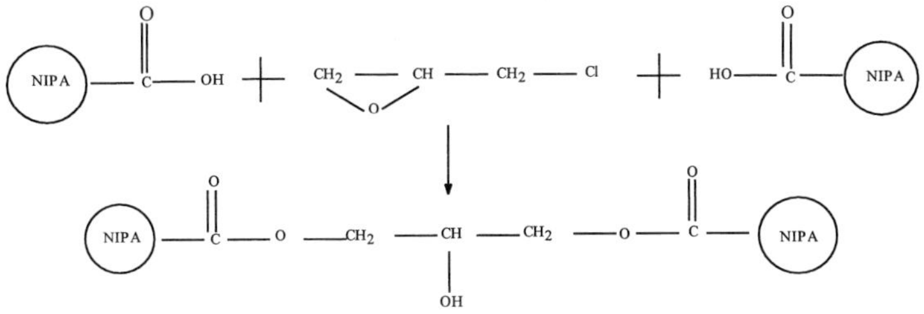

NIPA—C(=O)—OH + CH_2—CH($-$O$-$)—CH_2—Cl + HO—C(=O)—NIPA

↓

NIPA—C(=O)—O—CH_2—CH(OH)—CH_2—O—C(=O)—NIPA

Scheme 2

NIPA—OH + CH_2=CH—SO_2—CH=CH_2 + HO—NIPA

↓ OH^-

NIPA—O—CH_2—CH_2—SO_2—CH_2—CH_2—O—NIPA

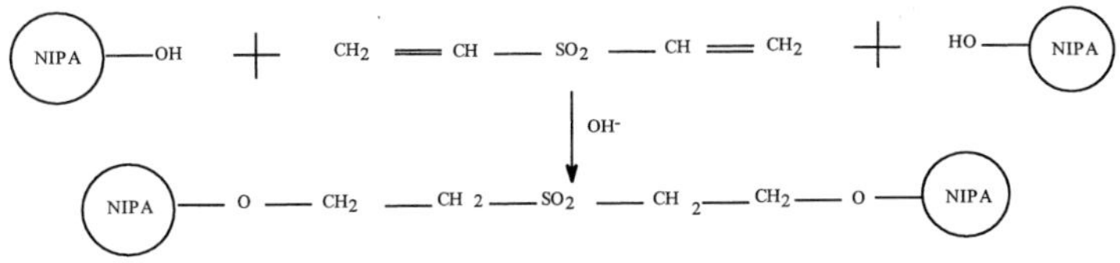

We have developed two schemes for covalently bonding nanoparticle assemblies in either organic or aqueous media. In the first scheme, the NIPA-AA nanoparticles in water were first precipitated by adding acetone. These concentrated nanoparticles were transferred to a mixture of acetone/epichlorohydrin (1:3 volume ratio) to form a short-range-ordered structure by steric and electrostatic repulsion. After incubation at 98 °C for 10 h, the nanoparticles formed a three-dimensional network. This network was immersed in acetone and finally in de-ionized water for thorough washing. In the second scheme, the NIPA-HEAc nanoparticles first self-assembled into different structural arrays at various concentrations in an aqueous solution (pH=12). Then, divinylsulfone (DVS) [12] was added to bond them at room temperature. The second scheme can be carried out directly in water and is particularly useful for hydrogels. Since the colloidal particles are linked together by covalent bonding, they cannot be re-dispersed into a solution, in contrast to well-known colloidal aggregates. It is noted that the crosslinkers can react not only with the functional groups of different nanoparticles but also with those of the same particles. It will be an interesting study to find the ratio of intra- versus inter-particle crosslinking.

Results and Discussion

Characterization of NIPA nanoparticles. Figure 1 shows a typical hydrodynamic radius distribution of NIPA nanoparticles in dilute dispersions at 25 and 40 °C respectively, where dynamic light scattering measurements were done at an angle of 20 degrees. [17] At temperatures lower than the phase transition temperature ($T_c \sim 34$ °C), the microgels are in their swelling state and are narrowly distributed. The average hydrodynamic radius ($<R_h>$) at 25 °C is 132 nm. At 40 °C, the nanoparticles shrink sharply ($<R_h> = 49.8$ nm with a narrower size distribution) but are fully dispersed.

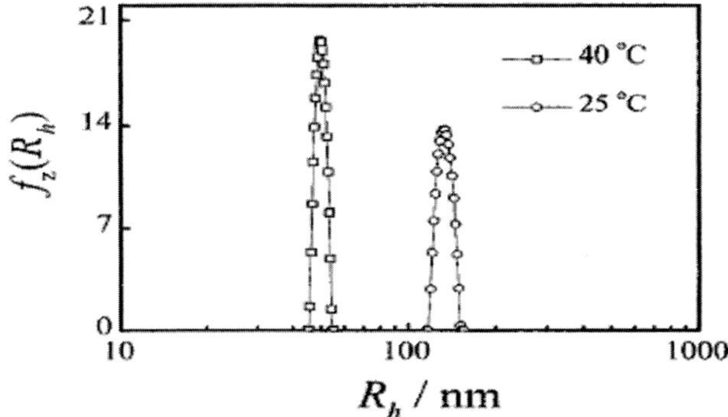

Figure 2. Hydrodynamic radius distributions ($f(R_h)$) of NIPA nanoparticle spheres in water at $T = 25.0$ °C (circles) and 40.0 °C (squares), respectively, where $C = 1.37 \times 10^{-5}$ g/g, and the scattering angle is 20°.[17]

Self-assembling of NIPA nanoparticles in water. As the concentration of NIPA nanoparticle dispersion increases at room temperature (18 to 22 °C, below T_c), all spheres can be close-packed, resulting in remarkable change in rheological and optical properties. From light scattering, optical and turbidity measurements, a schematic "phase diagram" is constructed for a very broad distribution of polymer concentrations ranging from 0.01 to 14 wt % as shown in Fig. 3. The assemblies exhibit different structures as the polymer concentration changes: the *liquid* phase (Fig. 3. (A-C)) where spheres are free to move around; the *crystal* phase (D-E) where the freedom of the spheres is limited and the particles can self-assemble into a long-range ordered state, and the *glass* phase (F-I) where the movement of the spheres is highly limited.

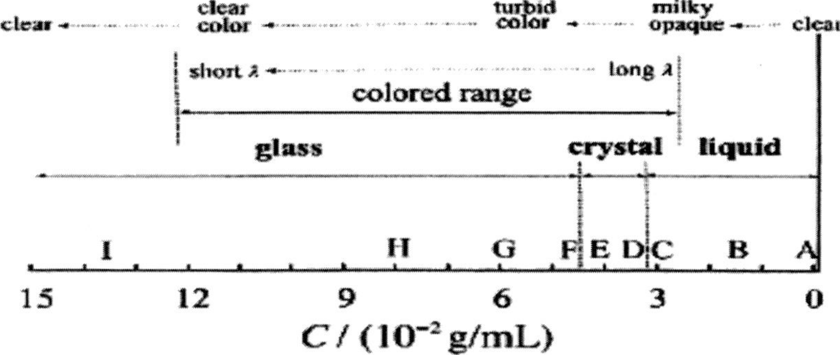

Figure 3. Schematic "phase diagram" of the NIPA colloidal water dispersions as a function of polymer concentration C (wt%).[17] The average hydrodynamic radius of NIPA spheres in water at 25 °C is 132 nm.

The corresponding optical appearance changes first from clear to opaque in dilute dispersion, then to pink, green, blue and blue-violet color in the concentrated dispersions, and eventually to transparent in highly concentrated dispersion. The

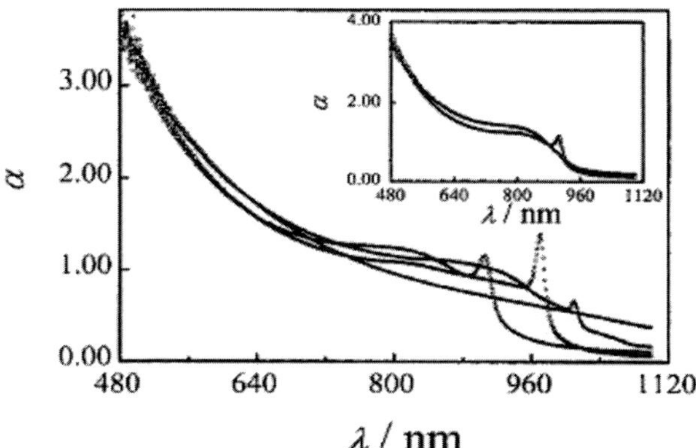

Figure 4. The turbidity is plotted as a function of wavelength for NIPA microgel dispersions at various C, where the average hydrodynamic radius of NIPA microgel spheres in dilute water dispersions at 25 °C is 216 nm. [17] The curves with a peak from left to right correspond to C=4.47x10^{-2}, 3.65x10^{-2}, and 3.21x10^{-2} g/g, respectively. The curve without a peak corresponds to C=1.99x10^{-2} g/g. The inset compares the turbidity curves before (with a peak) and after shear melting (without a peak) for the sample of C=4.47x10^{-2} g/g.

formation of crystals around a concentration of about 3 ~5 wt % adds iridescent patterns to the background colors.

The optical appearance of the NIPA nanoparticle dispersions can be better revealed by turbidity measurements as shown in Figure 4 for polymer concentration between ~ 3 and ~ 5 wt%. Here NIPA microgel spheres have not only strong interaction but also enough freedom to form large colloidal crystals at room temperature. These crystals are easy to observe due to their iridescent patterns. The turbidity exhibits either a sharp peak or a sudden increase with decrease of wavelength, as shown in Figure 4. The iridescent patterns originate from Bragg diffraction. Constructive interference occurs if Bragg condition: $2nd\sin\theta = m\lambda$, is satisfied, where d, θ, n, λ, m are the lattice spacing, the diffraction angle, the refractive index of the gel medium, the wavelength of light in vacuum and an integer, respectively. These crystals are not strong and can be easily destroyed by shaking the dispersion as shown in the inset of Figure 4, where turbidity curves are compared before and immediately after shear melting of the crystals.

We consider that the color of the dispersions originates from interference that relates to the ordered inter-particle distance. Let us define λ_c as the transition wavelength corresponding to the peak or the middle height of the shoulder in the turbidity curve. From Figure 4 and other measurements (not listed), λ_c as a function of polymer concentration was measured and shown in Figure 5. Here, we have converted the

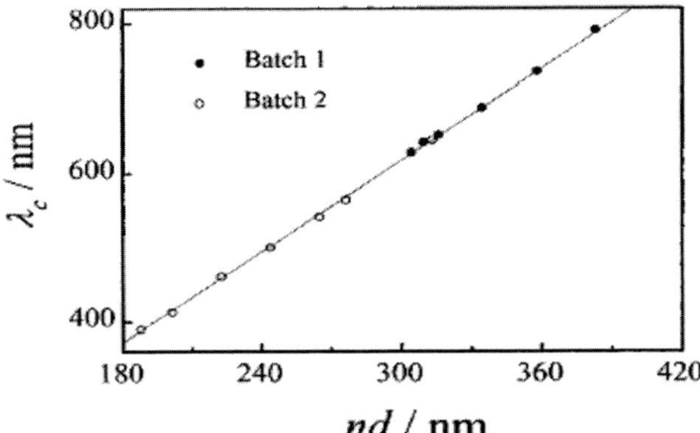

Figure 5. Influence of the inter-particle light distance nd on λ_c, at which the turbidity exhibits a peak or a shoulder-shape increase with the decrease of wavelength.[17] Two Batches of samples are combined together: Batch 1 (solid circles), R_h = 132 nm and Batch 2 (open circles), R_h = 216 nm at 25 ^{0}C. The line represents the least square fitting of λ_c (nm) = 2.04nd + 4.1 where n is the refractive index of the microgel.

concentration of the samples into the inter-particle distances by considering that the microgel spheres form face-centered cubic structured as revealed by small angle neutron scattering. [26] The increase in weight fraction of microgel spheres reduces the inter-particle distance so that λ_c shifts to a lower wavelength. Because of suitable size range of our samples, λ_c can be tuned to cover whole visible light wavelength range. It is also noted that λ_c is larger for dispersions with the same weight fraction but larger particle size.

The least square linear fit of λ_c with inter-particle optical path length (*nd*) in Figure 5 leads to the relation: λ_c (nm) = 2.04nd + 4.1. Here we have taken into account of the refractive index (n) change with the weight fraction of microgel spheres using weight additive method, $n = 1.6C + 1.33(1\text{-}C)$, where 1.6 and 1.33 are refractive indexes of PNIPA and water respectively, and C is the concentration. λ_c well satisfies the Bragg diffraction condition of $2nd\sin\theta = m\lambda_c$ at $\theta = 90°$ and $m = 1$.

The properties of NIPA nanoparticle networks. The covalent linkages between nanoparticles lead to the remarkable thermal stability of the crystal structure. Figure 6 shows the temperature dependence of a crystal hydrogel in a tube after polymerization with Scheme 2. The iridescent pattern (left panel) at room temperature became invisible when the sample was heated to 50 °C, at which point the gel became cloudy due to phase separation (right panel). When the sample at 50 °C was cooled back to 21 °C, the pattern reappeared within 10 s, and this process was reversible.

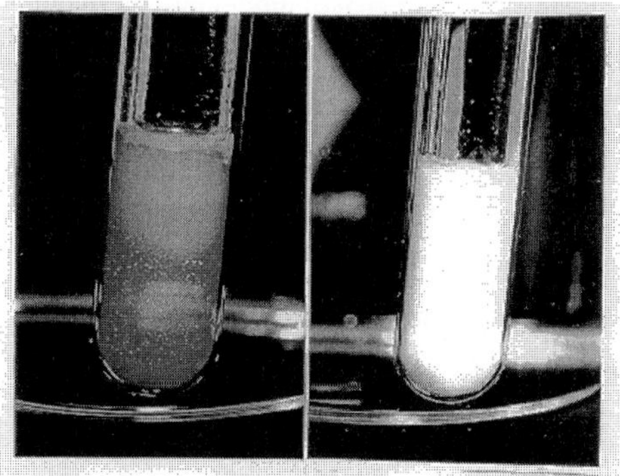

Figure 6. The NIPA crystal hydrogels made according to Scheme 2.[16] The sample in the tube is transparent and exhibited an iridescent pattern at 21 °C (left) until the temperature was raised to 50 °C (right), at which point the pattern became invisible. As the sample at 50 °C was cooled back to 21 °C, the patterns reappeared within 10 s.

In contrast, a non-crosslinked nanoparticle assembly with the same concentration was completely disrupted as the temperature was raised to 30 °C. It required about 1000 times longer for a non-crosslinked assembly to re-assemble into a crystal structure. It is apparent that disturbed nanoparticles in a crosslinked assembly are able to return to their equilibrium crystalline positions quickly through restoring forces provided by the network elasticity. The fast response rate of the crystal hydrogels could be a major advantage for developing sensor or display technologies over conventional colloidal crystal arrays.

It is noted from the left panel in Figure 6 that the crystal hydrogel in water displays an iridescent color with good transparency and no sedimentation (without adding an index-matching or a density-matching fluid). This is because the building blocks are hydrogel nanoparticles that contain up to 97 wt% water; both the refractive index and the density of the particles are nearly matched with those of the surrounding water.

Conclusion

Along the avenue paved by the late Professor Tanaka and coworkers on NIPA gels, we have synthesized NIPA gel opals by covalently bonding self-assembled gel nanoparticles. The major steps consist of (1) synthesizing monodisperse polymer gel nanoparticles, (2) self-assembling them into a 3D network, and (3) covalently bonding these particles. Covalent bonding contributes to the mechanical and thermal stability of the assemblies, while self-assembly provides the bulk hydrogels with crystal structures which diffract light, resulting in colors. The polymer nanoparticle networks could be used as a directly observable (in 3D real space) model system to study a wide variety of fundamental phenomena such as crystallization, phase transition, and fracture mechanics. Creating such nanostructured materials is also of importance in technological applications including controlled drug delivery, biomaterials, sensors, devices, bio-adhesives, and displays.

Acknowledgment This work is supported by the National Science Foundation under Grant No. DMR-0102468, the U.S. Army Research Office under Grant No. DAAD 19-01-1-0596. I thank Drs. J. Gao and X. H. Lu for their contributions to this project.

[1] T. Tanaka, *Phys. Rev. Lett.* **1978**, 40, 820.
[2] T. Tanaka, I. Nishio, S. T. Sun, S. Ueno-Nishio, *Science* **1982**, 218, 467.
[3] S. Hirotsu, Y. Hirokawa, T. Tanaka, *J. Chem. Phys.* **1987**, 87, 1392.
[4] M. Shibayama, T. Tanaka and C. C. Han, J. Chem. Phys. **1992,** 97, 6829; J. Chem. Phys. **1992**, 97, 6842.
[5] G. Liao, Y. Xie, K. F. Ludwig, R. Bansil, and P. Gallagher, Phys. Rev. E **1999**, 60, 4473.
[6] Y.L. Yin and F. Stanley and R.K. Prud'homme, In Polyelectrolyte Gels, R. Harland and R.K. Prud'homme, eds., ACS Symposium Series, **1992**, 480.
[7] J.P. Baker, H.W. Blanch and J.M. Prausnitz, Polymer **1995**, 36, 1061.
[8] R. A. Siegel, B. A. Firestone, *Macromolecules* **1988**, 21, 3254.
[9] Y. Osada, H. Okuzaki, H. Hori, *Nature* **1992**, 355, 242.
[10] G. Chen, A. S. Hoffman, *Nature* **1995,** 373, 49.
[11] Y. Osada, and J. P. Gong, *Adv. Mater.* **1998**, *10,* 827.
[12] D. C. Harsh, S. H. Gehrke, *J. Controlled Release* **1991**, 17, 175.
[13] Z. Hu, X. Zhang, Y. Li, *Science* **1995**, 269, 525.
[14] M. J. Snowden, M. J. Murray and B. Z. Chowdry, *Chemistry & Industry* **1996**, 531.
[15] Z. Hu, X. Lu, J. Gao, and C. Wang, *Advanced Materials* **2000**, 12, 1173.
[16] Z. Hu, X. Lu and J. Gao, *Advanced Materials* **2001,** 13, 1708.
[17] J. Gao and Z. Hu, *Langmuir* **2002**, 18, 1360
[18] A. S. Hoffman, *J. Control. Release* 6, 297 (1987).
[19] J. M. Weissman, H. B. Sunkara, A. S. Tse, S. A. Asher, *Science* **1996**, 274, 959 .
[20] J. H. Holtz, S. A. Asher, *Nature* **1997**, 389, 829.
[21] E. Sokolov, F. Yeh, A. R. Khokhlov, V. Y. Grosberg and B. Chu, *J. Phys. Chem. B* **1998**, 102, 7091.
[22] R. H. Pelton, P. Chibante, *Colloids and Surfaces* **1986**, 20, 247.
[23] C. Wu and S. Zhou, *Macromolecules* **1996**, *29*, 1584.
[24] H. Senff, W. Richtering, *J. Chem. Phys.* **1999**, 111, 1705.
[25] J. D. Debord and L. A. Lyon, *J. Phys. Chem.* **2000**, *104*, 6327.
[26] T. Hellweg, C.D. Dewhurst, E. Bruckner, K. Kratz, and W. Eimer, *Colloid Polym. Sci.* **2000**, *278*, 972.

Unresolved Issues in the Preparation and Characterization of Thermoresponsive Microgels

Robert Pelton

Department of Chemical Engineering, McMaster University, Hamilton, ON, L8S 4L7, Canada
E-mail: peltonrh@mcmaster.ca

Summary: The preparation and properties of colloidal thermal responsive microgels based on poly(N-isopropylacrylamide) are described with an emphasis on the unresolved issues which will impair commercial implementation. The main difficulties are: 1) removal of sol, surfactants and other impurities; 2) obtaining colloidally stable particles less than 50 nm in diameter; 3) synthesis of concentrated, colloidally stable microgel dispersions; 4) accurate measurement of water content; and, 5) control of microgel particle morphology.

Introduction

Poly(N-isopropylacrylamide), PNIPAM, is the cornerstone of many current approaches to the development of controllable materials for a variety of potential applications. This popularity stems from the fact that PNIPAM is water soluble at room temperature but phase separates when heated above 32 °C.[1] Thus temperature change is the trigger causing dramatic changes in the properties of PNIPAM solutions.

Most of the published studies involve PNIPAM hydrogels manifest temperature sensitivity as a temperature-dependent degree of swelling. The temperature corresponding to the most rapid change in swelling is called the volume phase transition temperature. The transition temperature can be raised by copolymerizing with acrylamide, or lowered by employing more hydrophobic monomers. Macroscopic PNIPAM gels were first described in the scientific literature by Tanaka.[2] Monodispersed colloidal microgels (these days we would call them nanogels) were first prepared by Pelton and Chibante in 1978.[3] Both macro and microgels can display nearly a ninety percent decrease in volume upon heating, giving spectacular changes in most properties. The number of PNIPAM gel papers seems to be growing geometrically, if not exponentially and there have been a number of excellent reviews.[4-8] The objective of this paper is to review the properties of PNIPAM thermoresponsive microgels and to highlight the most important unresolved issues. These have been largely ignored by most researchers including ourselves. Many of these issues are also relevant to macrogel preparation and characterization.

 DOI: 10.1002/masy.200450306

An Overview of Microgel Preparation and Characterization

N-isopropylacrylamide (NIPAM) is the major building block for temperature sensitive microgels. With a structure close to acrylamide, many of the properties of NIPAM are similar to those of acrylamide. In aqueous solution, it undergoes rapid free radical polymerization to give high molecular weight polymers at rates similar to that of acrylamide.[1-11] The first published account of PNIPAM-based microgels described a "surfactant-free emulsion polymerization" of aqueous NIPAM and methylene-bis-acrylamide (BA).[3] This recipe was essentially the same as that used to prepare monodisperse, surfactant-free polystyrene latex.[12] The polymerizations were conducted at 60 to 70 °C in order to generate free radicals by the decomposition of the persulfate initiator. However, elevated temperature was also required so that growing PNIPAM chains phase separated to form colloidal particles - some authors call this procedure "precipitation polymerization". The original procedure has been reproduced by many others.[13-19]

This simple polymerization procedure can produce remarkably uniform particles. Figure 1 shows a transmission electron micrograph of typical microgel particles. The TEM sample was prepared by placing a dilute drop of aqueous particles onto the sample grid and allowing it to dry. The surface tension forces pulled the swollen particles into a closely packed array of spheres which then dehydrated in vacuum to give disks. The original dimensions of the swollen microgels can be estimated from the center-to-center spacing of the ordered disks in Figure 1.

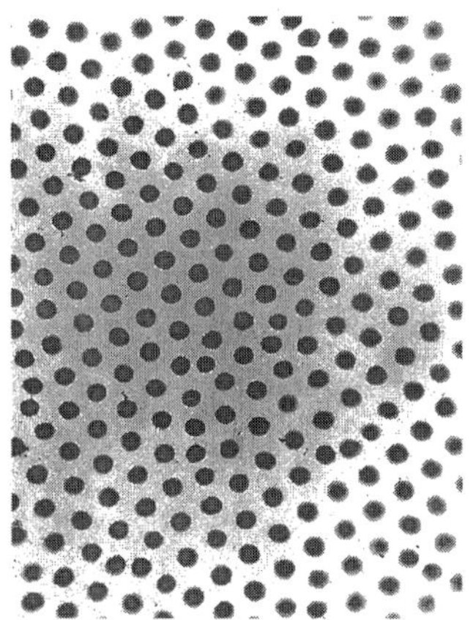

Figure 1. Transmission electron micrograph of PNIPAM microgels. Published with permission. [3]

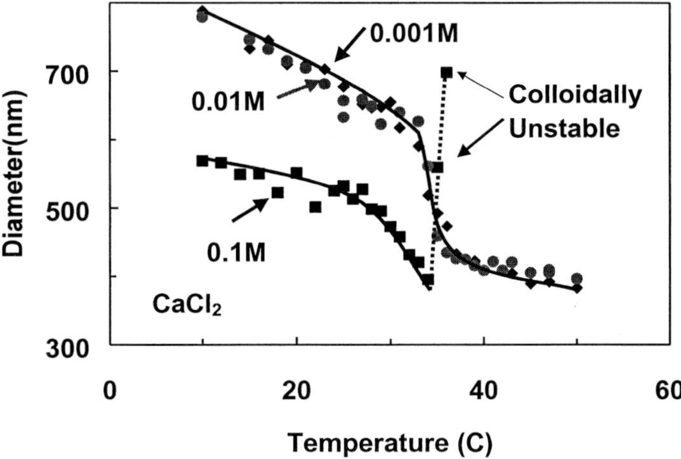

Figure 2. The diameter of PNIPAM microgel spheres as a function of temperature and $CaCl_2$ concentration. The particles display a volume phase transition at ~ 35 °C which is close to the lower critical solution temperature of linear PNIPAM. Adapted from Pelton et al. [29]

Temperature sensitive microgel swelling is most easily demonstrated by dynamic light scattering (DLS). Figure 2 shows that particle diameters can change by more than a factor of two over a narrow temperature range. The microgels whose behaviors are shown in Figures 1 and 2 are generally considered to be homogenous in composition.

Much work has also been reported on the preparation of core-shell gel particles. Core-shell gels consist of a water-insoluble latex particle coated with a gel layer. The composite particles have properties characteristic of both the core and the shell. The core dominates light scattering (turbidity) behavior, whereas colloid stability is determined by the hydrogel shell. The first reported preparation of PNIPAM-coated core-shell particles was by Pelton[20] who described both the one-shot surfactant free preparation of polystyrene/PNIPAM gels and the grafting of PNIPAM onto existing latex particles. Unlike the homogenous microgel preparations, these polymerizations are conducted below the LCST so that PNIPAM does not phase separate. Similar polystyrene-co-PNIPAM surfactant free latexes are obtained when starting with either NIPAM monomer or with PNIPAM. This reflects the fact that NIPAM polymerizes much more quickly than does styrene in aqueous media. Thus, most of the NIPAM is converted to PNIPAM at extremely low styrene conversions.

Makino et al.[21] refined Pelton's procedure with a two-stage approach. The first stage was a one-shot surfactant free polymerization with styrene and NIPAM. In the second stage, the surface gel layer was expanded by adding more NIPAM monomer and a nonionic water-soluble free radical initiator.

Zhu and Napper employed a two-stage procedure for the preparation of polystyrene-PNIPAM core-shell particles.[22] In the first stage, styrene was added slowly to an aqueous PNIPAM solution, presumably at room temperature, in the presence of a redoxinitiator. The objective of this stage was to produce a soluble NIPAM-co-styrene copolymer; however, the polymer from this stage was not isolated. In the second stage, the remainder of the styrene was added quickly to yield latex.

Duracher and coworkers[23] investigated the preparation and properties of cationic polystyrene-PNIPAM core shell particles. Positively charged groups were introduced by both cationic initiator and aminoethylmethacrylate. The cationic monomer gave smaller particles; however, a variety of particle morphologies was obtained, depending upon the polymerization conditions.

Publications describing new microgel synthesis schemes are published nearly every month. The above summary is meant to be neither a comprehensive nor an up-to-date account of these activities. However, in spite of scores of publications, there remain a number of unresolved issues regarding the preparation and characterization of microgels. These are now considered.

Unresolved Issues

1. Microgel Cleaning

Microgel polymerizations can yield a significant fraction of sol (i.e. linear or slightly branched polymer). Centrifugation, followed by decantation and redispersion in clean water is currently the only effective cleaning procedure. However, this is a tedious procedure requiring extended ultracentrifugation, because microgel densities are close to that of water. Membrane processes and ion-exchange beds, commonly used to generate clean polystyrene latex[24], are not effective with PNIPAM microgels – membranes do not remove high molecular weight sol components and nonionic sol components will not be captured in an ion-exchange resin bed.

Surfactant contamination is related to the cleaning issue. Most microgel preparations involve surfactants such as sodium dodecyl sulfate (SDS) to facilitate the formation of uniform particles. Furthermore, most ionic surfactants change the swelling/temperature response of PNIPAM gels. For example, Figure 3 shows that the presence of SDS increases microgel swelling and attenuates the temperature sensitivity. Thus, removing surfactant is important. There have been at least two published SDS binding isotherms,[25,27] one of which is shown in Figure 4. It is generally assumed that surfactant binding is reversible which, in turn, means that it is possible to remove bound surfactant by washing. The complete reversibility of SDS binding has not been proved for SDS or other surfactants. In other words, the extent to which surfactant can be removed from microgels has not been proved. In summary, microgel cleaning is tedious in a research laboratory and an unsolved problem for commercial scale production.

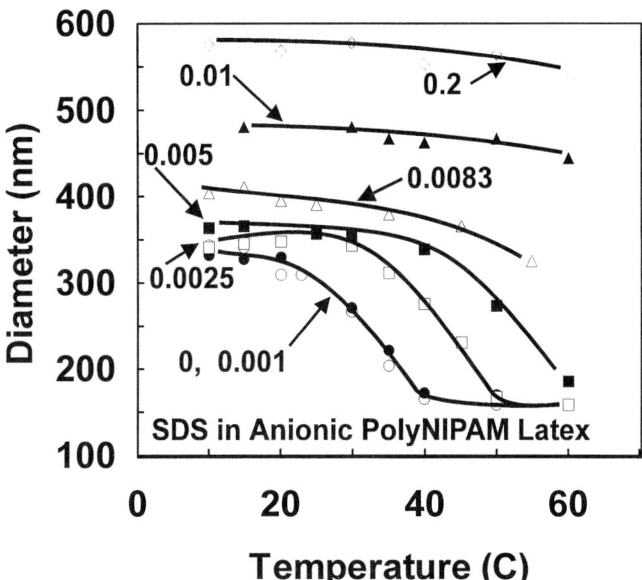

Figure 3. Diameter of PNIPAM microgel spheres as functions of temperature and the molar concentration of SDS. From Tam et al.; used with permission. [26]

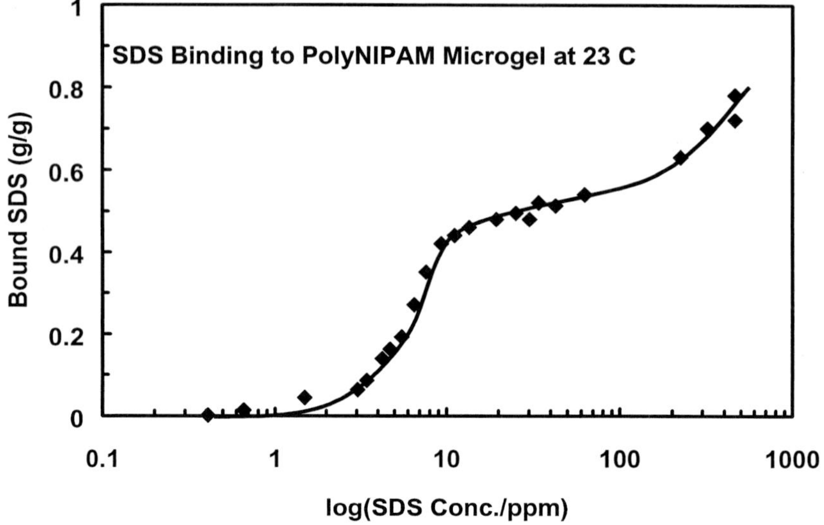

Figure 4. The amount of SDS bound on PNIPAM microgel as a function of the equilibrium SDS concentration. Adapted from Mears et al. [27]

2. Very small particles

It is difficult to prepare small microgels by surfactant-free methods because there is simply not enough available charge to stabilize high concentrations of small particles. The surface charge density of a PNIPAM microgel latex was found to be 3.8 μeq/g [28] which is about two orders of magnitude lower than that of a corresponding surfactant free polystyrene latex.[12] Surfactant will give smaller particles; however, colloidally stable, clean, 10~20 nm microgels have yet to be prepared. Indeed, such small species would not normally be called microgels.

3. High volume fraction synthesis

Commercial latex dispersions typically contain 30 to 60 wt% polymer whereas most microgel recipes produce less than 5 wt%. This is a significant limitation for anything other than the very highest value added applications. In my opinion, this limitation can be eliminated by using high surfactant concentrations; however, the cleaning issue remains a problem.

4. Accurate measurement of water content

Temperature dependent swelling with water is perhaps the most interesting feature of

PNIPAM gels. In the case of macroscopic gels, it is relatively easy to measure water contents; one simply weighs a piece of gel before and after drying. Most microgel swelling is reported as average particle diameters or volumes measured by dynamic light scattering.[29] However, DLS measurements alone do not give absolute water contents. More information is required to relate particle volume to polymer or water content of the microgel. Some have avoided this problem by reporting swelling volume or diameter ratios calculated by comparing particle size to a reference particle size. However, if the average mass of polymer per gel particle is known, then the diameters are easily converted to absolute measures of swelling. A few approaches have been applied to the measurement of absolute water content in microgels. McPhee et al.[28] centrifuged PNIPAM microgels to give an ordered, packed bed of particles, which was iridescent, indicating ordered packing. The water content of the bed was calculated from the bed mass before and after drying.

Wu's group was the first to employ intensity light scattering to measure the water content of uniform microgels[30,31] based on the weight average molecular weight and particle radius of gyration. This is the most direct method; however, it does require careful determination of the microgel optical properties.

Another approach is to measure the particle size at a temperature above the VPTT and assume a water content based on macrogel data, such as that given by Dong and Hoffman.[32] For example, consider a microgel with a particle diameter at 45 °C of 100 nm and a corresponding diameter of 200 nm at 20 °C. If we assume that the mass concentration of water in the particles at 45 °C is 25%, then the corresponding concentration of water in the particles at 20 °C is 0.8912. Note that this calculation is based upon the assumptions that: the density of PNIPAM is 1269 kg/m3, the PNIPAM density is independent of temperature over this range; and, that there is no excess volume of mixing. If the actual water content is 30% instead of 25% at 45 °C (i.e. an error of 20%), the corresponding water content of the particles at 20 °C is 0.8996. Thus, a large (20%) error in the estimate of the water content at 45 °C corresponds to only a 0.94 % error in the estimated water content at 20 °C. It seems therefore that this is a robust method for estimation of microgel water content in swollen gels.

5. Microgel particle morphology

Most of the reported microgel preparations are batch copolymerizations of NIPAM and N-methylenebisacrylamide (BA) crosslinker. Figure 5 shows the consumption of NIPAM and BA crosslinker as functions of time and temperature[10]. BA is consumed more quickly than NIPAM, indicating that the particles are unlikely to have a uniform composition. Indeed, it seems reasonable to speculate that there exists a zone of relatively high crosslink density in each particle. Since polyBA is more hydrophilic than PNIPAM at elevated temperature, one might further speculate that at least part of the high crosslink density zone is situated on the particle/water interface.

For applications requiring very well controlled gel morphologies, it might be better to completely decouple the growth of PNIPAM linear chains from the crosslinking reactions. For example, Hiratani et al. described gel crosslinking with reversible disulfide bonds.[33]

PNIPAM microgel polymerizations are usually initiated with ionic initiators such as potassium persulfate which lead to the incorporation of electrically charged groups in the gel network. Although many experimental and theoretical studies have addressed the electrokinetic properties of PNIPAM microgel dispersions, it has not been possible to measure the topographical distribution of electrical charges in PNIPAM gels. There is no reason to presume that the charges are only located near the particle surfaces.

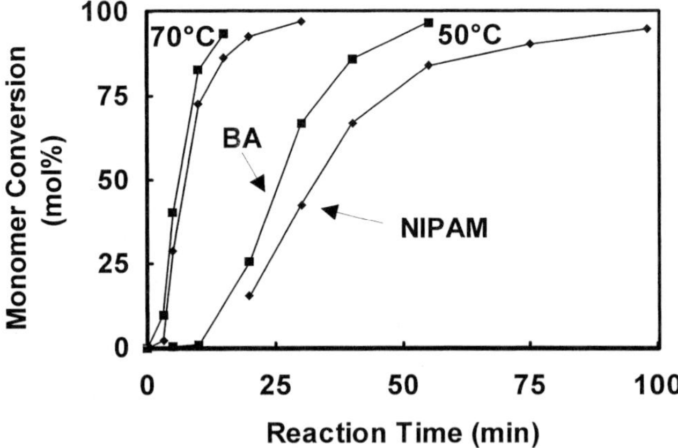

Figure 5. Monomer consumption versus time during PNIPAM microgel preparation. Adapted from Wu et al. [10] The crosslinking monomer, BA, polymerized at a faster rate than did NIPAM.

Concluding Remarks

The many current research efforts focused on PNIPAM microgels are likely to lead to promising applications. For example, Zhibing Hu has described the assembly of microgels into macrogels with nonlinear optical properties.[34] Highlighted in this paper are engineering difficulties which will have to be overcome before microgels can be prepared on a commercial scale.

[1] M. Heskins, J. E. Guillet, *J. Macromol. Sci., Chem.* **1968,** *A2(8)* 1441.
[2] M. Shibayama, T. Tanaka, *Adv. Polym. Sci.* **1993**, *109*, 1.
[3] R.H. Pelton, P. Chibante, *Colloids and Surfaces*, **1986,** *20*, 247.
[4] R. Pelton, X. Wu, W. McPhee, K. C. Tam, in: "Colloidal Polymer Particles", J.W. Goodwin, R. Buscall (Eds.), Academic Press, **1995**, 81.
[5] H. Kawaguchi, in: "Biomedical Functions and Biotechnology of Natural and Artificial Polymers", M.Yalpani (Eds.), ATL Press, 1996, 157.
[6] H. Kawaguchi, in: "Microbeads, Microcapsules & Liposomes", R. Arshady (Eds.), STN Books, London, 1998, 233-248.
[7] M. J. Murray , M.J. Snowdon, *Adv. Colloid Interface Sci.* **1995,** *54*,73.
[8] R. Pelton, Adv. Colloid Interface Sci., **2000**, *85*,1.
[9] W.C. Wooten, R.B. Blanton, H.W. Coover, Jr., *J. Poly. Sci.* **1957,** *XXV*,403.
[10] X. Wu, R.H. Pelton, A.E. Hamielec, D.R. Woods, W. McPhee, *Colloid and Polym. Sci.*, **1994,** *272,* 467.
[11] G. Bokias, A. Durand, D. Hourdet, *Macromol. Chem. Phys.* **1998,** *199*, 1387.
[12] J.W. Goodwin, R.H. Ottewill, R. Pelton, G. Vianello, D. Yates, *British Polymer Journal* **1978,** 10, 173.
[13] M.J. Snowden, B. Vincent, J. *Chem. Soc, Chem. Commun.* **1992**, 1103.
[14] H. Kawaguchi, K. Fujimoto, Y. Mizuhara, *Colloid and Polymer Sci.* **1992,** *270*, 53.
[15] M. Mielke, R. Zimehl, *Prog. Colloid Polym. Sci.* **1998,** *111*, 74.
[16] D.M. Öle Kiminta, P.F. Luckham, S. Lenon, *Polymer*, **1995,** *36(25)*, 4827.
[17] C. Wu, S. Zhou, *J. Polym. Sci., Part B, Polym. Phys.* **1996**, *34*, 1597.
[18] J.M. Weissman, H.B. Sunkara, A.S. Tse, S.A. Asher, *Science* **1996,** *274*, 959.
[19] K. Kratz, W. Eimer, *Ber. Bunsenges Phys. Chem.*, **1998,** *102,* 848.
[20] R.H. Pelton, *J Polym. Sci.,* **1988,** *26,* 9.
[21] K. Makino, S. Yamamoto, K. Fujimoto, H. Kawaguchi, H. Oshima, *J. Colloid Interface Sci.* **1994,** *166*, 251.
[22] P.W. Zhu, D.H. Napper, *Phys. Rev. E* **1994,** *50(2)*, 1360.
[23] D. Duracher, F. Sauzedde, A. Elaissari, A. Perrin, C. Pichot, *Colloid Polymer Sci.,* **1998,** *276*, 219.
[24] M. C. Wilkinson, J. Hearn, P. A. Steward, *Adv. Colloid Inter. Sci.*, **1999,** *81*, 77.
[25] T.Gilanyi, I. Varga, R. Meszaros, G. Filipcsei, M. Zrinyi, *Langmuir* **2001**, *17*, 4764.
[26] K.C. Tam, S. Ragaram, R.H. Pelton, *Langmuir* **1994,** *10*, 418.
[27] S. J. Mears, Y. Deng , T. Cosgrove, R. Pelton, *Langmuir* **1997,** *13*, 1901.
[28] W. McPhee, K.C. Tam, R. Pelton, *J. Colloid Interface Sci.* **1993,** *156*, 24.
[29] R.H. Pelton, H.M. Pelton, A. Morphesis, R.L. Rowell, *Langmuir* **1989,** *5*, 816.
[30] C. Wu, S. Zhou, *J. Macromol. Sci. – Phys* **1997** B36(3), 345.
[31] S. Peng, C Wu *Macromolecules* **2001** 34, 6795.
[32] L. Dong, A.S. Hoffman, *J. Controlled Release*, **1990,** *30*, 21.
[33] H. Hiratani, C. Alvarez-Lorenzo, J. Chuang, O. Guney, A.Y. Grosberg, T. Tanaka, *Langmuir*, **2001,** *17*, 4431.
[34] Z. Hu, X. Lu, J. Gao, *Adv. Mater.* **2001**, *13*, 1708.

Mechanical and Dielectric Studies of Carrageenan Sols and Gels

Makoto Takemasa,[1] Akio Chiba,[1] Munehiro Date*[2]

[1] Department of Applied Physics, School of Science and Engineering, Waseda University, 3-4-1 Okubo, Shinjuku-ku, Tokyo, 169-8555 Japan
[2] Kobayasi Institute of Physical Research, 3-20-31 Kokubunji, Tokyo, 185-8533, Japan

Summary: The temperature dependence of the dynamic shear modulus, strain optical coefficient, DC conductivity, and complex dielectric spectrum of κ- and ι-carrageenan aqueous solutions with K, Ca, Cs, and Na were measured in order to clarify the formation process of the cross-linking region and the gel network structure. From the correlation analysis between the shear modulus and the strain optical coefficient, we found that the stress inducing unit orientation increases with decreasing temperature, which strongly suggests that the branching number in a cross-linking region increases with decreasing temperature, which depends on counterion species. In terms of the correlation parameters, an increasing scheme of the branching number depends on counterion species. Just below the coil-helix transition temperature, dielectric relaxation arises, with relaxation time ~100μs and relaxation strength $\sim 10^3$.Dielectric relaxation can be assigned to the counterion fluctuation in the parallel direction to the helical axis. The fluctuation distance of the counterion estimated from the relaxation time increases sharply in the initial stage of gelation and gradually reaches a constant value. We concluded that the longitudinal length of the aggregated region increases sharply at the initial state of gelation while the number of helical molecules bundled in a cross-linking region increases successively with decreasing temperature.

Introduction

Carrageenans are natural polysaccharide extracted from red algae. Their primary structures are based on an alternating disaccharide repeating unit of 1,3-linked β-D-galactose and 1,4-linked 3,6-anhydro-α-D-galactose. These aqueous solutions with specific cations form physically cross-linked thermo-reversible polyelectrolyte gels. The macroscopic properties of carrageenan gels are strongly affected by the concentration and species of cations. Although a number of studies on the gelation mechanism have been performed over the last 3 decades using various methods, such as X-ray diffraction,[1] electron microscopy,[2] and light scattering,[3,4,5] both the formation process of the crosslinks and the gel network structure remain uncertain, and a detailed model has not been constructed. Undoubtedly, the structure of the branching and the cross-link

 DOI: 10.1002/masy.200450307

points are essential features governing the macroscopic properties of carrageenan gels. The purpose of this study is to clarify the formation process and growth. Measurement f multiple properties, has provided a physical insight into the cross-linking mechanism. We investigate responses to mechanical deformation, i.e. the dynamic shear modulus and the strain optical coefficient, and electrical properties, i.e. DC conductivity and dielectric properties.

The response to mechanical deformation provides information about the cross-link structure and the polymers connecting between cross-linking points. From the correlation analysis between shear modulus and strain optical coefficient, we estimate the optical anisotropy of polymer chains and the stress required to induce orientation, and discuss the increasing scheme of the branching number in a cross-linking region.

On the other hand, the electrical properties are sensitive to the change of binding nature of the counterions around the aggregated region of highly charged helical molecules and the abundance of freely movable counterions. According to the Manning's counterion condensation theory, bound counterions are classified into two types; loosely and tightly bound counterions, of the relative abundance of which depends on charge density along the polymer axis. The charge density of the carrageenan chain increases with decreasing temperature due to the temperature induced coil to helix transition and the aggregation of helices. Then, the number of the bound counterions increases, and the binding state of counterions changes in the sol-gel transition.

Linear polyelectrolyte solutions exhibit two dielectric relaxation processes appearing at the low frequency (~kHz) and the high frequency (~MHz) in the semi-dilute concentration region,[6,7] and it is suggested that these relaxations arise from the counterion fluctuations in the directions parallel and perpendicular to the polymer chain axis, in other words, tightly and loosely bound to polyions,[8,9] respectively. We investigated the formation and the longitudinal length of the high charge density region by using the low frequency dielectric relaxation analysis. Under the assignment of the relaxation process, the fluctuation distance can be obtained from the relaxation time τ,[10] which is expected to reflect the size of the high charge density region. The fluctuation distance of counterions reflects the size of the associated region of helical molecules. It is expected that a new physical insight into the cross-linking region of aggregated carrageenan helices will be obtained by investigating the nature of free and bound counterions.

Experiments

Sample preparation: κ- and ι-carrageenan from *Eucheuma cottonii*, were purchased from Sigma Chemical Co. Ltd.. We prepared the potassium, sodium, calcium, and cesium-form κ- and ι-carrageenan. Purification was performed by dialysis, and the carrageenans were de-ionized by passage through Amberlite IR-120 (H^+) cation exchange resin, and neutralized with KOH, NaOH, CsOH, or $Ca(OH)_2$ aqueous solutions for each cation-form solution, and then freeze-dried. The cation content was confirmed by ion chromatography, and the sulphur content was estimated by inductively coupled plasma spectrometry.

Apparatus: In this study, we particularly focused on the two groups of property, the responses to the mechanical deformation, and electrical properties. We have custom-built apparati for the measurement of complex shear modulus, strain optical coefficient, and dielectric relaxation at low frequencies (~kHz).[11,12] The special feature of the apparatus measuring dynamic shear modulus is its ability to make an accurate measurements over the ~7 decades in a concentric cylindrical cell applying a small amplitude of the strain.[11] The strain optical coefficient can be measured separately from the rotation of the optical azimuth of the gelled sample.[11] Using the other custom built apparatus, we can obtain the dielectric spectrum at the low frequency (~kHz) with such quality that the separation of the relaxation process can be successfully done even in samples exhibiting high DC conductivity.[12]

Measurements of the dynamic shear modulus and the strain optical coefficient were performed in the frequency range 1.0^{-2}-10^{0} Hz and the temperature range 2°C-60°C. All the measurements were performed under the same thermal condition. The sample was cooled from 60°C to 2°C at a rate of 0.5°C/min, and kept at 2°C for 120 minutes, and then heated from 2°C at a rate of 0.5°C/min.

Results and Discussion

Figure 1 shows the typical temperature dependence of the multiple properties for 2.0wt% Cs-form κ-carrageenan aqueous solution; dynamic shear modulus (G' and G''), the optical transmission (I/I_0), the strain optical coefficient ($\Delta n/\Delta\gamma$), the DC conductivity (σ), the dielectric relaxation time (τ) of the low frequency (~kHz) process, and the dielectric relaxation strength ($\Delta\varepsilon$), where I is the intensity of the transmitted light through the cell with sample and the I_0 is the intensity without sample. All the properties showed a characteristic change at the onset temperature T_o at which a marked change starts in the cooling process. Since the temperature T_{cross} at which G' crosses with G'' depends on frequency and does not agree with the gelation temperature estimated with the commonly accepted method proposed by Winter and Chambon,[13,14] we think that T_o does not correspond to the sol-to-gel transition temperature, but neither corresponds to the coil-to-helix conformational transition temperature. For precise estimation of the gelation temperature T_{gel} using a mechanical method, measurement of the frequency dependence of G^* at a fixed temperature is employed. Under the condition of constant cooling rate, the temperature T_{cross} at which G' crosses with G'' becomes a guidepost of the sol to gel transition temperature T_{gel}. The temperature at which both the mechanical properties increase significantly lies very close to the coil helix transition temperature, which is evaluated by dc conductivity just corresponding to the optical rotation result.[18)] As shown in Figure 1, T_{cross} lies very close to T_o, which indicates that T_{gel} in this study lies very close to the coil-helix transition temperature T_{CH} under this experimental condition.

In the cooling process, both the responses to the mechanical deformation, $\Delta n/\Delta\gamma$, G', and G'' increase with decreasing temperature, which apparently shows that the formation of the gel network, such as an increase of the number of the cross-link points and an increase of the rigidity of the polymers connecting between cross-link points, successively proceeds. On the other hand, the decrease of the DC conductivity means a decrease of the amount of the freely movable counterions, and the increase of the dielectric relaxation strength with the relaxation time ~100 μs means an increase of the amount of the bound counterions to polyions fluctuating along the polymer axis.

We have carefully checked that a frequency dependence of G^* and $\Delta n/\Delta\gamma$ observed just below the sol-gel transition temperature does not affect the conclusion of the growth process of the gel network.[11] The data of G^* and $\Delta n/\Delta\gamma$ shown in Figure 1 were measured at a constant frequency 0.5Hz. Since we confirmed that almost no apparent frequency dependence was observed for G^* and $\Delta n/\Delta\gamma$ far below the gelation temperature T_{gel}, we consistently used the data at a constant frequency of 0.5Hz.

Mechanical properties: In the case of a rubber-like network, the temperature and frequency dependence of viscoelasticity offers information about the formation of cross-link points and the number of polymer chains in the gels. In contrast, in the case of gels

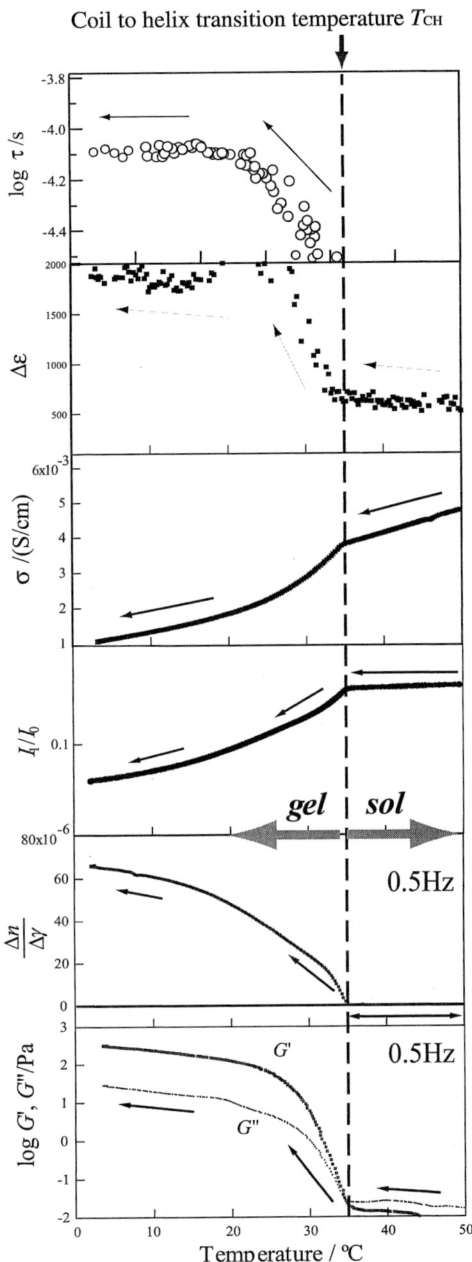

Figure 1 A typical temperature dependence of the relaxation time τ, the dielectric relaxation strength Δε of the dielectric relaxation process observed at the low frequencies (~kHz), DC conductivity σ, the optical transparency I_1/I_0, the strain optical coefficient $\Delta n/\Delta\gamma$, and the shear modulus G', G'' of 2.0wt% Cs-form κ-carrageenan. (The figures of I_1/I_0, $\Delta n/\Delta\gamma$, and G^* are reconstructed with permission from Fig.7 in *Macromolecules*. 2001, *34*, 7427-7434.)

made of rigid segments, these properties are affected by the structure of the cross-link region, and the topological structure of the gel network. However, it is difficult separate these effects using only the viscoelastic experiments. Moreover, the viscoelasticity of the κ-carrageenan solutions does not exhibit significant frequency dependence far below the gelation temperature. On the other hand, the investigation of the strain optical coefficient offers information about the anisotropy of the refractive index of the polymer chain and the polymer chain orientation caused by mechanical deformation of gel network. Combining *G'* with Δn/Δγ, we have obtained new information about the growth process of the gel network through the change in the orientation of the polymer chains and the optical anisotropy of molecules during gelation. Figure 2 shows the correlation between log *G'* and log Δn/Δγ, as deduced from the data of the temperature dependence of *G'* and Δn/Δγ, eliminating temperature. As an important finding in this

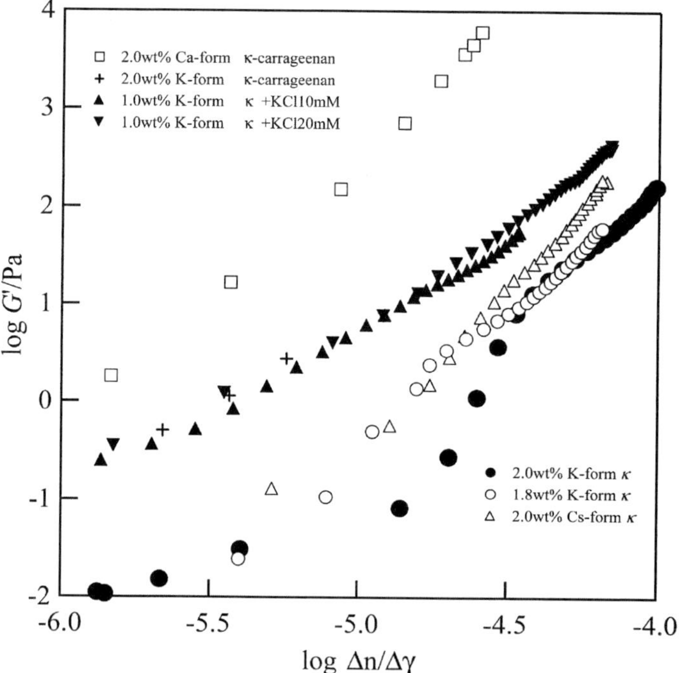

Figure 2 The correlation between the strain optical coefficient and the storage modulus. Upper right region corresponds to lower temperature. (+Ca-form κ 2.0wt%, □ Cs-κ 2.0wt%, ▪K-κ 1.5wt%, ●K-κ 1.8wt%, ○ K-κ 2.0wt%,▲ K-κ 1.0wt%+KCl10mM, ▼ K-κ 1.0wt%+KCl20mM) (Reproduced with permission from Figure 12, *Macromolecules* **2001,** *34,* 7427-7434. Copyright 2001 Am. Chem. Soc.)

study, a linear relation between log G' and log $\Delta n/\Delta\gamma$ was observed for all the samples at the temperatures far below the gelation temperature T_{gel}. The linear relation is formulated as,

$$G' = \left(\frac{d\sigma}{d\gamma}\right) = \alpha\left(\frac{d\Delta n}{d\gamma}\right)^{\beta}. \tag{1}$$

The two parameters α and β characterizing the correlation between G' and $\Delta n/\Delta\gamma$ have an important physical meaning. α corresponds to the optical anisotropy of the segment, and β characterizes the growth scheme of the gel network, reflecting the stress required for inducing the unit stress.[11] The value of β is larger than unity for all samples, which indicates that the stress required for inducing unit orientation increases with decreasing temperature over a wide temperature range. The topological structure of gel network made of rigid segment such as helical molecules should largely determine the mechanical properties. The change in topological structure of the gel network is explained by an increase of the branching number in a cross-link region. An increase of the branching induces an increase of the number of the polymer chain acting as a brace. Based on the commonly accepted model of a crosslink structure for κ-carrageenan gel, which indicates that the branching is an interhelical association,[15] we conclude that the parameter β is the characteristic parameter connected with the increasing scheme of the branching number in a cross-linking region.

Logα represents the offset of the vertical axis at log($\Delta n/\Delta\gamma$)=0 in Figure 2, and strongly depends on the value of β. Since the values of $\Delta n/\Delta\gamma$ are much smaller than 10^0, the value of α reflects the optical anisotropy expanded by the value of β. For the more adequate evaluation of optical anisotropy in eq 1, we use the value of the horizontal offset α' defined as,

$$\log G' = \beta\left(\frac{d\Delta n}{d\gamma} - \alpha'\right), \qquad \alpha' = \frac{\log\alpha}{\beta}. \tag{2}$$

As listed in Table 1, the values of α' with different cation species (Cs, K, and Ca-form κ-carrageenan) are quite different. We think that the difference of α' is dominated by the effects of cations bound to helical molecules, rather than the change in the optical anisotropy during conformational transition of the polymer chain. This indicates that the structure inside the cross-link region and the conformation of the chains connecting

Table 1. The correlation exponents α and β in eq 1 estimated from the data in Figure 2 by a least squares fitting procedure. (Reproduced with permission from Table 2, *Macromolecules* 2001, *34,* 7427-7434. Copyright 2001 Am. Chem. Soc)

sample	polymer concentration(wt%)	logα	β	$-\frac{\log a}{b}$
κ K-form + KCl 10mM	1.0	8.8	1.63	-5.42
κ K-form + KCl 20mM	1.0	9.0	1.63	-5.51
ι K-form	2.0	8.5	1.54	-5.50
κ K-form	1.8	12.1	2.48	-4.88
κ K-form	2.0	12.8	2.65	-4.82
κ Ca-form	2.0	16.6	2.83	-5.89
κ Cs-form	2.0	16.7	3.46	-4.84

cross-link regions change successively in the low temperature region.

We found that in the case of K-form ι-carrageenan, the value of the exponent β is much smaller than that for K-form κ-carrageenan and without added salt it is rather close to unity. This suggests that the growth mechanisms of κ- and ι-carrageenan are quite different. Piculell *et al.* pointed out the possibility that aggregation is essential for the gelation of κ-carrageenan, but is not essential for that of ι-carrageenan.[16] We think that the difference of the values of β for κ- and ι-carrageenan is caused by the difference in cross-linking mechanism, i.e., the branching number in the cross-link region increases during gelation in the case of κ-carrageenan gel, but hardly increases in the case of ι-carrageenan gel.

Light scattering studies focusing on change of the molecular weight during coil-helix transition and aggregation of helices showed a continuous increase of the molecular weight,[5,17] indicating an increase of the number of helices associated in a cross-link region. We think that our results are consistent with those results.

***The electrical properties*:** Knowledge about the formation of the cross-linking region obtained from the correlation analysis between *G'* and $\Delta n/\Delta\gamma$ is reinforced by the estimation of the longitudinal length of the aggregated region using low frequency dielectric relaxation measurements. We focused on the dynamics of counterions bound to the high charge density region reflecting the cross-linking region in relation to the conformational transition.

Figure 3 shows the temperature dependence of the dielectric dispersion (ε') spectrum for

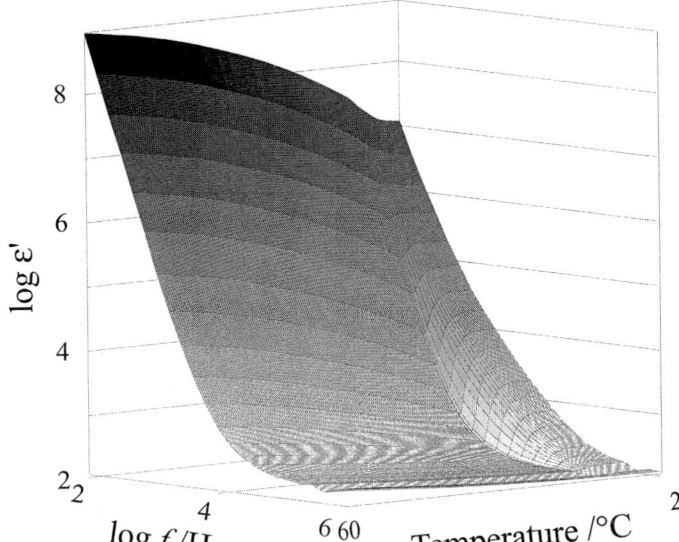

Figure 3 The temperature dependence of the dielectric dispersion ε' spectrum of 1.0wt% Ca-form κ-carrageenan. (Reproduced with permission from 2002**,** *35,* 5595. Copyright 2001 Am. Chem. Soc)

1.0wt% Ca-form κ-carrageenan solutions in the frequency range 10^2-10^6Hz in the temperature range 2°C-60°C. The shape of the ε' spectrum shows the characteristic change at the coil-helix transition temperature, and is quite different below and above the coil to helix transition temperature T_{CH}. Since in the case of the sample with non-gelling promoting cation of Na, which does not show the coil-helix transition at this ionic concentration,[18] such a behavior was not observed, the increase of ε' near the 10kHz must be connected with the coil-helix transition.

Figure 4 shows that the ε' and ε'' spectrum of 1.0wt% Ca-form κ-carrageenan solution at 60°C and 5°C, corresponding to the coil and helical state, respectively. It is seen that ε'' obeys ε'' $\sim 1/\omega$ over the wide frequency range, 10^0-10^6Hz, which means that the dominant effect in this range is DC conductivity. An increase in ε' with decreasing temperature at the frequencies lower than 10^3Hz is due to the electrode polarization effect ε_{ep}^*, which is represented by the empirical function,[19]

$$\varepsilon_{el}^{*} = \frac{\varepsilon_{ep}}{(i\omega)^{\gamma}}, \tag{3}$$

where ω is the angular frequency and ε_{ep} and γ the electrode polarization parameter. We can well reproduce the data at sol state (60°C) by using the effects of DC conductivity and electrode polarization to reproduce the experimental data,

$$\varepsilon^{*} = \varepsilon_{\infty} + \frac{1}{\frac{i\omega}{\sigma} + \frac{(i\omega)^{\gamma}}{\varepsilon_{ep}}}, \tag{4}$$

where ε_{∞} is the instantaneous permittivity and σ the DC conductivity. In contrast, we cannot reproduce the data with eq 4 at 5°C. Only an addition of the Cole-Cole relaxation term gave an excellent fit, i.e.

$$\varepsilon^{*} = \frac{1}{\frac{1}{\varepsilon_{\infty} + \frac{\Delta\varepsilon}{1+(i\omega\tau)^{\beta}} + \frac{\sigma}{i\omega}} + \frac{1}{\frac{\varepsilon_{ep}}{(i\omega)^{\gamma}}}}, \tag{5}$$

where $\Delta\varepsilon$ is the relaxation strength defined as ε_0-ε_{∞}, τ the relaxation time, and β is the distribution parameter of the relaxation time in the Cole-Cole dielectric relaxation

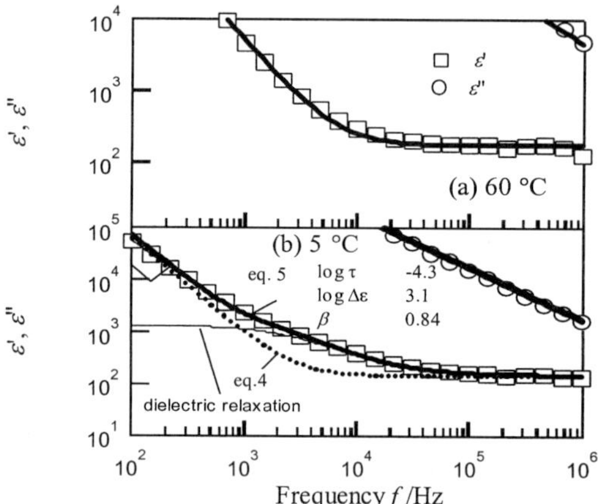

Figure 4 The temperature dependence of the dielectric dispersion ε' and loss ε'' of 1.0wt% Ca-form κ-carrageenan at (a) 60°C and (b) 5°C. The thick solid, the dotted, and the thin solid lines represent the best fit curve with eq 5, the fitted curve with eq 4, and the relaxation term in eq 5, respectively.

function. It is confirmed that the electrode polarization parameters are nearly constant in the entire temperature range.[12] The residual resulting from the fitting procedure with eq

3 is much larger than that with eq 4, especially at the temperatures lower than T_{CH}. We therefore used eq 4.

As shown in Figure 1, the dielectric relaxation parameters, τ and $\Delta\varepsilon$ of Cs-form κ-carrageenan estimated by the least square fitting procedure show the characteristic change at T_{CH}. σ decreases sharply just below the T_{CH}. This means that the sharp decrease of the σ is caused by a decrease of the amount of counterions contributing to σ. This is a strong evidence that the dielectric relaxation arises due to a change of the binding nature of the counterions linked with the coil to helix transition.

According to the Manning's counterion condensation theory,[20] the bound counterion to the polyion is classified as loosely and tightly bound counterions, in terms of the dimensionless charge density parameter ξ defined as

$$\xi = \frac{e^2}{\varepsilon bkT}, \tag{6}$$

where b is the charge separation distance, e the elementary of electric charge, ε the dielectric constant of solvent, k the Boltzmann constant, and T the temperature. The tightly bound counterions arise when ξ is higher than $1/z$, where z is the counterion valence.[20]

The charge separation distance b of the helix state is shorter than that of the coil state, and the helical content increases with decreasing temperature below T_{CH}. For κ-carrageenan aqueous solutions, the values of ξ for the coil and helical state, estimated from the chemical structure of the monomers and the X-ray data.[1,21] are ~0.68 and 1.65, respectively. The value of ξ in the helical part in the polymer chains exceeds unity. An increase of the amount of tightly bound counterions with decreasing temperature is anticipated.

As mentioned in the introduction, low frequency dielectric relaxation is ascribed to the fluctuation of counterions tightly bound to polyions along the polyion axis.[8,9] The relaxation time τ for the process is proportional to d^2/μ, where μ is the mobility of the counterion and d is thc fluctuation distance of the counterion.[10] Based on the cross-linking mechanism of carrageenan gels, a side-by-side aggregation of helical molecules,[2,15] it is expected that the fluctuation distance d of tightly bound counterions estimated from the relaxation time of the low frequency process will reflect the size of the high charge density region. In other words, d reflects the longitudinal length of the aggregated region of helices.

We think that the cross-linking and the gel network structure are dominated by whichever process, growth of the length of helices or aggregation of helices, precedes in the initial stage of gelation, since $\Delta\varepsilon$ increases sharply just below T_{CH}, and gradually reaches a constant value at low temperatures. At the initial stage of gelation, in other words, just below T_{CH}, some short helices appear in a chain. When the aggregation rapidly occurs just below T_{CH} and the gel network is formed before the growth of the long helices, the longitudinal length of the aggregated region is difficult to increase due to the constraint by the other chains in the network. The long helices can be formed when the growth of the long helix occurs before the helices take part in forming the gel network.

The length of the rigid segment, that is, the helical molecule, is important for enlarging the stress for the orientation. In terms of the dielectric relaxation time, the longitudinal length of the helices and their aggregated region sharply increase in the initial stage of gelation and gradually reaches a constant value at low temperatures. We think that this result reinforces the suggestion that the enlargement of the stress inducing polymer orientation is caused by the increase of the number of helical molecules bundled in a cross-linking region.

[1] N. S. Anderson, J.W. Campbell, H. M. Harding, D. A. Rees, J. W. B. Samuel, *J. Mol.Biol.*, **1969**, *45*, 85.
[2] A. M. Hermansson, *Carbohydr. Polym.*, **1989**, *10*, 163.
[3] R. A. Jones, E. J. Staples, A. Penman, *J. Chem. Soc. Perkin Trans. I*, **1973**, 1608.
[4] I. T. Norton, D. M. Goodall, E. R. Morris, D. A. Rees, *J. Chem. Soc., Faraday Trans. I*, **1983**, *79*, 2475.
[5] K. Bongaerts, H. Reynaers, F. Zanetti, S. Paoletti, *Macromolecules*, 1999, 32 (3), 683.
[6] S. Uemura, R. Hayakawa, Y. Wada, *Biophys. Chem*, **1980**, *11*, 317.
[7] F. van der Touw, M. Mandel, *Biophys. Chem. 2*, **1974**, 231.
[8] K. Ito, A. Yagi, N. Ookubo, R. Hayakawa, *Macromolecules*, **1990**, *23*, 857.
[9] N. Okubo, Y. Hirai, K. Ito, and R. Hayakawa *Macromolecules*, **1989**, *22*, 1359.
[10] F. Oosawa, Polyelectrolytes, Marcel Dekker, New York, **1971**.
[11] M. Takemasa, A. Chiba, M. Date, *Macromolecules*, **2001**, *34*, 7427.
[12] M. Takemasa, A. Chiba, M. Date, *Macromolecules* **2002,** *35,* 5595.
[13] H. H. Winter, F. Chambon, *J. Rheol.* **1986**, 30, 367.
[14] F. Chambon, H. H. Winter, *J. Rheol.* **1987**, 31, 683.
[15] E. R. Morris, D. A. Rees, G. Robinson, *J. Mol. Biol.* **1980**, *138*, 349.
[16] L. Piculell, *"Food Polysaccharides and Their Applications*"; Stephen, A. M., Marcel Dekker, New York, 1995, p205.
[17] V. Meunier, T. Nicolai, D. Durand, *Macromolecules*, **1999**, *32*, 2610.
[18] C. Rochas, M. Rinaudo, *Biopolymers*, **1980**, *19*, 1675.
[19] J. F. Johnson, R. H. Cole, *J. Am. Chem. Soc.*, **1951**, *73*, 4536.
[20] G. S. Manning, *J. Chem. Phys.* **1969**, *51*, 924.
[21] S. Arnott, W. E. Scott, D. A. Rees, C. G. A. McNab, *J. Mol. Biol.* **1974**, *90*, 253.

Deuterium Substitution and Fluorescence Studies on Polymer Hydrogels and Complexes

Hideaki Shirota,*[1] *Kazuyuki Horie**[2]

[1] Department of General Systems Sciences, Graduate School of Arts and Sciences, University of Tokyo, 3-8-1 Komaba, Meguro-ku, Tokyo 153-8902, Japan
[2] Department of Organic and Polymer Materials Chemistry, Faculty of Technology, Tokyo University of Agriculture and Technology, 2-24-16 Naka-cho, Koganei-shi, Tokyo, 184-8588 Japan

Summary: Deuterium isotope effects on swelling kinetics and volume phase transition in typical polymer hydrogels (poly(*N*-isopropylacrylamide) and polyacrylamide gels) are discussed. Deuterium substitutions affect on the swelling kinetics and volume phase transition of the polymer hydrogels. The slower swelling kinetics of hydrogels in D_2O than in H_2O arises mainly from the high viscosity of the medium. The deuterium isotope effect on the swelling-shrinking curve of hydrogels would come from the different polymer-solvent interaction. The microenvironments of hydrogels studied by solvatochromic fluorescence probe are compared with the bulk state. The zipper-type hydrogen-bonding inter-polymer complexes (poly(acrylic acid)-polyacrylamide and poly(acrylic acid)-poly(*N*-acryloylglycineamide)) are also investigated and show the huge isotope effect on the phase separation temperature.

1. Introduction

Phase transitions and critical phenomena in polymer gels have been paid great attentions by theoretical and experimental scientists and engineers. Volume phase transition of polymer gel was theoretically predicted by Dusek and Patterson in 1968.[1] They showed the possibility of a discontinuous volume change of polymer gels based on the analogy of the coil-globule transition of linear polymers in solution.[2] Tanaka experimentally discovered the discontinuous volume change of a partially ionized polyacrylamide (PAAM) gel against the continuous change of the solvent composition in a mixture of acetone and water in 1978.[3] After this discovery, the generality of volume phase transition in polymer gels has been well recognized and the study of this topic has become more extensive.[4,5]

This article is dedicated to the late Prof. Toyoichi Tanaka.

 DOI: 10.1002/masy.200450308

The critical phenomenon (swelling-shrinking curve) of a polymer gel can be expressed by Flory-Rehner equation:[6-8]

$$\Pi = -\frac{k_B T}{V_s}[\phi + \ln(1-\phi) + \chi\phi^2] + \nu k_B T\left[\frac{1}{2}\left(\frac{\phi}{\phi_0}\right) - \left(\frac{\phi}{\phi_0}\right)^{\frac{1}{3}}\right] \tag{1}$$

where Π is the net osmotic pressure, k_B is the Boltzmann constant, T is the absolute temperature, V_s is the molar volume of the solvent, ν is the number of cross-links per unit volume, χ is the Flory's interaction parameter, and ϕ and ϕ_0 are the network volume fractions at the equilibrium and reference states, respectively. The first term of the right side denotes the mixing free energy and the second term of the right side denotes the elastic free energy. Although χ is very complex,[9-12] χ is often assumed to be linearly dependent on ϕ for the simplicity. [13-15]

$$\chi = \chi_1 + \phi\chi_2 \tag{2}$$

where

$$\chi_1 = \frac{\Delta H - T\Delta S}{k_B T} \tag{3}$$

and χ_2 is a constant. ΔH and ΔS are the mixing enthalpy and entropy of the polymer solution system. χ is a very important parameter, because polymer-polymer and polymer-solvent interactions are included in this factor.

The interactions in polymer solutions are comprised of the following four fundamental molecular interactions: (i) van der Waals interaction, (ii) hydrophobic interaction, (iii) electrostatic interaction, and (iv) hydrogen-bonding interaction. Here, we focus on the hydrogen-bonding interaction among these four fundamental molecular interactions. Although the bonding energy of a hydrogen bond is not large in comparison with that of a covalent bond, it plays a key role in many chemical and biological systems.[16-19] It would be expected that the hydrogen bonds have a great role in phase transitions of polymer gel and polymer solution, especially aqueous systems.

Deuterium substitution is a useful method to investigate both the static and dynamic features of condensed phases.[20] Because of the heavier mass of deuterium, the zero-point vibrational energy level of the deuterated hydrogen bond is lower than that of normal hydrogen bond. As the result, the hydrogen bond of deuterium is more stabilized than that of hydrogen, and the hydrogen-bonding network is more stiffened.[21] The differences between the properties of deuterated and undeuterated hydrogen bonding systems arise from the stabilized hydrogen-bonding interaction for deuterium in comparison with hydrogen. For example, the viscosity of D_2O is higher than that of H_2O and the melting and boiling points of D_2O are higher than those of H_2O.[22]

The deuterium isotope effects on the phase separation of nonaqueous polymer solutions have been well studied.[20] According to the extensive studies of the deuterium isotope effects of polystyrene solutions by Van Hook and co-workers, the region of the miscibility becomes large by the increase of the H/D ratio in the solvent and the decrease of the H/D ratio in the polymer.[23-25] In contrast to nonaqueous polymer systems, the deuterium isotope effects in polymer hydrogels have not been much extensively studied. As far as we know, the deuterium isotope effect of water on the volume phase transition temperature of a polymer hydrogel (poly(*N*-isopropylacrylamide) (PNIPAM) gel) was firstly observed by Shibayama et al. for the neutron scattering study.[26] However, the deuterium isotope effects on the whole swelling-shrinking state of hydrogels, the origin of the deuterium isotope effect in hydrogels, and the generality of the deuterium isotope effect in hydrogels were open questions.

On the other hand, fluorescence probe has been widely used to investigate the microscopic behavior of polymer systems. Especially, the dansyl group is well used to investigate the microscopic feature of the phase transition, phase separation, and conformational change in polymer systems, because its fluorescence shows the solvatochromism and it is relatively easy to synthesize the labeled polymer.[27-41] The microenvironment behavior on the phase transition of the polymer systems is not directly correlated with the bulk behavior in some cases.

In this review article, we focus on the deuterium isotope effects on the swelling kinetics[42] and volume phase transition[43-45] of typical polymer hydrogels, PNIPAM and PAAM gels, and the phase separation of the zipper-type hydrogen-bonding inter-polymer complexes, poly(acrylic acid) (PAA)-PAAM and PAA-poly(*N*-acryloylglycineamide) (PAG), in solution.[46] Since the deuterium substitutions change the hydrogen-bonding interactions in aqueous polymer systems, we can observe which state and process of polymer hydrogels and hydrogen-bonding inter-polymer complexes are strongly influenced by hydrogen bonds.

2. Swelling Kinetics of Polymer Hydrogel

To design polymer hydrogels as soft materials such as drug delivery systems, it is very important to understand the kinetics of swelling and shrinking processes. Tanaka and co-workers[47,48] and Peters and Candau[49,50] showed that the characteristic time of swelling process τ of spherical, cylindrical, and disk PAAM gels in water is proportional to the square of the size of the polymer gel:

$$\tau = \frac{[l(\infty)/(2\pi)]^2}{\alpha D} \tag{4}$$

where $l(\infty)$ is the diameter of a spherical or a cylindrical gel or the thickness of a disk gel at the equilibrium state, D is the cooperative diffusion coefficient, and α is the shape factor. The values of α in a spherical gel, a long cylindrical gel, and a disk gel are 1, 2/3, and 1/3, respectively.[48] In this section, we discuss the deuterium isotope effect of the medium on the swelling kinetics of polymer hydrogels.[42]

Figure 1 shows the time dependence of the diameter, l, of (a) PNIPAM gel in H_2O and in D_2O and (b) PAAM gel in H_2O and in D_2O at 298 K. The exponential fit functions are also shown in Figure 1. The initial diameters of the sample hydrogels correspond to the as-prepared state at 289 K of each hydrogel. The swelling kinetics factors are summarized in Table 1. Since the shape of sample gels is short cylindrical, D is tentatively estimated by using eq. (4) with $\alpha = 1$.

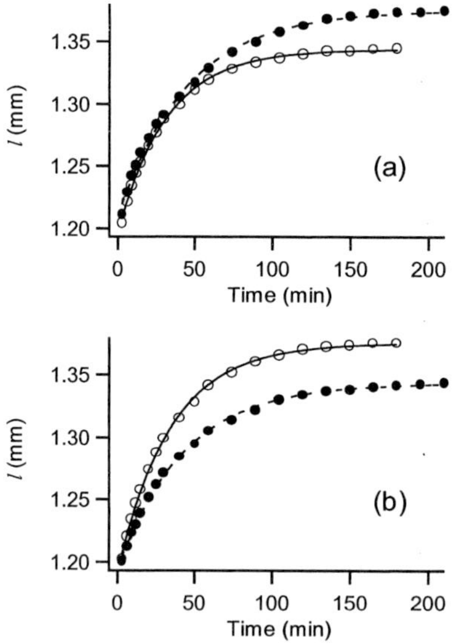

Figure 1. Comparison of the time dependence of the diameter l for (a) PNIPAM gel in H_2O (open circles) and D_2O (filled circles) and (b) PAAM gel in H_2O (open circles) and D_2O (filled circles). Exponential fit curves are also shown by solid line (H_2O) and broken line (D_2O).

The interesting points to be noted from Figure 1 and Table 1 are followings. (i) The characteristic time constant τ in swelling kinetics of the polymer hydrogels in D_2O is larger than that of the polymer hydrogels in H_2O, and the cooperative diffusion coefficient D of the polymer hydrogels in D_2O is smaller than that of the

Table 1. Characteristic times of swelling kinetics τ, diameters at the infinite time $l(\infty)$, and diffusion coefficient D of PNIPAM and PAAM gels in H_2O and D_2O.

Hydrogel	Solvent	τ (min)	τ_{D2O}/τ_{H2O}	$l(\infty)$ (mm)	$l_{D2O}(\infty)/l_{H2O}(\infty)$	D (cm^2/s)	D_{D2O}/D_{H2O}
PNIPAM	H_2O	32.0		1.344		2.38×10^{-7}	
	D_2O	44.8	1.40	1.375	1.023	1.78×10^{-7}	0.748
PAAM	H_2O	34.7		1.375		2.30×10^{-7}	
	D_2O	43.3	1.25	1.343	0.977	1.76×10^{-7}	0.765

polymer hydrogels in H_2O. (ii) The infinite-time (equilibrium state) diameter $l(\infty)$ of PNIPAM gel in D_2O is larger than that of PNIPAM gel in H_2O. In contrast to PNIPAM gel, the $l(\infty)$ of PAAM gel in D_2O is smaller than that of PAAM gel in H_2O. The cooperative diffusion coefficient D is expressed as,[51]

$$D = \frac{k_B T}{6\pi \eta \xi} \tag{5}$$

where k_B is the Boltzmann constant, T is the absolute temperature, η is the viscosity of the solvent, and ξ is the correlation length. As shown in eq. (5), D depends on the viscosity of the medium. The viscosities of H_2O and D_2O at 298 K are 0.890 mPa·s[22] and 1.095 mPa·s,[52] respectively. The inverse of the viscosity ratio between H_2O and D_2O ($(\eta_{D2O}/\eta_{H2O})^{-1}$) is, thus, 0.813. When the viscosity ratio is compared with the diffusion constant ratio, it would be clear that the slower swelling kinetics of hydrogels in H_2O than in D_2O arises mainly from the different viscosities between H_2O and D_2O. Since the polymer-solvent interaction is changeable by deuterium substitutions of the sources of hydrogel and solvent molecules, the residue of the deuterium isotope effect on the swelling kinetics can arise from the different microfrictions of deuterated and undeuterated systems.

Wang and Wu have studied the deuterium isotope effects of the average radius of gyration and the hydrodynamic radius of the coil-state linear PNIPAM in water using dynamic light scattering measurement.[53] Interestingly, both the average radius of gyration and the hydrodynamic radius of linear PNIPAM at about 298 K in D_2O are about 1.1 times larger than those in H_2O. This means that ξ of linear PNIPAM in D_2O is about 1.1 times longer than that in H_2O. If we include the deuterium isotope effects of η and ξ into eq. (5), the deuterium isotope effect on D of PNIPAM gel observed can be well reproduced. Kujawa and Winnik suggested that the hydration volume of PNIPAM in D_2O is larger than that of PNIPAM in H_2O from the pressure perturbation calorimetry study.[54] It would be possible that the deuterium isotope effect on the radiuses arises from the different interactions of PNIPAM and H_2O and PNIPAM and D_2O.

Another interesting point to be noted that the deuterium isotope effect on the diameter of hydrogels at the equilibrium state ($l(\infty)$). This deuterium isotope effect in PNIPAM gel shows the opposite feature to that in PAAM gel: Although the $l(\infty)$ of PNIPAM gel in D_2O is larger than that of PNIPAM gel in H_2O, the $l(\infty)$ of PAAM gel in D_2O is smaller than that of PAAM gel in H_2O.

PNIPAM in water has a lower critical solution temperature (LCST) and PAAM in water has an upper critical solution temperature (UCST) at near ambient temperature. A polymer solution system showing a UCST feature is characterized by positive values of ΔH and ΔS in a polymer-solvent interaction parameter, while one showing an LCST feature is characterized by negative values of these quantities (Table 2). In other words, the features of polymer-solvent interaction in PNIPAM and PAAM are different: PNIPAM dissolves into water enthalpically and PAAM dissolves into water entropically. The mixing free energy ΔG (= $\Delta H - T\Delta S$) is negative in the *swollen* polymer hydrogel.

Deuterium isotopic substitutions give stabilization and structural effects on hydrogen-bonding systems.[21] In the hydrogen-bonding systems, the stabilization effect gives more negative ΔH, and the structural effect gives more negative ΔS by deuterium isotopic substitutions. The present result shows PNIPAM gel in D_2O is larger than in H_2O and PAAM gel in D_2O is smaller than in H_2O. Namely, the ΔG of PNIPAM in D_2O is more negative than that in H_2O and the ΔG of PAAM in D_2O is more positive than that in H_2O (Table 2). The dominant factor in the deuterium isotope effect on the swelling ratio of polymer hydrogel could thus be ΔH for LCST polymer and ΔS for UCST polymer for the driving force of mixing, though the deuterium isotopic substitutions should affect on both ΔS and ΔH of a polymer-solvent interaction parameter.

Table 2. Mixing parameters for UCST and LCST polymers and the isotope effects.

	ΔH (ΔH_D–ΔH_H)	ΔS (ΔS_D–ΔS_H)	ΔG (ΔG_D–ΔG_H)
LCST (H → D)	– (negative)	– (negative)	– (negative)
UCST (H → D)	+ (negative)	+ (negative)	– (positive)

3. Volume Phase Transition of Polymer Hydrogel

After Tanaka's discovery of the volume phase transition phenomenon of polymer gel,[3] the generality of volume phase transition in polymer gels was recognized not only by changing the composition of the solvent but also by temperature, ionic and pH changes, light irradiation, and electric field.[4,5] Here, we discuss the deuterium isotope effects of polymer hydrogel on the volume phase transition of polymer hydrogels triggered by temperature[43,44] and solvent composition.[45] Further, the bulk behavior of the volume phase transition of polymer hydrogel based on the size change are also compared with the microenvironment behavior observed by the fluorescence probe linked chemically with polymer hydrogels.[44,45]

3.1. Temperature Dependence

PNIPAM gel (and liner PNIPAM) in water is well known that this hydrogel shows the volume phase transition at about 307 K (the phase separation and coil-globule transition in the case of linear PNIPAM in water).[55-59] In D_2O, about 0.7 K increase of the transition temperature of PNIPAM gel in comparison with in H_2O have been observed.[26,43,44] The cloud point (or coil-globule transition temperature) of linear PNIPAM in D_2O is also about 1 K higher than that of liner PNIPAM in H_2O.[44,53,54,60-62]

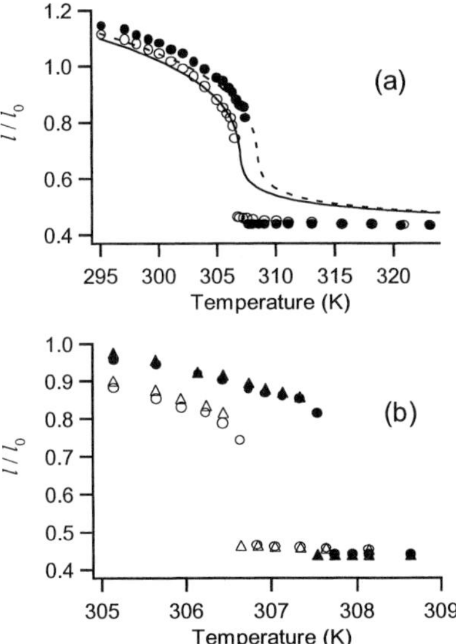

Figure 2. (a) Comparison of the swelling-shrinking curves of PNIPAM gel obtained by experiment (symbols) and prediction using the modified Flory-Rehner model (lines). Open circles and solid line show PNIPAM gel in H_2O, and filled circles and broken line show PNIPAM gel in D_2O. (b) Comparison between the heating (circles) and cooling (triangles) processes of the swelling-shrinking curves.

Figure 2 shows the swelling-shrinking curve of PNIPAM gel in H_2O and in D_2O against temperature. Besides the transition temperature, it becomes clear that (i) the swelling ratio at the swollen state of PNIPAM gel in D_2O is larger than that in H_2O, (ii) the swelling ratio at the shrunken state of PNIPAM gel in D_2O is equivalent to that in H_2O, and (iii) the magnitudes of the hysteresis of the transition temperatures (the difference between the transition temperatures in the heating process and the cooling process) are not affected by isotopic substitutions. These features have observed in PNIPAM gels with the several polymer concentrations.[44]

In order to understand the origin of the deuterium isotope effect on the volume phase transition of PNIPAM gel in water, we have examined the experimental result by a simple modified Flory-Rehner model proposed by Shibayama and co-workers.[15] The advantage of this model in comparison with the original Flory-Rehner model is that the swollen state is well reproducible, because the effect of the entanglements, which work as quasi-cross-links, is incorporated into the original Flory-Rehner model.[6-8,15] In this model, the number of cross-links per unit volume ν in eq. (1) is replaced by

$$\nu \frac{C_0}{C_{0,\mathrm{ref}}} \tag{6}$$

where $C_{0,\mathrm{ref}}$ is the lowest initial monomer concentration at which a uniform gel can be formed (0.3 M for PNIPAM).[14] As shown in Figure 2, the deuterium isotope effects of the swollen state and the transition temperature are qualitatively reproduced the experimental results. The parameters for PNIPAM gel in H_2O used were reported by Hirotsu,[14] and the three parameters, i.e., ΔH, ΔS, and χ_2, for PNIPAM gel in D_2O were determined by the fit to the experimental data.[43,44]

The different parameter between the simulation results of PNIPAM gels in H_2O and D_2O is only ΔH in χ, except for the molar volume of the solvent. The values of ΔH of PNIPAM gels in H_2O and D_2O are -1.246×10^{-20} J and -1.252×10^{-20} J, respectively. The change in the other possible parameters, χ_2 and ΔS, did not reproduce the experimental results.[43,44] We thus conclude that the deuterium isotope effect on the swelling-shrinking curve of PNIPAM gel would arise from ΔH in polymer-solvent interaction parameter χ. Namely, the solvent-polymer interaction in aqueous PNIPAM systems becomes stronger by deuterium substitutions because of the stabilized hydrogen-bonding interaction.

On the other hand, the deuterium isotope effects on the swelling ratio at the shrinking state have not been observed within experimental error. If the inter-polymer and/or intra-polymer hydrogen-bonding interaction is the dominant interaction to determine the shrinking state of the polymer hydrogel, the deuterium isotope effects on the swelling ratio of the polymer hydrogel at the shrinking state might be observable. Namely, the hydrophobic interaction is the dominant interaction to collapse PNIPAM gel. Maeda et al. also suggested that the globule state of linear PNIPAM is dehydrated around the hydrophobic groups of the polymer by IR experimental result.[60]

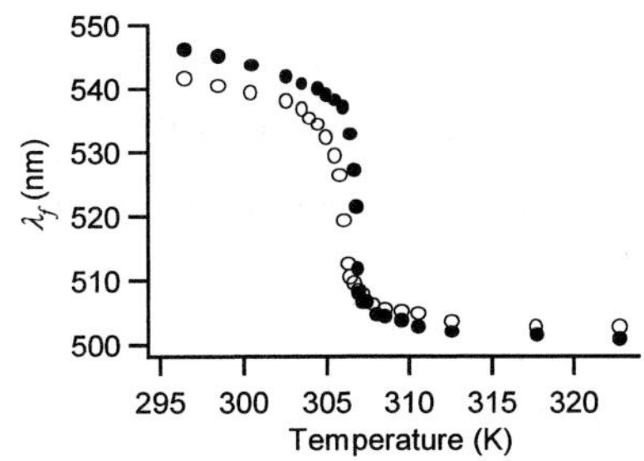

Figure 3. Temperature dependence of the fluorescence peak wavelength λ_f of the dansyl-group-labeled PNIPAM gel in H_2O (open circles) and D_2O (filled circles).

It may be interesting to compare the deuterium isotope effect of the swelling-shrinking curve of the bulk PNIPAM gel with that of the microscopic aspect of PNIPAM gel. Figure 3 shows the temperature dependence fluorescence peak wavelength λ_f of dansyl group chemically bonded with PNIPAM gel in H_2O and D_2O. Dansyl group shows a solvatochromism: λ_f becomes longer wavelength with higher polarity of the medium.[45] Therefore, it can observe the conformational change of polymer by using the fluorescence probe. The dramatic change of λ_f at about 310K, thus, indicates the microscopic (conformational) change of the polymer corresponds to the volume phase transition. It is clear from the comparison between Figures 2 and 3 that the overall features of the deuterium isotope effects in both macroscopic and microscopic observations are qualitatively similar. Aqueous linear PNIPAM solutions have been also shown the similar features.[44]

3.2. Solvent-Composition Dependence

The deuterium isotope effects on the swelling-shrinking curves of PNIPAM and PAAM gels against the solvent composition of the binary mixtures H_2O-acetone (and D_2O-acetone) and H_2O-methanol (and D_2O-CH_3OD) have been investigated.[45] Figure 4 shows the swelling-shrinking curves to the solvent composition of (a) PNIPAM gel and (b) PAAM gel in H_2O-methanol and D_2O-deuteriomethanol. Since PNIPAM gel has isopropyl groups in the side groups of the polymer, the selling-shrinking curve of the hydrogel in some aqueous binary mixtures shows a recurring feature to the solvent composition. The notable points from Figure 4 are followings. (i) The swelling ratio of PNIPAM gel in D_2O is larger than that in H_2O at the low organic solvent composition region. (ii) In contrast to PNIPAM gel, the swelling ratio of PAAM gel in D_2O is smaller than that in H_2O at the low organic solvent composition region. (iii) There is no isotope effect in the swelling ratio of both PNIPAM and PAAM gels at high organic solvent composition region. Acetone-water mixtures also show the similar deuterium isotope effect on the swelling-shrinking curve in both PNIPAM and PAAM gels.[45] The result in the low organic solvent composition region is similar to the results of the diameter at the swollen state of PNIPAM and PAAM gels in pure H_2O and D_2O discussed previously (*vide supra*).[42] The experimental result in the high organic solvent composition region indicates that the hydrophobic interaction (both polymer-polymer and polymer-solvent interactions) does not or very weakly affect on the swelling ratio (volume) of the polymer hydrogels by deuterium substitutions. (Note that the hydrogels for deuterated samples were rinsed and immersed in D_2O before put into the mixtures of organic solvent and D_2O. Therefore, hydrogens of amide groups of PNIPAM or PAAM gel exchange to deuteriums.) The swelling ratio of PNIPAM gel in pure water at the shrinking state in high temperature region (> 310 K) also showed no deuterium isotope effect on the swelling ratio of the hydrogel.[43,44]

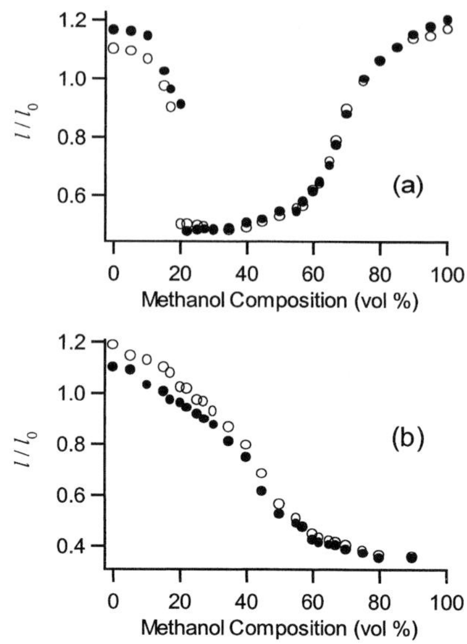

Figure 4. Methanol composition dependence of the swelling ratio l/l_0 for (a) PNIPAM gel and (b) PAAM gel in methanol-H_2O (open circles) and CH_3OD-D_2O (filled circles).

Figure 5 shows the fluorescence peak wavelength λ_f dependence of fluorescent-labeled (a) PNIPAM gel and (b) PAAM gel on the solvent composition of H_2O-methanol or D_2O-deuteriomethanol. The deuterium isotope effect in the microenvironment of the polymer hydrogels is also observed only in the low organic solvent composition region. However, the feature of the deuterium isotope effects of the swelling ratio and microenvironment of PAAM gel in the low organic solvent composition region is different from that in PNIPAM gel. λ_f of the fluorescent-labeled PNIPAM gel in D_2O is *longer* than that in H_2O and the swelling ratio of PNIPAM gel in D_2O is *larger* than that in H_2O. In contrast, the fluorescence peak of the fluorescent-labeled PNIPAM gel in D_2O is *longer* than that in H_2O, though the swelling ratio of PAAM gel in D_2O is *smaller* than that in H_2O.

The studies of aqueous PNIPAM or PAAM solutions and the polymer hydrogels by the molecular dynamics simulation and Raman and IR spectroscopies showed that the existence of the differently interacting water molecules.[60-70] Tamai et al. classified the water molecules of polymer solution in three categories: (i) around the hydrophilic group, (ii) around the hydrophobic group, and (iii) the bulk region.[63,64] Both the static and dynamic features of the hydrogen bonds in the three different regions were different from each other. It can be expected that the balance of the amplitudes of the three regions should determine the swelling ratio of the polymer hydrogel, because the solvent-polymer interaction parameter is related to the ensemble of the interaction strengths of solvent molecule and a part of polymer. In contrast to the macroscopic region, the information of the microscopic environment probed by fluorescent groups might come from one of the three categories in polymer solutions.

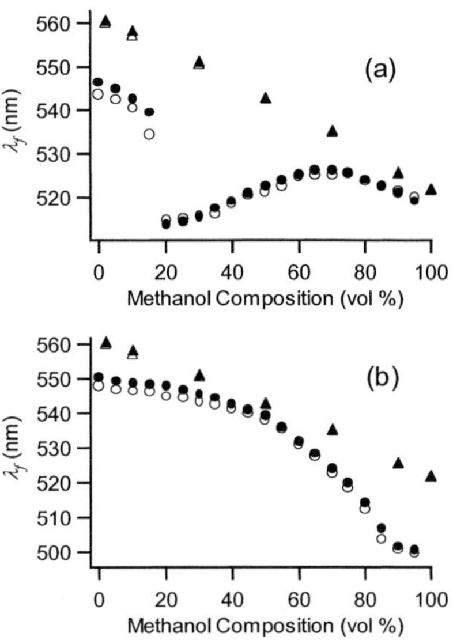

Figure 5. Methanol composition dependence of the fluorescence peak wavelength λ_f for dansyl-group-labeled (a) PNIPAM gel and (b) PAAM gel in methanol-H_2O (open circles) and CH_3OD-D_2O (filled circles). For the comparison, methanol composition dependence of λ_f for dansyl-group-labeled monomer in methanol-H_2O (open triangles) and CH_3OD-D_2O (filled triangles) are also shown.

4. Phase Separation of Inter-Polymer Complexes in Solution

Poly(acrylic acid) (PAA) forms cooperative inter-polymer hydrogen bonds with PAAM or the other polymers with amino groups as side groups only at low temperature, but dissociates at a certain temperature.[71-77] This phenomenon is so-called “zipper effect“. When an inter-polymer complex is formed due to cooperative inter-polymer hydrogen bonds, the binary polymer solution becomes turbid arising from the aggregations of the inter-polymer complexes due to the dehydration of the polymers. By utilizing the polymer-polymer interactions through the cooperative inter-polymer hydrogen bonds, a temperature-responsive polymer gel forming an interpenetrating polymer network has been prepared and studied by several groups.[77-81] We discuss the deuterium isotope effect on the phase separation of the zipper-type hydrogen-bonding inter-polymer complexes in solution in this section.[46]

Figure 6 shows the temperature dependence of the light transmittance at 500 nm of PAA-PAAM in H_2O and D_2O. For the reference, the temperature dependence of turbidity of PAA and PAAM in D_2O are also shown in Figure 6. Amazingly, the phase separation temperature of PAA-PAAM in D_2O is about 16 K higher than that of PAA-PAAM in H_2O. The other zipper-type hydrogen-bonding inter-polymer complex, PAA-poly(*N*-acryloylglycineamide) (PAG) complex, also shows the large deuterium isotope effect on the cloud point (about 5 K).[46]

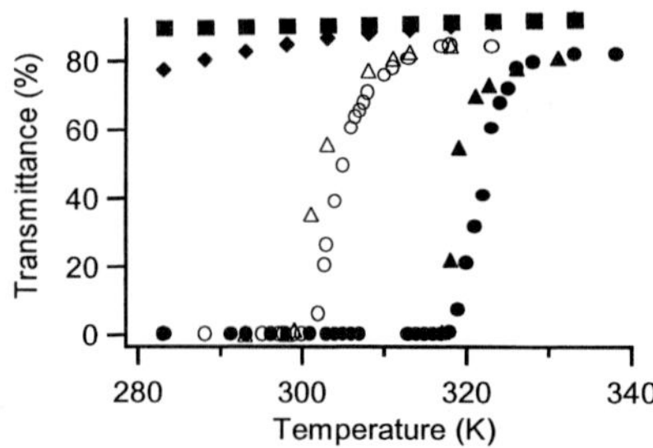

Figure 6. Temperature dependence of the light transmittance at 500 nm for PAA-PAAM in H_2O (open circles for heating process and open triangles for cooling process) and D_2O (filled circles for heating process and filled triangles for cooling process). Single component polymer solutions are also shown (filled squares denote PAA in D_2O and filled diamonds denote PAAM in D_2O).

These enormous deuterium isotope effects on the phase separation of the zipper-type hydrogen-bonding inter-polymer complex solutions have not been observed for other aqueous macromolecular solution systems, such as PNIPAM, cellulose, polypeptides, and proteins (about 1–2 K in these systems).[26,43,44,53,54,60,61,82-86] Since cooperative inter-polymer hydrogen bonds play a more significant role for the phase separation of the zipper-type polymer system than for the other simple aqueous polymer systems, it is possible to observe such huge isotope effects in the hydrogen-bonding inter-polymer systems. The smaller isotope effect on the phase separation temperature of PAA-PAG than PAA-PAAM could arise from the stronger hydration around PAG due to the extra hydrophilic groups present in PAG.

The interaction parameter χ in binary polymer solution systems is more complex than that in single polymer solutions. Although the phase state of single polymer solution can be expressed by one χ parameter, it is necessary to consider three interaction parameters in the case of polymer-polymer solutions: polymer(A)-solvent χ_{AS}, polymer(B)-solvent χ_{BS}, and so-called segregation factor χ_{AB} arising from polymer(A)-polymer(B) interaction.[8,51] From the estimation of the stabilization energies of formic acid dimer, formic amide dimer, and formic acid - formic amide dimer by MP2/6-31G(d,p)-level calculation, the molecular mechanics stabilization energy of formic acid - formic amide dimer is the largest among these three dimers.[87] We can easily imagine that since the inter-polymer hydrogen bonds has a great role to determine the phase state of hydrogen-bonding inter-polymer complex solution, the deuterium isotope effect on the phase separation of the hydrogen-bonding inter-polymer solution should mainly arise from the segregation factor. Although the polymer-solvent interaction factors can be changed by deuterium substitutions like in PNIPAM, PAAM, and other simple aqueous polymers, the contribution to the deuterium isotope effect of the polymer-solvent interactions on the phase separation of hydrogen-bonding inter-polymer complexes should be minor in comparison with the contribution of the polymer-polymer interaction.

5. Concluding Remarks

In this article, we have reviewed the deuterium isotope effects on the swelling kinetics and volume phase transition in the typical polymer hydrogels: PNIPAM and PAAM gels. The deuterium isotope effects of the phase separation of the hydrogen-bonding inter-polymer complexes have also discussed. Regarding the swelling kinetics of polymer hydrogels, the difference of the viscosities of H_2O and D_2O is the main origin of the isotope effect on the cooperative diffusion constant in the swelling process. The difference of the volume phase transition and phase separation of the polymer hydrogels in H_2O and D_2O arises from the different polymer-solvent interaction parameters. The deuterium isotope effect on the phase separation temperature in zipper-type hydrogen-bonding inter-polymer complex solution, PAA-PAAM in water, is as large as 16 K. In the case of zipper-type hydrogen-bonding inter-polymer complex, the polymer-polymer interaction could give large contribution to the phase separation rather than the polymer-solvent interactions. From the fluorescence studies, the microscopic features of the swelling-shrinking curves of polymer hydrogels have been observed. The deuterium isotope effects of the bulk and microenvironment in the polymer hydrogels are not always consistent. The difference between bulk and microenvironment of the polymer hydrogels may arise from the microheterogeneity of the polymer hydrogel systems.

Acknowledgement: We are grateful to our co-workers, Ms. Noriko Endo, Ms. Nozomi Kuwabara, and Mr. Kazuya Ohkawa (University of Tokyo), for their significant contributions to the works. We also thank to Dr. Hiroshi Ushiyama (University of Tokyo) for fruitful discussions regarding the inter-polymer complexes. The works were partially supported by *Ministry of Education, Science, Sports and Culture* of Japan.

[1] K. Dusek, D. Patterson, *J. Polym. Sci. Part A-2* **1968**, *6*, 1209.
[2] O. B. Ptitsyn, A. K. Kron, Y. Y. Eizer, *J. Polym. Sci. Part C: Polym. Symp.* **1968**, *16*, 3509.
[3] T. Tanaka, *Phys. Rev. Lett.* **1978**, *40*, 820.
[4] K. Dusek, *Responsive Gels: Volume Transitions I, Vol. 109*, Springer-Verlag, Berlin Heidelberg, 1993.
[5] K. Dusek, *Responsive Gels: Volume Transitions II, Vol. 110*, Springer-Verlarg, Berlin Heidelberg, 1993.
[6] P. J. Flory, J. Rehner, Jr., *J. Chem. Phys.* **1943**, *11*, 521.
[7] P. J. Flory, *J. Chem. Phys.* **1950**, *18*, 108.
[8] P. J. Flory, *Principles of Polymer Chemistry*, Cornell University Press, Ithaca, NY, 1953.
[9] H. Inomata, S. Goto, S. Saito, *Macromolecules* **1990**, *23*, 4887.
[10] R. Moerkerke, R. Koningsveld, H. Berghmans, K. Dusek, K. Solc, *Macrmolecules* **1995**, *28*, 1103.
[11] H. Schafer-Soenen, R. Moerkerke, H. Berghmans, R. Koningsveld, K. Dusek, K. Solc, *Macromolecules* **1997**, *30*, 410.
[12] R. Moerkerke, F. Meeussen, R. Koningsveld, H. Berghmans, W. Mondelaers, E. Schacht, K. Dusek, K. Solc, *Macromolecules* **1998**, *31*, 2223.
[13] B. Erman, P. J. Flory, *Macromolecules* **1986**, *19*, 2342.
[14] S. Hirotsu, *J. Chem. Phys.* **1991**, *94*, 3949.
[15] M. Shibayama, Y. Shirotani, H. Hirose, S. Nomura, *Macrmolecules* **1997**, *30*, 7307.
[16] I. D. Kuntz, W. Kauzmann, *Adv. Protein Chem.* **1974**, *28*, 239.
[17] G. A. Jeffrey, W. Saenger, *Hydrogen Bonding in Biological Structures*, Springer-Verlag, Berlin, 1991.
[18] J. A. Rupley, G. Careri, *Adv. Protein Chem.* **1991**, *41*, 37.
[19] C. L. Perrin, J. B. Nielson, *Annu. Rev. Phys. Chem.* **1997**, *48*, 511.
[20] G. Jancso, P. N. Rebelo, W. A. Van Hook, *Chem. Rev.* **1993**, *93*, 2645.
[21] G. Nemethy, H. A. Scheraga, *J. Chem. Phys.* **1964**, *41*, 680.
[22] *CRC Handbook of Chemistry and Physics*, 74 ed., CRC Press, Boca Raton, Fl, 1993.
[23] J. Szydlowski, W. A. Van Hook, *Macrmolecules* **1991**, *24*, 4883.
[24] M. Luszczyk, L. P. N. Rebelo, W. A. Van Hook, *Macrmolecules* **1995**, *28*, 745.
[25] M. Luszczyk, W. A. Van Hook, *Macrmolecules* **1996**, *29*, 6612.
[26] M. Shibayama, T. Tanaka, C. C. Han, *J. Chem. Phys.* **1992**, *97*, 6829.
[27] K. Horie, H. Ushiki, F. W. Winnik, *Molecular Photonics: Fundamentals and Practical Aspects*, Kodansha, Tokyo and WILEY-VCH, Weinheim, 1999.
[28] G. Weber, *Biochem. J.* **1952**, *51*, 155.
[29] I. M. Klotz, H. F. Fiess, *Biochim. Biophys. Acta* **1960**, *38*, 57.
[30] T. Iio, Y. Iwashita, H. Watanabe, *Bull. Chem. Soc. Jpn.* **1972**, *45*, 2206.
[31] T. Iio, *Bull. Chem. Soc. Jpn.* **1973**, *46*, 335.
[32] U. P. Strauss, G. Vesnaver, *J. Phys. Chem.* **1975**, *79*, 1558.
[33] U. P. Strauss, G. Vesnaver, *J. Phys. Chem.* **1975**, *79*, 2426.
[34] K. Horie, I. Mita, J. Kawabata, S. Nakahama, A. Hirao, N. Yamazaki, *Polym. J.* **1980**, *12*, 319.
[35] Y. Hu, K. Horie, H. Ushiki, *Macrmolecules* **1992**, *25*, 6040.
[36] Y. Hu, K. Horie, H. Ushiki, F. Tsunomori, T. Yamashita, *Macrmolecules* **1992**, *25*, 7324.
[37] Y. Hu, K. Horie, H. Ushiki, *Polym. J.* **1993**, *25*, 651.
[38] T. Torii, T. Yamashita, K. Horie, *Biopolymers* **1994**, *34*, 101.
[39] M. Asano, T. Yamashita, K. Horie, *Polym. Gels Networks* **1995**, *3*, 281.
[40] M. Asano, F. W. Winnik, J. Yamanaka, K. Horie, *Macrmolecules* **1995**, *28*, 5861.
[41] K. Horie, M. Kuriyama, *Rep. Prog. Polym. Jpn.* **1997**, *40*, 159.
[42] H. Shirota, K. Horie, *Chem. Phys.* **1999**, *242*, 115.
[43] H. Shirota, N. Endo, K. Horie, *Chem. Phys.* **1998**, *238*, 487.
[44] H. Shirota, N. Kuwabara, K. Ohkawa, K. Horie, *J. Phys. Chem. B* **1999**, *103*, 10400.
[45] H. Shirota, K. Ohkawa, N. Kuwabara, N. Endo, K. Horie, *Macromol. Chem. Phys.* **2000**, *201*, 2210.
[46] N. Endo, H. Shirota, K. Horie, *Macromol. Rapid Commun.* **2001**, *22*, 593.
[47] T. Tanaka, D. J. Fillmore, *J. Chem. Phys.* **1979**, *70*, 1214.
[48] Y. Li, T. Tanaka, *J. Chem. Phys.* **1990**, *92*, 1365.
[49] A. Peters, S. J. Candau, *Macromolecules* **1986**, *19*, 1952.
[50] A. Peters, S. J. Candau, *Macromolecules* **1988**, *21*, 2278.
[51] P. D. de Gennes, *Scaling Concepts in Polymer Physics*, Cornell University Press, Ithaca, NY, 1979.
[52] N. Matsunaga, A. Nagashima, *J. Phys. Chem. Ref. Data* **1983**, *12*, 933.
[53] X. Wang, C. Wu, *Macrmolecules* **1999**, *32*, 4299.
[54] P. Kujawa, F. W. Winnik, *Macrmolecules* **2001**, *34*, 4130.
[55] J. Eliasaff, A. Silberberg, *J. Polym. Sci.* **1959**, *41*, 33.

[56] M. Heskins, J. E. Guillet, *J. Macromol. Sci., Chem.* **1968**, *A2*, 1441.
[57] H. G. Schild, *Prog. Polym. Sci.* **1992**, *17*, 63.
[58] M. Shibayama, T. Tanaka, *Adv. Polym. Sci.* **1993**, *109*, 1.
[59] S. Hirotsu, *Adv. Polym. Sci.* **1993**, *110*, 1.
[60] Y. Maeda, T. Higuchi, I. Ikeda, *Langmuir* **2000**, *16*, 7503.
[61] Y. Maeda, T. Nakamura, I. Ikeda, *Macromolecules* **2001**, *34*, 1391.
[62] Y. Maeda, T. Nakamura, I. Ikeda, *Macrmolecules* **2001**, *34*, 8246.
[63] Y. Tamai, H. Tanaka, K. Nakanishi, *Macromolecules* **1996**, *29*, 6750.
[64] Y. Tamai, H. Tanaka, K. Nakanishi, *Macrmolecules* **1996**, *29*, 6761.
[65] T. Terada, Y. Maeda, H. Kitano, *J. Phys. Chem.* **1993**, *97*, 3619.
[66] Y. Maeda, N. Tsukida, H. Kitano, T. Terada, J. Yamanaka, *J. Phys. Chem.* **1993**, *97*, 13903.
[67] M. Shibayama, M. Morimoto, S. Nomura, *Macrmolecules* **1994**, *27*, 5060.
[68] S.-Y. Lin, K.-S. Chen, L. Run-Chu, *Polymer* **1999**, *40*, 2619.
[69] Y. Maeda, T. Higuchi, I. Ikeda, *Langmuir* **2001**, *17*, 7535.
[70] O. Ramon, E. Kesselman, R. Berkovici, Y. Cohen, Y. Paz, *J. Polym. Sci.: Part B: Polym. Phys.* **2001**, *39*, 1665.
[71] E. A. Bekturov, L. A. Bimendina, *Adv. Polym. Sci.* **1981**, *41*, 99.
[72] E. Tsuchida, K. Abe, *Adv. Polym. Sci.* **1982**, *45*, 1.
[73] Y. Osada, *J. Polym. Sci., Polym. Chem. Ed.* **1979**, *17*, 3485.
[74] K. Abe, M. Koide, E. Tsuchida, *Macrmolecules* **1977**, *10*, 1259.
[75] V. Y. Baranovsky, A. A. Litmanovich, I. M. Papisov, V. A. Kababov, *Eur. Polym. J.* **1981**, *17*, 969.
[76] D. J. Eustace, D. B. Siano, E. N. Drake, *J. Appl. Polym. Sci.* **1988**, *35*, 707.
[77] H. Katono, H. Maruyama, K. Sanui, N. Ogata, T. Okano, Y. Sakurai, *J. Controlled Release* **1991**, *16*, 215.
[78] F. Ilmain, T. Tanaka, E. Kokufuta, *Nature* **1991**, *349*, 400.
[79] H. Katono, K. Sanui, N. Ogata, T. Okano, Y. Sakurai, *Polym. J.* **1991**, *23*, 1179.
[80] H. Sasase, T. Aoki, H. Katono, K. Sanui, N. Ogata, R. Ohta, T. Okano, Y. Sakurai, *Makromol. Chem., Rapid. Commun.* **1992**, *13*, 577.
[81] T. Aoki, M. Kawashima, H. Kitano, K. Sanui, N. Ogata, Y. Sakurai, *Macrmolecules* **1994**, *27*, 947.
[82] F. W. Winnik, *J. Phys. Chem.* **1989**, *93*, 7452.
[83] F. W. Winnik, *Macromolecules* **1991**, *20*, 2745.
[84] C. Luan, D. W. Urry, *J. Phys. Chem.* **1991**, *95*, 7896.
[85] G. I. Makhatadze, G. M. Clore, A. M. Gronenborn, *Nat. Struct. Biol.* **1995**, *2*, 852.
[86] R. Guzzi, L. Spotelli, C. L. Rosa, D. Milardi, D. Grasso, *J. Phys. Chem. B* **1998**, *102*, 1021.
[87] H. Ushiyama, work in progress.

Macromol. Symp. **2004**, *207*, 95-104

Structural and Optical Properties of Poly(4-vinyl pyridine)/Pyridine Gels

Evgenia Vaganova, Shlomo Yitzchaik*

Institute of Chemistry and the Farkas Center for Light Induced Processes, The Hebrew University of Jerusalem, Jerusalem 91904, Israel

Summary: A study of photoinduced processes in pyridine-based gels is presented. Poly(4-vinyl pyridine) dissolved in pyridine with water addition forms a photosensitive material. Under irradiation at the main absorption at 250 nm, a new intense absorption at 360 nm and an emission at 515 nm appeared. Upon continuation of radiation, together with an increase in the intensity of the new absorption, a prolonged tail through the whole visible range is observed. A new far red-shifted emission arises in correlation with the absorption spectrum changes. The presence of lower energy aggregates leads to polymer morphology changes. Short-range aggregates enhance polymer-polymer interactions. Long range aggregates have a tendency to crystallize. The photoinduced morphology changes was studied by transmission electron microscopy (TEM) of the polymer thin film and differential scanning calorimetry (DSC) of the gel. Phase separated species, micelles-like forms with domain size over 200 nm, and nanocrystals with average size 20-30 nm are demonstrated.

Introduction

The presence of the lone pair of sp^2 electrons on the pyridine nitrogen atom influences the photophysical behavior of the molecule. A comparatively weak base with pK_a= 5.21 and hydrogen bond basicity pK_{HB}=1.86,[1] pyridine turns into a strong base when in the photoexcited state. According to semi-empirical calculations[2] the difference in proton affinity between the ground and excited states amounts to 12 kcal/mol. In the presence of water, under UV-irradiation from the main absorption at 254 nm, pyridine is known to undergo photoisomerization to a Dewar pyridine. As the reaction continues, 5-amino-2,4-pentadienal (AP) is produced. AP is not stable and reverts in the dark to pyridine with water elimination.[3-7]

Ten years after its discovery,AP was synthesized.[8] It exists as yellow crystals with maximum absorption at λ_{max}=339 nm (absorption spectrum taken in 0.02 N HCl solution) and extinction coefficient ε=34400. These yellow crystals are also unstable and rapidly turn brownish at room temperature upon exposure to air.[8]

 DOI: 10.1002/masy.200450309

In pyridine-based polymers the presence of the unshared electron pair of the pyridine N results in a strong flexibility of the optical properties. In poly(*p*-pyridine) and poly(*p*-pyridyl vinylene), additional (n,π*) states result from the promotion of a lone-pair electron from the nitrogen heteroatom to the π* backbone. The optical properties of these polymers were shown to be morphology-dependent. Photoluminescence (PL) of the polymers was red-shifted in thin films compared to in solution, a result that was explained by aggregate formation, where the ground state and excited state wavefunctions are delocalized over several chains. The practical application of these polymers as a symmetrically configured light-emitting device (device which emits light under forward and reverse d.c. bias) was shown.[9]

Recently an unusual novel photosensitive material composed from poly(4-vinyl pyridine)/pyridine (P4VPy/Py) mixture was observed.[10-13] Under UV-irradiation the mixture gels. The photogenerated gel has altered optical properties, especially multicolor emission.[10] FTIR studies of the material reveal the formation of pyridinium ion attached to the polymer backbone immediately after polymer dissolution [13] and hydrogen bond formation during photoinduced gelation.[10]

Here we present the study of the structural and optical properties changes of P4VPy/Py/H_20 mixture caused by irradiation at 250 nm. The P4VPy/Py/H_20 mixture, initially a viscous solution, turns to the chemically crosslinked gel, which is difficult to dissolve even in polar solvents. A new absorption band centered at 360 nm, and intense green and red PL at 515 nm and 640 nm, respectively, accompany the structural changes. We have carried out optical measurements, transmission electron microscopy (TEM), and differential scanning calorimetry (DSC) to study the optical and structural properties of the photoinduced gel. By considering the evolution of the optical properties and comparing the latter with the structural investigations on the nanometer scale, we concluded that under UV-irradiation at 250 nm several photoprocesses occur. One of these is the pyridine ring opening reaction. Another one is the formation of aggregates. The aggregates of the initial stages of irradiation have agreen emission at 515 nm. Upon prolongation of the irradiation another type of the aggregates appears. These aggregates exhibit red emission at 640 nm. Comparison of the parameters of the red (640 nm) and green (515 nm) PL such as: lifetime – 2 ns and 5.5 ns, Stokes losses – 0.09 eV and 0. 29 eV; full width at half maximum (FWHM) – 0.3 eV and 0.55 eV it is suggested that these changes are due to the more rigid and well-defined structures of the red emitting centers.

TEM pictures illustrate modification of the gel. Micelle-like structures with size >200 nm and nanocrystals of size 20-30 nm form under irradiation. The nanocrystals exhibit an electron diffraction pattern with Bragg peaks. A phase transition to the crystal phase can also be seen in a narrow temperature range with T_m=113°C in the calorimetry diagram of the irradiated gel.

The results presented conclude that the nanocrystals mostly involve red emitting centers and the semisolid micelles mostly contain the green emitting centers.

The photoproduct, we believe AP, is much more stable in the viscous media than in a pyridine-water solution. With prolongation of irradiation the concentration of photoproduct increases. The increasing concentration of the photoproduct occurs simultaneously with improvement of the rigidity of the system. Thus we can conclude that AP is an efficient crosslinker in the system.

Experimental Techniques

For this work, poly (4-vinyl pyridine) (P4VPy) with a molecular weight of 50000 (Polyscience Corporation) and gelatin (Aldrich) were used. P4VPy was carefully dried in a vacuum oven (10^{-3} torr), at 90°C for one week before use. P4VPy was dissolved in pyridine-d_5 or in pyridine-h_5 (Py), pK_{HB}=1.86.[5] Both kinds of pyridine were anhydrous (water < 0.003%) and obtained from Aldrich. The ratio between free solvent molecules and polymer side-chain groups was 1:1 in all solutions. Water used was triply distilled with pH=7.0. The water concentration was varied: 0.02, 3, and 50 wt %. Py-d_5 was used for the preparation of the gel with water concentration 0.02 and 3 wt %. Polymer dissolution was carried in a dry box under a N_2 atmosphere. Spectroscopic measurements were performed immediately after preparation of the mixture. Samples were placed in a 2.5-mm quartz cuvette (Starna). For TEM investigation the composition (with water concentration 3 wt %) was dissolved in pyridine so that the P4VPy concentration was 1 wt %. The polymer solution was deposited on 400-mesh copper grids, and then irradiated.

Absorption spectra were recorded on a Shimadzu UV-3101PC scanning spectrophotometer. Excitation and PL spectra were taken on a Shimadzu RF-5301PC spectrofluorimeter. Data were collected at right angles to the excitation beam. UV-irradiation at 250nm (0.5 mW/cm^2) was accomplished by a Xenon short arc lamp (Ushio) inside the Shimadzu RF-5301PC. Periodically, the irradiation was interrupted, and UV-VIS and photoluminescence measurements were performed. The resolution of the emission and excitation spectra was 1 nm; the resolution of the absorption spectra was 2 nm. PL lifetime measurements were performed by phase modulation with a ISS K-2 unit. The gelation was evaluated visually, depending on the ability of the solution to flow or not. The transmission electron microscope Tecnai-12 (110kV) (Philips) was applied for structural study at the nanometer scale. Differential scanning calorimetry DSC 822 (Mettler Toledo) was used for calorimetry measurements. The CS Chem3D CS ChemOffice Software was used for modeling.

Photophysics of the P4VPY/PY Composition

Figure 1 presents the measured absorption and emission spectra of the composition with water addition in ratio 1:1:0.3 (3 wt %) before and after irradiation at 250 nm for 60 min. The initial absorption spectrum of the gel is almost identical to that of pyridine (curve 1), the PL spectrum is dark blue centered at 450 nm by excitation at 380 nm (curve 3). After irradiation a new absorption centered at 356 nm is observed (curve 4). In conjunction with the absorption spectra change a new green emission peaking at 515 nm (curve 6) by excitation at 468 nm appears in PL spectra.

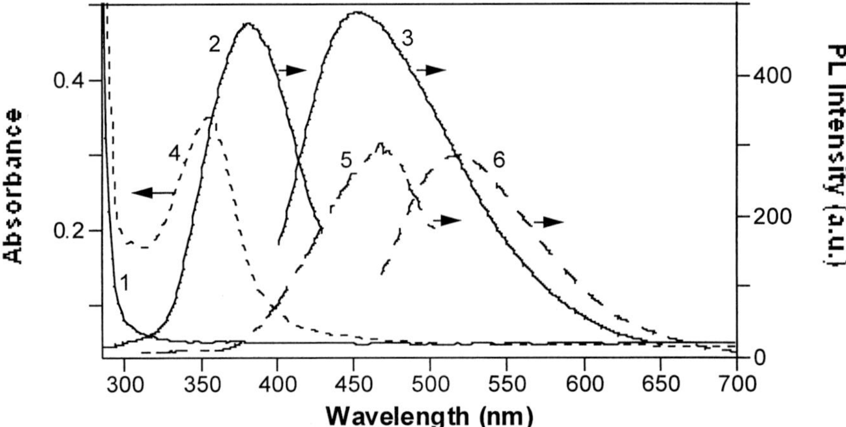

Figure 1. Absorption spectra (1,4); emission (3,6) and excitation (2,5) spectra before irradiation (1,2,3 - solid lines) and after UV-irradiation with 250 nm for 60 min (4,5,6 - dashed lines), for the P4VPy/Py/water composition (mol ratio 1:1:0.3).

The appearance of the absorption at 356 nm can be explained by the presence of some amount of the pyridine open form - AP as a result of the photochemical reaction. As was mentioned above, it is well established that pyridine in the presence of water undergoes a ring opening reaction under excitation at 250 nm.[3-7]

To elucidate the structure of the green emitting centers, a model system - gelatin/water/pyridine gel at different pH's (5.0, 6.0, 7.0) was studied. The gel was sandwiched between two quartz slides and irradiated with 250 nm wavelength for 30 min.

Figure 2 shows the absorption spectra of the gelatin gel (at pH 5.0, 7.0) before irradiation (1); and absorption (2), excitation (3), and emission (4) spectra after irradiation. As can be seen, the intensity of the pyridine absorption at 250 nm is decreased, and simultaneously an intense new absorption at 360 nm is rising. Moreover, in the absorption spectrum of the gel at pH=5.0 an additional shoulder at 460 nm can clearly be observed.

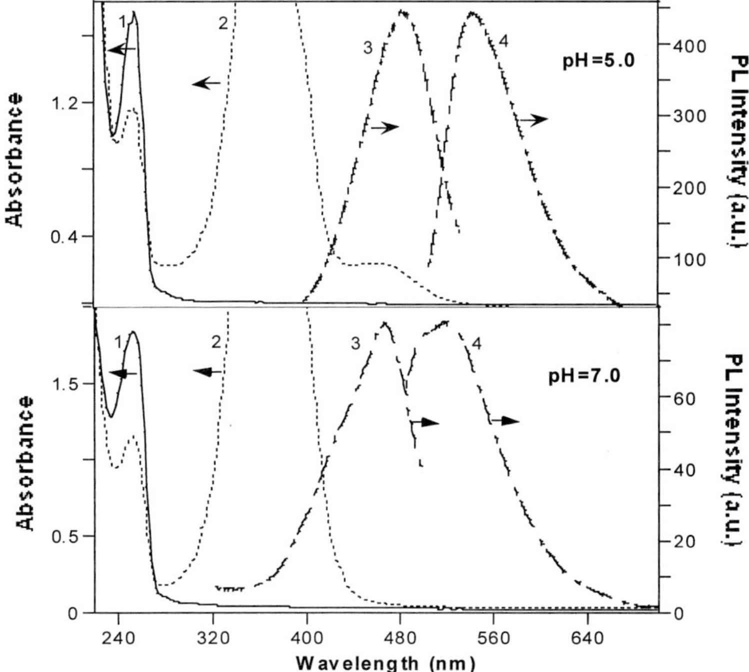

Figure 2. Absorption spectra before irradiation (1); absorption (2), excitation (3) and emission (4) spectra after UV-irradiation with 250 nm for 30 min of the gelatin/Py/water composition with different pH: 5.0 (pH=5.0) and 7.0 (pH=7.0).

The initial the gel practically has no emission. Nevertheless, under irradiation a green emission similar to the one observed in the P4VPy/Py/water (Fig.1) is registered (Fig.2).

The intensity and the peak position of the green emission is observed to be dependent on pH, as shown by Fig.2. The peak position is red-shifted from 517 nm to 542 nm and the intensity increases upon decreasing pH from 7.0 to 5.0.

It is important to note that the appearance of the identical green emission as a result of UV-irradiation in P4VPy/Py gel was observed earlier.[10] Transition energy of the green PL on the transition energy of the initial blue PL depends on the following empirical equation:

$$E_G^{PL} = E_B^{PL} - \Delta_B/2 + C \text{ (eV)} \qquad (1)$$

where E_B^{PL} is the energy of blue PL (eV); E_G^{PL} is the energy of green PL (eV); Δ_B denotes the Stokes losses of the blue PL (eV) and C is a constant in the first approximation. (Note that C is a function of the molecular rearrangement in the excited and ground states; in our case it is in the range 15-50 meV).[10] Regarding the classical characterization of the dimer given by Kasha[14] a similar equation describes the "dimer" emitting center structure, where the building block is the emitting center with blue PL.

We now consider the results of the green PL study. Under irradiation in the investigated pyridine-content systems a new aggregate with a radiative energy transition at 2.4 eV (460 nm-520 nm) appears. The appearance of the aggregates occurs as in the P4VPy-based gel as well as in the gelatin gel. In the gelatin gel with low pH the absorption at 460 nm is clearly present in the absorption spectrum after irradiation, which demonstrates a high concentration of the photoinduced aggregates.

At this stage we do not have enough information about the detailed structure of the aggregate. However, flexibility of acid-base properties of pyridine in the excited state, namely, the increase of pyridine basicity in the excited state,[2] and the high concentration of photoinduced aggregates in the gelatin gel at pH=5.0, similar to the "dimer" structure of the emitting center, suggest the following model. A green emitting center appears as a result of photoinduced bonding of two closely located species, which were present in the system before irradiation. An initial building block we believe is the linear pyridine-pyridinium (protonated pyridine) aggregate.[10] Complexation of pyridine-water due to hydrogen bonding was extensively investigated.[15,16]

Continuation of UV-irradiation with 250 nm of the P4VPy/Py samples containing water for 300 min led to an increase of the intensity of the new absorption at 360 nm and the appearance of the prolonged tail throughout the whole visible range (Fig.3). Simultaneously the new intense red emission at 640 nm by excitation at 610 nm is accompanied by a change in the absorption (Fig. 3).

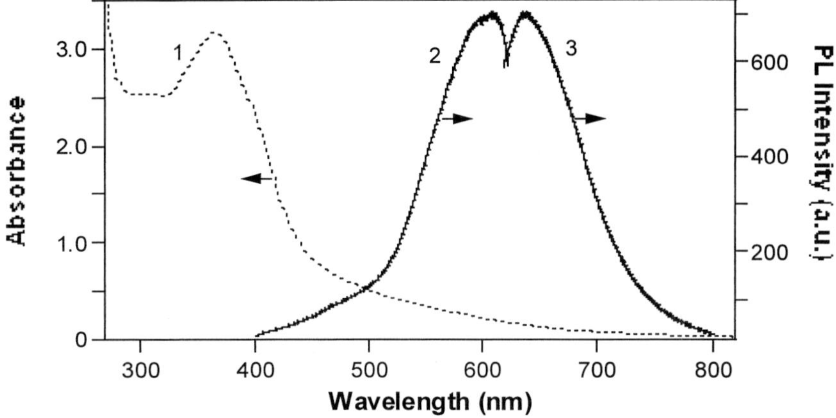

Figure 3. Absorption (1); excitation (2) and emission (3) spectra measured after irradiation of P4VPy/Py/water composition (mol ratio 1:1:10) with 250 nm for 300 min.

The red emission has significantly different properties compared to the green one. Comparing the parameters such as: lifetime – 2 ns and 5.5 ns, Stokes losses – 0.09 eV and 0. 29 eV; FWHM – 0.3 eV and 0.55 eV for red and green PL respectively, argue for a more rigid and well-defined structure of the red emitting centers.

TEM and Differential Scanning Calorimetry Studies

Together with formation of the luminescent aggregates, changes were also observed in the polymer morphology. TEM and calorimetry studies were applied for the polymer morphology investigation.

Transmission electron microscopy (TEM) was carried out on samples obtained from the thin gel film before and after irradiation at 250 nm for 60 min. The TEM image of the gel before irradiation is presented in Fig.4a in two scales: with low and high magnification (Fig.4a, including insert). A homogeneous structure of the polymer film is observed.

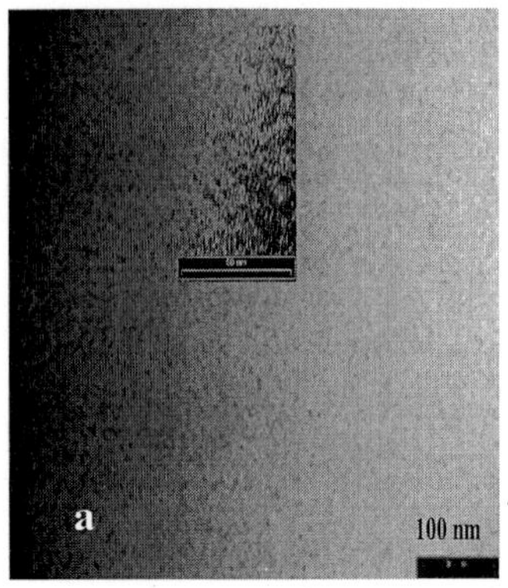

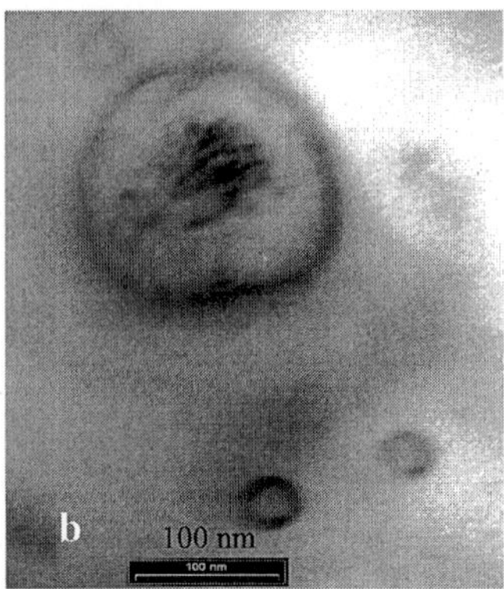

Figure 4. TEM images of the thin gel film before (a) and after UV-irradiation (b) by 250 nm for 60 min.

In support of the results of the optical study (formation of two different types of the aggregates) two new kinds of molecular packing are observed after irradiation (Figs. 4b, 5, 6). The typical region of the irradiated gel film with lower magnification is presented in Fig. 4b.

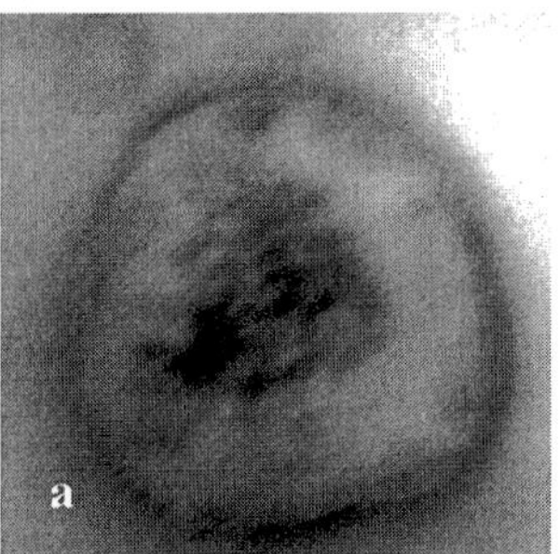

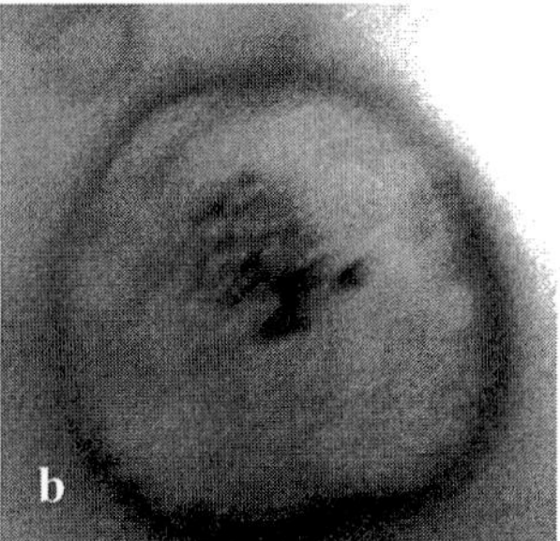

Figure 5. TEM images of the same micelle, taking one (a) after another (b) with the difference in the registration time 2 min.

The micelles with the size in the range of 200 nm (Fig.4 b; 5 a,b), and the nanocrystals with the average size 20-30 nm (Fig. 6) are two kinds of the phase-separated structures, which characterize the gel after UV-irradiation. Interesting behavior of the micelles is shown in Fig. 5a,b. The surface of the micelle has a non-homogeneous structure. The dark spots on the surface belong to the regions with higher electron density. The topography of the darkest areas was constantly changing during the period of the time of the image registration. We suggest that this observation reflect the isomerization of the inter- or intrachain polymer crosslinking.

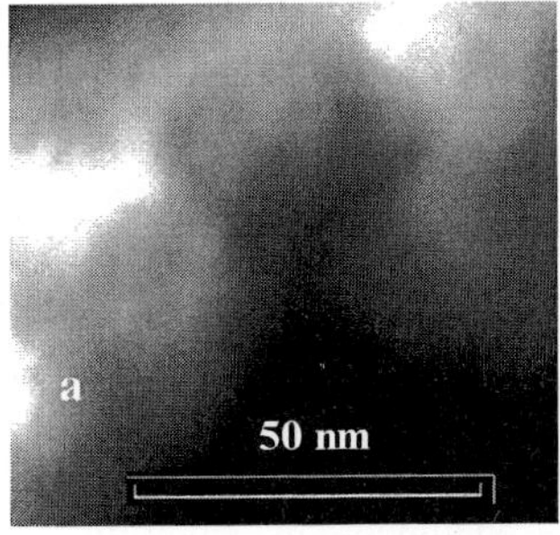

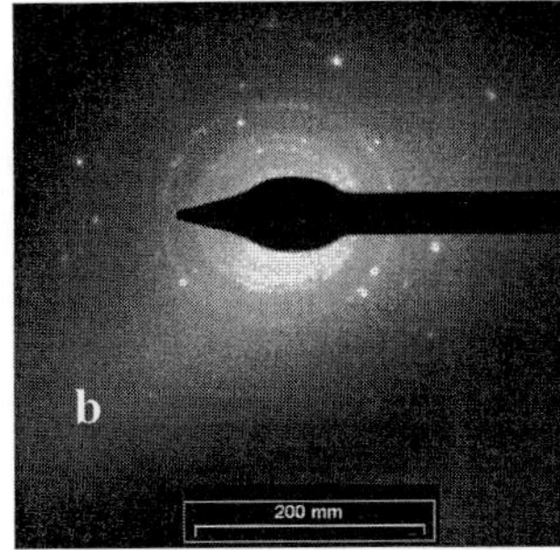

Figure 6. TEM image of a typical nanocrystal (a) and electron diffraction pattern of the same area (b).

The electron diffraction pattern of the typical crystal, shown in Fig. 6b, clearly indicates the concentric ring diffraction pattern and Bragg spots.

Crystalline phase formation in the gel during irradiation with 250 nm also has been confirmed by scanning calorimetry study. Figure 7 presents the calorimetry diagram of P4VPy/Py/water gel before and after UV-irradiation for 300 min in a bulk (the PL and absorption spectra of the same sample are presented in Fig. 3). The sharp peak in the diagram of the irradiated sample demonstrates crystal phase appearance after irradiation.

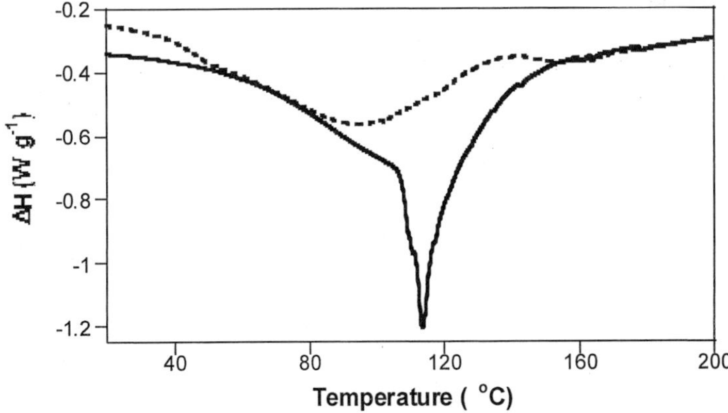

Figure 7. DSC diagram of the gel before (dashed line) and after irradiation with 250 nm for 300 min (solid line) taken at a scan rate of 10^0C/min.

Conclusion

The photophysics of the pyridine-based system P4VPy/Py shows changes in optical properties resulting from the photochemical reaction in the system. The efficiency of the reaction is strongly dependent on the quantity of water in the system. With increasing water content new lower energy aggregates result as the reaction is occurring. The first stage aggregate has a structure similar to the dimer. This photoinduced "dimer" has absorption at 460 nm, and green emission.

Upon continuation the irradiation optical properties change significantly. Longer chain aggregates with lower energy ground and excites states reveal themselves in the prolonged tail throughout the whole visible range and intense red emission.

Short chain and long chain aggregates change the morphology of the gel. Short range aggregates enhance polymer-polymer interaction. Long range aggregates have a tendency towards crystallization. The photoinduced morphology changes were studied by TEM and DSC. Phase separated species of micelles-like form with size over 200 nm, and nanocrystals with average size 20-30 nm are demonstrated. Based on the results of the experiments, and a comparing of our results with the known properties of the aminopentadienal as the product of the open ring reaction, at this stage we can suggest that aminopentadienal acts as the effective crosslinker in the system.

Acknowledgement: E.V. gratefully thank the Israel Ministry of Absorption for financial support.

[1] M. Berthelot, C. Laurence, M. Safar, F. Besseau, *J. Chem. Soc. Perkin Trans.* **1998**, *2*, 283.
[2] E.Vaganova, S. Yitzchaik, L. Shapiro, M. Sigalov, V. Khodorkovsky, *Adv. Mater.* **2000**, *12,22*, 1669.
[3] J. Joussot-Dubien. *Tetrahedron Letters* **1967**, *44*, 4389.
[4] J. Joussot-Dubien, J. Houdard-Pereyre. *J. Bull. Soc. Chim. Fr.* **1969**, 2619.
[5] H. Freytag, *Chem. Ber.* **1936**, *69B*, 32.
[6] J.C. Andre, M. Niclause, J. Joussot-Dubien, X. Deglise, *J. of Chemical Education* **1977**, *54*, 387.
[7] K.E. Wiltzbach, D.J. Rausch, *J. Am. Chem. Soc.* **1970**, *92*, 2178.
[8] D. Reinehr, T. Winkler,. *Angew. Chem. Int. Ed. Engl.* **1981**, *10*, 881.
[9] A.J. Epstein, J.W. Blatchford, Y.Z. Wang, S.W. Jessen, D.D. Gebler, L.-B. Lin, T.L. Gustafson, [10] H.-L.Wang, Y. W. Park, T.M. Swager, A. G. MacDiarmid, , *Synth. Metals* **1996,** *78*, 253.
[10] E.Vaganova, G. Meshulam, Z. Kotler, M. Rozenberg, S Yitzchaik, *J.of Fluorescence* **2000**, *10*, 81.
[11] E. Vaganova, M. Rozenberg, S. Yitzchaik, *Acta polym.* **1998**, *49*, 632.
[12] E. Vaganova, M. Rozenberg, S. Yitzchaik, *Chemistry of Materials* **2000**, *12*, 261.
[13] M. Rozenberg, E. Vaganova, S. Yitzchaik, *New Journal Chemistry* **2000**, 24, 332.
[14] M. Kasha, H.R. Rawis, El-Bayoumi, *A. Pure Appl. Chem.* **1965,** *11*, 371.
[15] S.Schliicker, R.K. Singh, B.P.Asthana, J. Popp, W. Kiefer, *J.Phys.Chem. A* **2001**,105, 9983.
[16] H.Takahashi, K.Mamola, E.K. Plyer, *J. of Molecular. Spectr.* **1966**, 21, 217.

Macromol. Symp. **2004**, *207*, 105-109

Image Analysis of the Spatial Distribution of Paramagnetic Pr^{3+} Ions in a Polymer Gel by a ^{1}H-Chemical Shift NMR Imaging Method

*Satomi Yokota, Akitsugu Sasaki, Yoshio Hotta, Yuji Yamane, Hideaki Kimura, Shigeki Kuroki, Isao Ando**

Department of Chemistry and Materials Science, International Research Center of Macromolecular Science, Tokyo Institute of Technology, Ookayama, Meguro-ku, Tokyo 152, Japan

Summary: ^{1}H-chemical shift NMR imaging patterns of a poly(methacrylic acid) gel containing water with paramagnetic praseodymium ions(Pr^{3+}) were successfully observed, in order to elucidate spatial distribution of Pr^{3+} ions in the gel. The ^{1}H chemical shift of water associated with Pr^{3+} ions in the gel moves largely downfield. By analyzing these experimental results, the immersion process of Pr^{3+} ions into the network of the polymer gel was spatially clarified. Further, it is shown that the chemical shift NMR imaging method is a useful means for determining the spatial distribution of paramagnetic metal ions in polymer gels.

Introduction

It has been demonstrated that the NMR imaging method is a useful means for obtaining non-destructively μm-scale spatial information on probe molecules in bulk matter.[1, 2] For this reason, the NMR imaging method has been successfully applied to polymer gel systems.[3-10] In particular, the response of polymer gels under the application of external stimuli, such as stress and electric field,[11,12] has been elucidated at the molecular level and at the macroscopic level by means of NMR imaging method. The target of those efforts was to understand the nature of polymer gels and to discover novel properties. Therefore, the introduction of new NMR methodology to polymer gels leads to new developments in gel science. In addition to the NMR imaging method, most recently, the ESR imaging method has been developed to identify the spatial distribution of radical species in polymer systems.[13,14] Thus, it can be said that both of these methods are useful means for images analysis of spatial distributions in polymer gels.

 DOI: 10.1002/masy.200450310

From such a background, we aim to elucidate the spatial distribution of paramagnetic praseodymium(Pr^{3+}) ions in a poly(methacrylic acid)(PMAA) gel containing water by using a ^{1}H-chemical shift imaging method, and to establish an NMR methodology for elucidating insight into the nature of polymer gels.

Experimental

Materials

Methacrylic acid (MAA) (Tokyo Kasei Kogyo Co., Japan) was distilled at 299 K under a pressure of 265 Pa. N,N'-methylenebis(acrylamide) (MBAA) (Wako Pure Chemical Industries, Japan) used as the crosslinking monomer, was recrystallized twice from an ethanol solution. $K_2S_2O_8$ (Wako Pure Chemical Industries), used as the polymerization initiator, was recrystallized from an aqueous solution. $Pr(NO_3)_3$ •$5H_2O$ as NMR shift reagent is obtained from Apache Chemicals(Illinois, USA)

The PMAA gel was prepared by radical polymerization of MAA(3.0 mol L^{-1}) and MBAA(1.0x10^{-2} mol/L) in an aqueous solution at 318 K for 24 h in a cylindrical tube with an inside diameter of ca. 5 mm. Then, the PMAA gel obtained was soaked in excess deionized water for 3 weeks to remove remaining monomers, linear polymers formed as a by-product, and initiator. The water was changed repeatedly. The degree of swelling of the polymer gel, which was defined by $q = M_{swollen}/M_{dry}$, was 12.4. The diameter of the cylindrical gel was 5.24 mm.

The PMAA gel obtained was placed in aqueous $Pr(NO_3)_3$•$5H_2O$ solution(40 mg/mL) for 3 min. The gel was shrunk to a degree of swelling of 3.3, with cylindrical diameter 3.55 mm.

Measurements

^{1}H chemical shift imaging measurements were carried out in a Bruker DSX 300 NMR spectrometer operating at 300 MHz with an imaging system at 300 K. In these experiments, the ^{1}H chemical shift images of water molecules in the gel were observed. Data processing for two-dimensional images was performed by the Fourier imaging method. In the ^{1}H NMR imaging experiments, the field gradient strengths of 100 G/cm used for the slice selection, phase-enclosing and read-out. The slice thickness was 1 mm. The data points numbered 256x256 for an image and 4000x4000x256 for a spectrum. The number of accumulation for an image was 1.

Results and Discussion

The ^{1}H chemical shift of aqueous $Pr(NO_3)_3$ solutions as a function of $Pr(NO_3)_3$ concentration: It is known that the ^{1}H chemical shift of molecules associated with paramagnetic Pr^{3+} ion, a lanthanide shift reagent, moves downfield with an increase in $Pr(NO_3)_3{\bullet}5H_2O$ concentration.[13,14] In Fig. 1 the dependence of the ^{1}H chemical shift of water in aqueous $Pr(NO_3)_3$ solutions as a function of $Pr(NO_3)_3$ concentration relative to pure water($\delta = 4.8$ ppm) is shown. It is shown that the ^{1}H chemical shift of water containing 60 mg/1mL moves largely downfield by about 2.5 ppm as compared with that of pure water. This shows that when Pr^{3+} ions are added into a PMAA gel and are spatially and heterogeneously distributed, water molecules at in the gel have different ^{1}H chemical shift values from place to place. Thus, it can be expected that the spatial distribution of Pr^{3+} ions in a PMAA gel can be determined by ^{1}H chemical shift imaging.

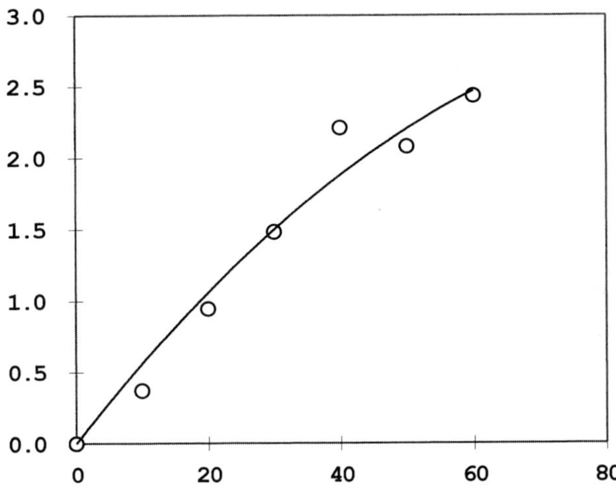

Figure 1. Plots of the ^{1}H chemical shift δ of water in aqueous $Pr(NO_3)_3$ solutions against $Pr(NO_3)_3$ concentration at 300 K.

^{1}H NMR images of water in a PMAA gel with $Pr(NO_3)_3{\bullet}5H_2O$: After a PMAA gel was placed in aqueous $Pr(NO_3)_3$ solution(40 mg/mL) for 3 min, the ^{1}H chemical shift NMR imaging measurements were carried out by placing the gel in an NMR probe. Conditions were set such that the gel consisted of two parts: an outside part containing Pr^{3+} ions, and an inner part containing no Pr^{3+} ions. In Fig. 2 one-dimensional ^{1}H NMR

spectrum of water in the gel and the ^1H chemical shift imaging patterns spliced transversely for the gel are shown. As seen from this figure, the one-dimensional ^1H NMR spectrum consists of some peaks, which arise from water interactions with Pr^{3+} ions, and without Pr^{3+} ions. The chemical shift range is large. Water having no interaction with Pr^{3+} ions appears at ca. 4.8 ppm, and on the other hand water appears in the range from 4.8 to 5.5 ppm. Probably, there exist network regions with different Pr^{3+} concentrations, as expected from Fig. 1.

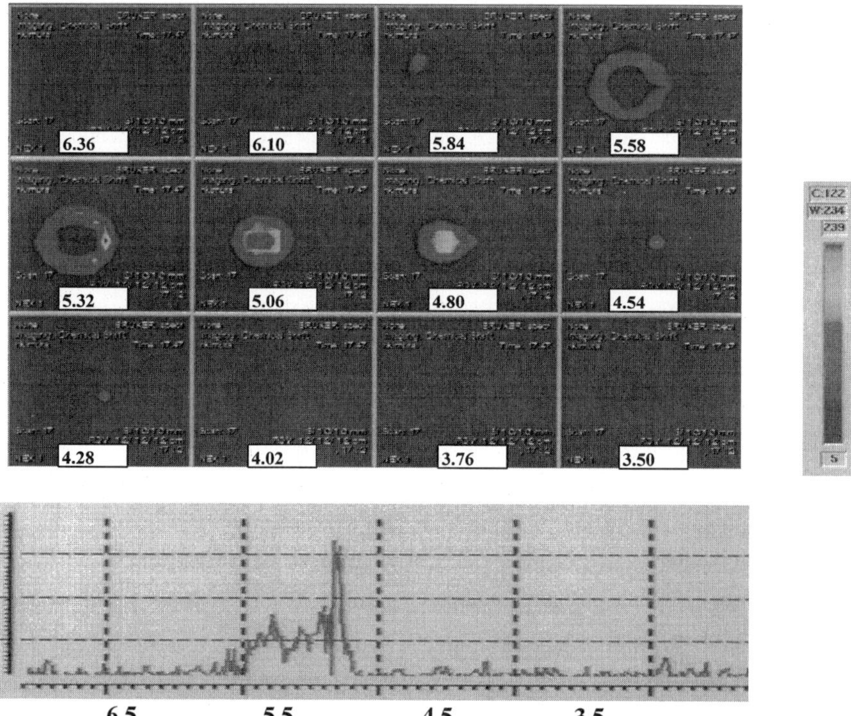

Figure 2. One-dimensional ^1H NMR spectrum of water in a PMAA gel with paramagnetic lanthanide Pr^{3+} ions and ^1H chemical shift imaging patterns of water in the gel spliced transversely to the gel.

If we look at the ^1H chemical shift imaging patterns transversely-sliced with a slice thickness of 1 mm at the central part of the gel (Fig.2), which are expressed as a function of ^1H chemical shift, we can understand that Pr^{3+} ions diffuse from the outside part to the inner part in the gel. For convenience, three-dimensional profiles are shown as a function of ^1H chemical shift in Fig. 3. The profile sliced at 4.8 ppm has an intense peak in the central part of the gel. The profile sliced at lower field than 4.8 ppm has a ring-like peak at the outside of the gel. From these ^1H chemical shift image profiles, the spatial distribution of Pr^{3+} ions in the gel is successfully elucidated. This shows that the spatial distribution of paramagnetic lanthanide metal ions in a polymer gel can be determined by ^1H chemical shift imaging.

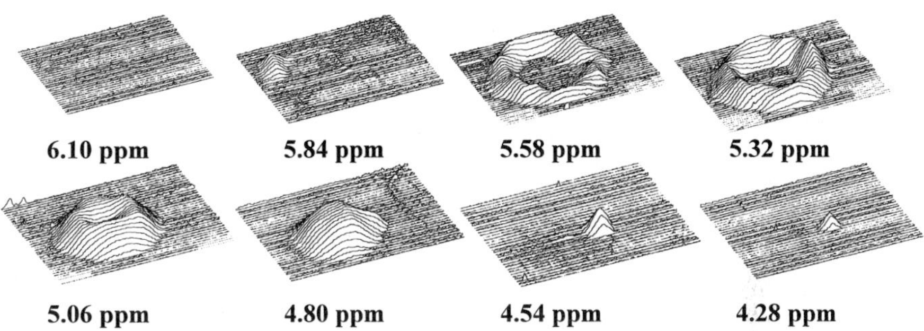

Figure 3. ^1H chemical shift imaging profiles of water in a PMAA gel with paramagnetic lanthanide Pr^{3+} ions sliced as a function of ^1H chemical shift δ.

We can conclude as follows. ^1H-chemical shift NMR imaging patterns of a poly(methacrylic acid) gel containing water with paramagnetic Pr^{3+} ions have been successfully observed and the spatial distribution of Pr^{3+} ions in the gel has been determined. Further, it has been shown that the chemical shift NMR imaging method is a useful means for the determining spatial distribution of paramagnetic metal ions in polymer gels.

[1] P.T. Callaghan, Principles of Nuclear Magnetic Resonance Microscopy, Oxford, 1991.
[2] B. Bluemich, NMR Imaging of Materials, Clarendon Press, 2000.
[3] H. Yasunaga, H. Kurosu and I. Ando, *Macromolecules*, 25, 6505 (1992).
[4] T. Shibuya, H. Yasunaga, H. Kurosu and I. Ando, *Macromolecules*, 28, 4377(1995).
[5] H. Kurosu, T. Shibuya, H. Yasunaga and I. Ando, *Polymer J.*, 28, 80 (1996).
[6] A. Yamazaki, Y. Hotta, H. Kurosu and I. Ando, *Polymer*, 28, 2082 (1997).
[7] A. Yamazaki, Y. Hotta, H. Kurosu and I. Ando, *Polymer*, 39, 1511 (1998).
[8] Y. Hotta, T. Shibuya, H. Yasunaga, H. Kurosu and I. Ando, *Polymer Gels and Networks*, 6, 1(1998).
[9] A. Yamazaki, Y. Hotta, H. Kurosu and I. Ando, *J. Mol. Structure*, 554, 47(2000).
[10] Hotta and I. Ando, *J. Mol. Structure*, 602/603, 165(2002).
[11] Tanaka, I. Nishino, S. Sun and S. Ueno-Nishio, *Science*, 218, 467(1982).
[12] Osada, *Adv. Polymer Sci.*, 82, 1(1987).
[13] R.E. Sievers, Nuclear Magnetic Resonance Shift Reagents, Academic Press, New York, 1973.
[14] G.L. La Mar, W.D. Horricks, Jr. and R.H. Holm(eds.), NMR of Paramagnetic Molecules, Academic Press, New York, 1973.

Water Diffusion in Methacrylate Based Copolymer Hydrogels of 2-Hydroxyethyl Methacrylate

David J.T. Hill, Mohammad Chowdhury, Phuong Y. Ghi, Naomi G. Moss, Andrew K. Whittaker*

Polymer Materials and Radiation Group, Department of Chemistry, University of Queensland, Brisbane, Australia 4072

Summary: The diffusion of water into cylinders of polyHEMA and copolymers of HEMA with THFMA, BMA and CHMA were studied over a range of copolymer compositions. The diffusion of water into the polymers was found to follow a Fickian, or case I mechanism. The diffusion coefficients of water were determined from mass measurements and NMR imaging studies. They were found to vary from $1.7 \pm 0.2 \times 10^{-11}$ $m^2 s^{-1}$ for polyHEMA at 37 °C to lower values for the copolymers. The mass of water absorbed at equilibrium relative to the mass of dry polymer varied from 52-58 wt% for polyHEMA to lower values for the copolymers.

Introduction

Poly(hydroxyethyl methacrylate), polyHEMA, and its copolymers with other methacrylate monomers have found numerous medical applications due to their well documented biocompatibility[1,2,3]. Applications of these copolymers include polymeric matrices for the controlled release of bioactive agents[4-6], intraocular and contact lenses[7] and soft cushioning materials for use in dentistry and surgery[8]. In controlled release systems, for example, the drug is initially incorporated within the core of a glassy polymer, and then, when body fluids migrate into the polymer plasticizing it, the trapped drug is allowed to diffuse out through the swollen hydrogel matrix. Thus the diffusion of water into the polymer is vital for release of the drug, and the water and drug diffusion kinetics are dependent on the nature of the polymer matrix.

The diffusion of water into a polymer matrix is influenced by: the polymer microstructure, the polarity of the copolymer segments, the glass transition temperature of the polymer, the flexibility of the polymer backbone, the molecular weight of the polymer, the crosslink density and interchain interactions, the degree of chain branching and the presence of bulky comonomer pendant groups. The kinetics of water sorption, as well as the equilibrium amount of water which can be absorbed by the matrix, are particularly dependent on the nature of the comonomers present, the comonomer

 DOI: 10.1002/masy.200450311

composition and the crosslink density.

Herein we review some of our work[9-14] on water diffusion into polyHEMA and its copolymers with tetrahydrofurfuryl methacrylate, THFMA, n-butyl methacrylate, BMA, and cyclohexyl methacrylate, CHMA. These investigations have included mass uptake measurements and NMR imaging studies of the water profile during its diffusion into the polymers. We also compare the sorption of water by polyHEMA with that for polyHEMA crosslinked with diethyleneglycol-dimethacrylate (DEGDMA) and with tetraethyleneglycol-dimethacrylate (TEGDMA)[11].

Experimental

Stabilized BMA and CHMA (Aldrich Chemical Company), THFMA (DAJAC Laboratories) and HEMA (Ubichem Ltd) were purified by chromatography using anhydrous Al_2O_3 (neutral, Merck) and distillation at reduced pressures (. 10^2 Pa). Only the middle fractions were used experimentally[12,13]. DEGDMA and TEGDMA (Aldrich Chemical Company) were purified by chromatography using anhydrous Al_2O_3[11]. The purity of the monomers was confirmed by NMR analysis.

Cylindrical samples of the polymers were prepared in cylindrical teflon moulds for the diffusion studies[11-13]. For the HEMA copolymers with THFMA, BMA and CHMA, the monomers were weighed into a 25 mL pyrex glass flask in the required mole ratios and 0.08 weight % BPO was added, based on the mass of the monomers. The mixtures were shaken until the BPO had dissolved, and then they were poured into the teflon moulds (diameters in the range 3.0 - 10 mm and length 20 - 40 mm). The polymerizations were forced to complete conversion of monomer to polymer in a vacuum oven using various temperature/time protocols[12,13] over the temperature range 40 ^{0}C - 120 ^{0}C. After polymerization, the polymer cylinders were removed from the teflon moulds, the absence of residual monomer was confirmed by FT-NIR analysis, and the ends were ground to a smooth, flat finish prior to the diffusion measurements.

Some polyHEMA cylinders were prepared using using ^{60}Co (-radiolysis at 25 ^{o}C, as were polyHEMA cylinders which were crosslinked with DEGDMA and TEGDMA[11]. Again, the absence of residual monomer was confirmed by FT-NIR analysis, and the ends were also ground to a smooth, flat finish prior to the diffusion measurements.

The copolymer cylinders were placed in distilled water in a water bath for which the temperature was controlled to $\pm$ 1 ^{0}C. The polymers were removed periodically from the bath, the surface water was removed, they were weighed, and an NMR image of the water diffusion front was obtained. The experimental details associated with the NMR imaging process have been reported elsewhere[14].

Results and Discussion

The distinguishing features of the monomers are the hydrophilic polar hydroxy group of HEMA; the bulky hydrophobic side-chains of BMA and CHMA, the less hydrophobic side-chain of THFMA and the crosslinking ability of the DEGDMA and TEGDMA. Studies on these series of copolymers provide information about the influence of the side-chains and the introduction of crosslinks on the water diffusion and equilibrium water contents of the polymers.

The nature of the copolymerizations of HEMA with the comonomers have been studied in detail[12,15,16] and it has been shown that they are close to ideal polymerizations. This means that there is relatively little tapering of the feed composition with conversion, and hence of the instantaneous copolymer composition, as the polymerizations proceed to high conversions.

Water Sorption by PolyHEMA.

The diffusion of water into a series of polyHEMA cylinders has been studied at 25 °C, 37 °C and 45 °C for polymers prepared by a range of polymerization protocols. A series of typical mass uptake data are demonstrated in Figure 1 for sorption at 37 °C. From Figure 1, it is possible to identify a change in the trend of the mass uptake at the point at which the glassy core disappears at a relative mass uptake of $\approx$ 0.6, and there is evidence for an over-shoot in the mass uptake just prior to equilibrium being established. These effects are related respectively to the dimensional changes that occur in the cylinders when the glassy core disappears, and the system becomes totally rubbery, and the slow relaxation of the polymer chains as equilibrium is approached.

The mass uptake of a penetrant into an infinite cylinder of radius *a* can be represented by equation 1 for a Fickian diffusion model[9]

$$\frac{M_t}{M_\infty} = 1 - \sum_{n=1}^{\infty} \frac{4}{a^2 \alpha_n^2} \exp(-D\alpha_n^2 t) \qquad (1)$$

where t is the time for which penetrant diffusion has occurred, M_t and M_4 are the mass uptakes at time t and at equilibrium, respectively, D is the diffusion coefficient, and α_n are defined in terms of the roots of the zero order Bessel function, $J_0(a\alpha_n) = 0$.

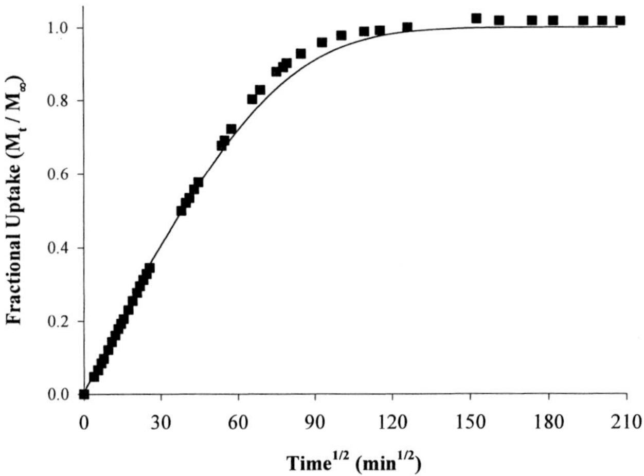

Figure 1. Water mass uptake data for a polyHEMA cylinder of diameter 1.0 cm measured at 37 °C. The solid line shows the fitted curve for equation 1 for data up to a relative mass uptake of 0.6, which is the point at which the glassy core disappeared.

The diffusion of water into polyHEMA cylinders has been found to be adequately represented by equation 1[12,13]. The best values of the diffusion coefficients for water, were determined by nonlinear least squares curve fits to equation 1 using only uptake data prior to the point of disappearance of the glassy core. A typical curve fit is shown in Figure 1. The diffusion coefficients obtained were found to be dependent on the details of the temperature/time protocols adopted during preparation of the cylinders. For example, for a range of polymerization protocols we have obtained values of D of

$1.7 \pm 0.2 \times 10^{-11}$ m^2 s^{-1}at 37 °C. The small variations in the values are believed to arise from the differences in the extent of chain entanglement and the crosslink density developed during polymerization. As expected, the mass uptake/time curves for cylinders varying in diameter from 3.0 to 10.0 mm prepared by the same polymerization protocols were found to be different. However, the values of D determined from the data fits to equation 1 were the same within the estimated experimental error of $\pm$ 0.2 x 10^{-11} m^2 s^{-1}.

The values of D obtained for a range of temperatures are shown in Figure 2, along with data reported by Gehrke et al.[17] for water sorption into flat sheets. The agreement between the data is good. The activation energy determined from the diffusion coefficients obtained herein was 12.9 $\pm$ 1.0 kJ mol^{-1}, which agrees well with the value of 11.9 kJ mol^{-1} reported by Gehrke and coworkers.

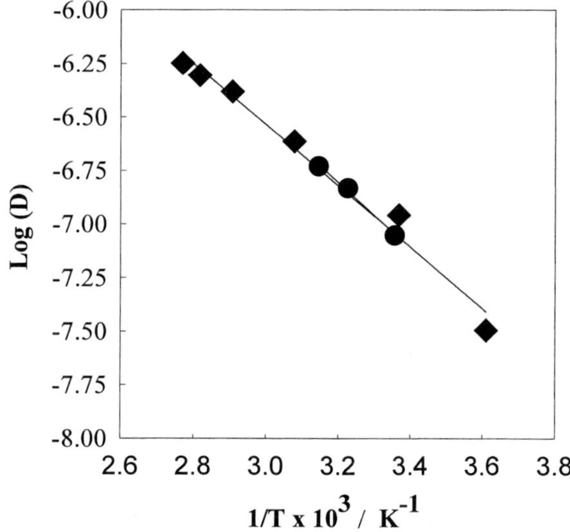

Figure 2. Plot of the logarithm of the diffusion coefficient versus reciprocal temperature for diffusion of water into polyHEMA. ♦ Gehrke et al.[17] , • This work.

The water uptake at equilibrium, S_4, expressed as a percentage of the original mass of the polymer, is:

$$S_4 = [(M_4 - M_0) / M_0] \times 100 \qquad (2)$$

where M_0 is the mass of the dry polymer cylinder at time zero and M_4 is the mass of the polymer cylinder plus water at time infinity.

The equilibrium water uptake by polyHEMA cylinders was also found to be dependent on the polymerization protocol adopted in their preparation, and values of S_∞ in the range 52 - 59 % have been obtained at 37 °C for cylinders prepared by thermal initiation using BPO[12,13]. The equilibrium water uptakes for the polyHEMA cylinders prepared by (-irradiation, 58 %, agree with these values within experimental error[11]. The studies on polyHEMA cylinders prepared by γ-radiolysis also showed that incorporation of the crosslinking monomers DEGDMA and TEGDMA at concentrations up to 0.5 wt % resulted in a decrease in the equilibrium water contents over that for polyHEMA, as demonstrated in Figure 3. This observation reflects the reduced swelling capability of polyHEMA matrices containing added crosslinker.

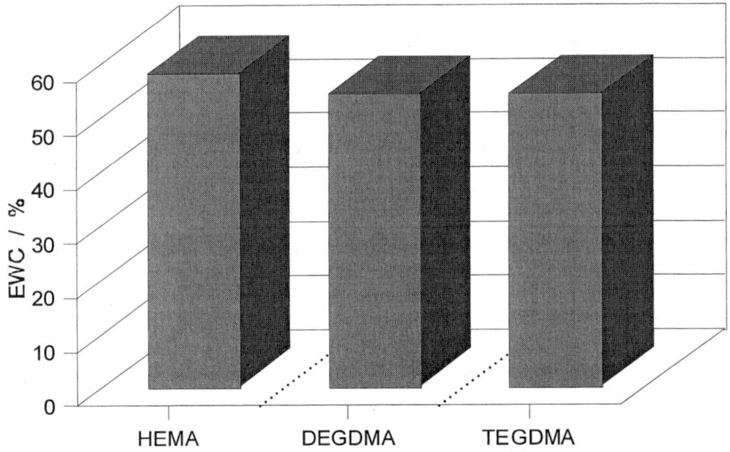

Figure 3. Equilibrium water contents (EWC) for polyHEMA and polyHEMA crosslinked with DEGDMA and TEGDMA (0.5 wt%) for water sorption at 37 °C.

Mass Uptake and Diffusion Coefficients for Copolymers.

Diffusion coefficients, have been obtained from mass uptake data for THFMA/HEMA[13], BMA/HEMA[12] and CHMA/HEMA[12] copolymers. Again, the mass uptake data were fitted to equation 1, using a least squares curve fitting procedure, in order to determine the best values for the diffusion coefficients. The results are shown in Figure 4 relative to the diffusion coefficient for polyHEMA.

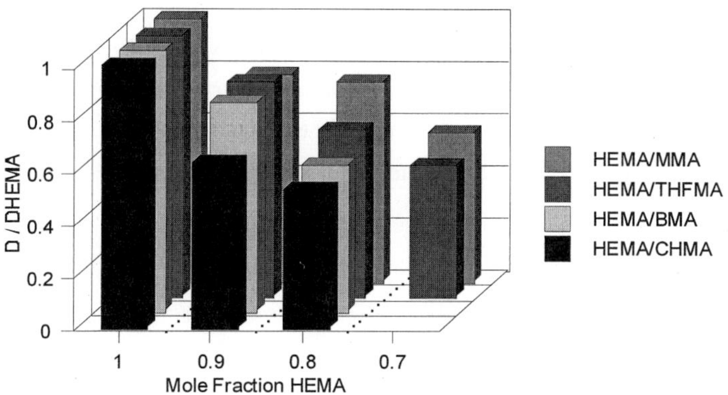

Figure 4. Diffusion Coefficients for HEMA copolymers relative to that for polyHEMA at 37 °C. The data for HEMA/MMA were calculated from the work of Franson and Peppas[18].

With increasing proportions of the comonomers, the diffusion coefficients decreased, reflecting the influence of the compositional microstructure on the diffusion of water into the cylinders. As the mole fraction of HEMA in the copolymers decreases, the fraction of HEMA units with two HEMA neighbours also decreases, and hence so do the average lengths of the HEMA sequences. These changes in the HEMA microstructure have the effect of inhibiting the diffusion of the polar water molecules through the copolymer matrices. The incorporation of small amounts of the non-polar side-chains of CHMA and BMA into polyHEMA were found to have a greater effect on the diffusion coefficient for water than does the incorporation of small amounts of the somewhat more polar side-chain of THFMA.

Franson and Peppas[18] have studied the diffusion of water into poly(hydroxyethyl methacrylate-co-methyl methacrylate), HEMA/MMA, at 37±0.5 °C. We have calculated the relative diffusion coefficients for these copolymers, using the data of Franson and Peppas, and they are also included in Figure 4. The diffusion coefficients for the MMA copolymers are larger than those for the other three copolymers at corresponding comonomer mole fractions. This presumably reflects the balance between the hydrophobic and hydrophylic natures of the groups in the side-chains of the various methacrylate comonomers.

Mass Uptake of Water at Equilibrium.

The equilibrium mass uptakes for the copolymer cylinders at 37 °C were found to decrease with increasing comonomer mole fraction, as demonstrated in Figure 5. Again, the data for the various comonomers reflect the changing hydrophobicity of the comonomer side-chains, as was observed for the diffusion coefficients.

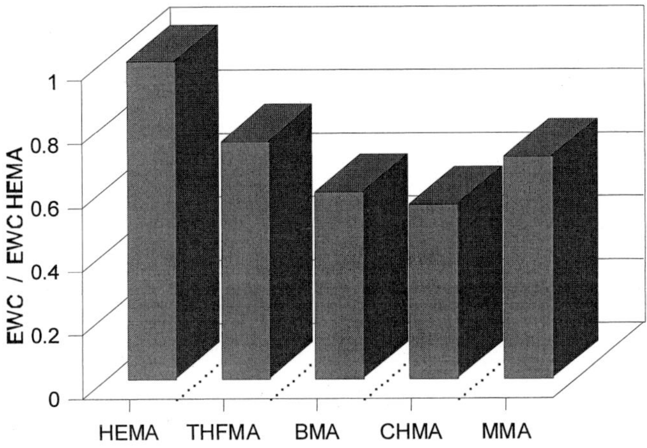

Figure 5. Equilibrium water contents, EWC, for a range of HEMA copolymer cylinders, F_{HEMA}= 0.8, of 10.0 mm diameter at 37 °C. The data for MMA/HEMA was taken from Franson and Peppas[18].

NMR Imaging of Water Diffusion.

Mass uptake measurements on cylinders which have imbibed water are bulk measurements, and, in order to obtain information about the processes controlling diffusion of the water into the polymer matrix, it is necessary to make assumptions about the diffusion mechanism, and then to assess the goodness of fit of each model to the uptake data, as outlined above. In addition, mass uptake measurements, being bulk data, are influenced by diffusion through the flat ends as well as through the curved cylinder walls. This introduces uncertainty, and in order to carry out an analysis of these data, it must be assumed that these end effects are negligible.

An alternative approach is to image the profile of the water diffusion front directly. This can be conveniently done in real time using NMR imaging of the water protons. Such an NMR profile is shown below in Figure 6 for water diffusion into a cylinder of BMA/ HEMA, with a mole fraction HEMA of 0.8.

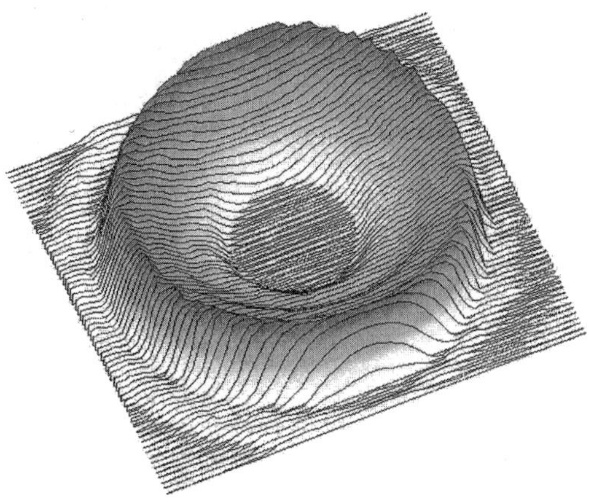

Figure 6. NMR image of the water diffusion front for a BMA/HEMA cylinder, F_{HEMA} = 0.8, at 37 °C. The diameter of the cylinder is 10 mm.

This image was obtained from a 3 mm slice through the centre of the cylinder, so it is uneffected by end effects. A cross section of this image profile is shown in Figure 7. The image and the cross section show that the maximum concentration of water in the cylinder is at the outer wall and that some of the central glassy core remains. The shape of the underlying profile appears to be Fickian-like, but there are clearly additional features on the walls of the profile. A similar cross section of a profile for a polyHEMA cylinder, see Figure 8, shows these additional features more clearly. We have assigned these features to water molecules located in cracks which are formed due to the tensile stresses generated in the glassy core at the interface between the glassy core and the surrounding, swelling rubbery region.

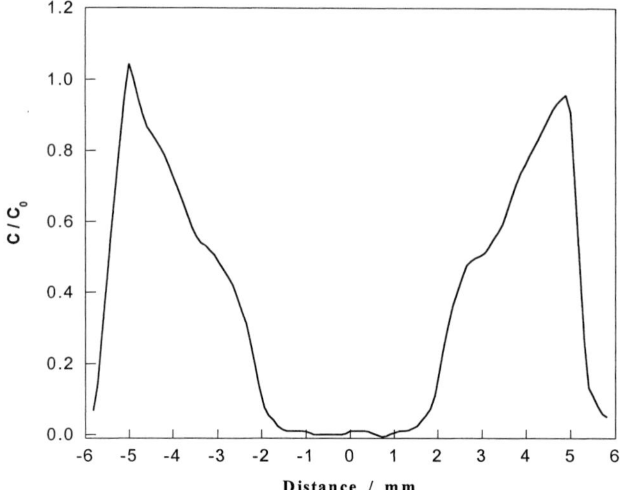

Figure 7. The cross section of an NMR image for water diffusion into a BMA/HEMA cylinder, $F_{HEMA} = 0.8$, at 37 °C. The profile was taken from the image shown in Figure 6. The symbol C represents the concentration of the water in the polymer matrix and C_0 the concentration of water at the surface of the cylinder.

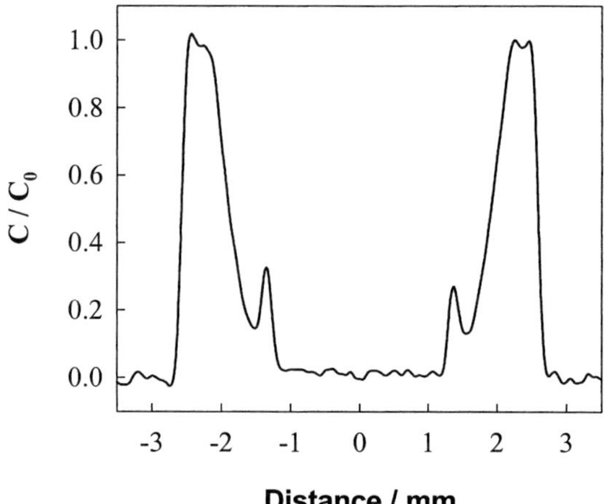

Figure 8. The cross section of an NMR image for water diffusion into a polyHEMA cylinder, diameter 5 mm, at 37 °C showing additional features at the edge of the glassy core. Fractional mass uptake 0.4. The symbol C represents the concentration of the water in the polymer matrix and C_0 the concentration of water at the surface of the cylinder.

As the diffusion front progresses through the cylinder, so too does the annulus of cracks. However, in the rubbery region behind the diffusion front the additional features are invisible because the plasticizing water molecules lower the Tg of the local matrix to below the measurement temparature. The greater mobility of the polymer chains therefore allow a repair process that can remove the remnants of the cracks. In the case of the BMA/HEMA cylinder, the presence of water in cracks is less evident than in polyHEMA cylinders, because of the inherently lower Tg of the matrix due to the presence of the bulky non-polar BMA side chains.

The diffusion coefficient for water can be determined by a curve fit of the profile of the diffusion front, such as those shown in Figures 7 and 8. For a Fickian model the variation of the water concentration with distance and time is given by:

$$\frac{C_{r,t}}{C_{0,\infty}} = 1 - \frac{2}{a}\sum_{n=1}^{\infty} \exp(-D\alpha_n^2 t)\frac{J_0(r\alpha_n)}{J_1(a\alpha_n)} \tag{3}$$

where t is the time for which penetrant diffusion has occurred, $C_{r,t}$ is the concentration of the penetrant at distance r at time t, C_0 is the constant surface concentration, α_n and $J_0(x)$ are defined as above, and $J_1(x)$ is the Bessel function of the first order.

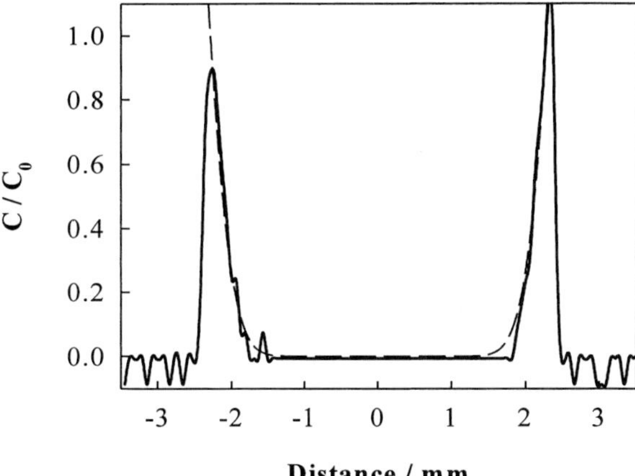

Figure 9. Concentration profile for diffusion of water into a polyHEMA cylinder, diameter 5 mm, at 37 °C showing a curve fit (----) to the profile based on equation 3. Fractional mass uptake 0.2. The symbol C represents the concentration of the water in the polymer matrix and C_0 the concentration of water at the surface of the cylinder.

Curve fits of the underlying water concentration profiles to equation 3 yielded values for the diffusion coefficients in agreement with those found from the mass uptake measurements within the experimental error. A typical curve fit to the profile of a polyHEMA cylinder is shown in Figure 9. For polyHEMA the best value of the diffusion coefficient, which fitted all of the profiles for fractional water uptakes from 0 to ≈0.95, was $1.5x10^{-11}$ m^2 s^{-1} [14]. This value compares well with the value of $1.7 \pm 0.2x10^{-11}$ m^2 s^{-1} obtained from the mass uptake study.

Conclusions

The diffusion of water into HEMA based polymers has been studied by mass uptake and NMR imaging. The value of the diffusion coefficient for water sorption into HEMA copolymers is lowered by the presence of a non-polar comonomer. By decreasing the proportion of HEMA in a copolymer, the average HEMA sequence length in the copolymer decreases, and so does the diffusion coefficient and equilibrium water content of the copolymer. For example the diffusion coefficient decreased significantly from $1.7 \pm 0.1x10^{-11}$ m^2s^{-1} for polyHEMA to $.9.1 \pm 0.1x10^{-12}$ m^2s^{-1} for CHMA/HEMA, $9.7 \pm 0.1x10^{-12}$ m^2s^{-1} for a BMA/HEMA and $1.3 \pm 0.1x10^{-11}$ m^2s^{-1} for a THFMA/ HEMA copolymer at compositions F_{HEMA}=0.8. The equilibrium water contents for the copolymers followed similar trends. The diffusion behaviour for water was found to be Fickian for the HEMA based copolymer systems. However, for polyHEMA and some of its copolymers with BMA, CHMA and THFMA, cracks were observed to appear in the glassy core adjacent to the diffusion front during water sorption, which allowed the formation of small water pools adjacent to the diffusion front.

Acknowledgements

The authors wish to acknowledge financial support for this work from the Australian Research Council. An ongoing collaboration with the Department of Biomaterials in Relation to Dentistry at Queen Mary and Westfield College, University of London is also gratefully acknowledged.

[1] Choudray, M.S. and Varma, I.K., *Macromol.Sci. Chem.*, **1983**, *A20,* 771.
[2] Wichterle, O. and Lim, D., *Nature*, **1960**, *185*, 177.
[3] Jeyanthi, R. and Pandurang Rao, K., *Biomaterials*, **1990**,*11*, 238.
[4] Peppas, N.A. and Moynihan, H.J., "Hydrogels in Medicine and Pharmacy", Vol II (N.A. Peppas, ed.), CRC Press, Boca Raton, **1987**, 49.
[5] Tighe, B.J., "Hydrogels in Medicine and Pharmacy", Vol III (N.A. Peppas, ed.), CRC Press, Boca Raton, **1987**, 53.
[6] Martin, A., Swarbrick, J. and Cammarata, A. (eds.) Physical Pharmacy 3rd Edition, Lea and Febiger, Philadelphia,**1983**, 579.
[7] Franklin, V.J., Bright, A.M. and Tighe, B., *TRIP*, **1993**, 1, 9.
[9] Parker, S. and Braden, M., *Biomaterials*, **1989**, *10*, 91.
[9] Ghi, P.Y., Hill, D.J.T., Maillet, D. and Whittaker, A.K., *Polym. Comm.*, **1997**, *38*, 3985.
[10] Hodge, R.M., Simon, G.P., Hill, D.J.T., and Whittaker, A.K., *J. Polym. Sci., Part B: Polym. Phys.*, **1998**, *36*, 463.
[11] Hill, D.J.T., Lim, M.C.H. and Whittaker, A.K., *Polym. Int.*, **1999**, *48,* 1046.
[12] Hill, D.J.T., Moss, N.G., Pomery, P.J. and Whittaker, A.K., *Polymer*, **2000**, *41*, 1287.
[13] Ghi, P.Y., Hill, D.J.T. and Whittaker, A.K., *J. Polym. Sci.*, **2000**, *38*, 1939.
[14] Ghi, P.Y., Hill, D.J.T. and Whittaker, A.K. *Biomacromolecules*, **2001**,*2*, 504.
[15] Whittaker, M.R., Hill, D.J.T., O'Donnell, J.H. and Pomery, P.J., *Polym. Gels and Networks*, **1995**, *4*, 85.
[16] Ghi, P.Y., Hill, D.J.T., O'Donnell, J.H., Pomery, P.J. and Whittaker, A.K., *J. Polym. Sci., Part A: Polym. Chem.*,**1995**, *3*, 1730.
[17] Gerhke, S.H., Biren, D. and Hopkins, J.J., J. *Biomater. Sci., Polymer, Edn*, **1994**, *6*, 375.
[18] Franson, N.M. and Peppas, N.A., *J. Appl. Polym. Sci.*, **1983**, *28*, 1299.

Thermoreversible Gelation Strongly Coupled to Polymer Conformational Transition

Fumihiko Tanaka

Department of Polymer Chemistry, Graduate School of Engineering, Kyoto University, Kyoto 606-8510, Japan

Summary: Thermoreversible gelation of polymers driven by the coil-to-helix transition in chain conformation is theoretically studied. For pairwise association of single helices, there are three fundamental types of self-assemblies as a result of competition between helix growth and helix association: Type I network (random coils connected by paired short helices), Type II network (helices connected by short random coils) and pairing (pairs of long helices without branching). Two distinct phase diagrams showing sol/gel transition and coil/helix transition are derived for weak and strong association.

Introduction

Most natural polymers undergo conformational transition preceding gelation. Activation of the particular functional groups on a polymer chain accompanied by a proper three dimensional conformation change is a necessary prerequisite for the interchain cross-linking. For instance, water-soluble natural polymers such as agarose and κ-carrageenan first change their conformation from the random coil state to a partially helical state, and then helical parts aggregate to form network junctions (Figure 1).[1-4] Recently, a similar two-step mechanism of gelation through coil-to-helix transition was confirmed for synthetic polymers with stereo-regularity.[5,6] It was found that, in solutions of syndiotactic poly(methyl methacrylate) in toluene, a fast intramolecular conformational change is followed by an intermolecular association leading eventually to gelation. The main purpose of this study is to derive possible phase diagrams in which two phase transitions, sol/gel transition and coil/helix transition, interfere.

Thermoreversible Gelation Induced by Coil/Helix Transition

A large body of experimental and theoretical study on the problem of single-chain coil/helix transition now exists,[7] but there is room for the theoretical study of gelation of many chains induced by coil/helix transition. We consider a single polymer chain with total number η statistical units from which j_ζ sequences of ζ contiguous chain

 DOI: 10.1002/masy.200450312

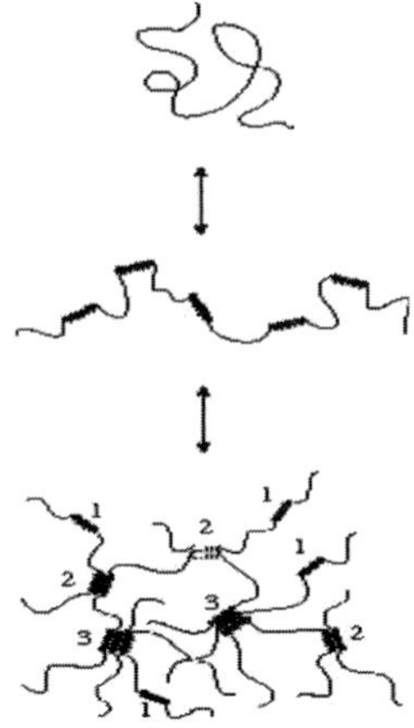

Figure 1. Thermoreversible gelation by association of partial helices on polymers. Pairwise or triple association of helices form network junctions connecting random chains. If a polymer chain carries many short helices, its functionality is high but association energy of each helix is small. On the contrary, if a polymer chain has small number of long helices, its functionality is low but association energy is large. Therefore, there should be a competition between helix growth and helix association. Network structure varies depending upon the length distribution of helices on a chain.

segments are selected for helix formation. The total number of $W(\eta,\{j\})$ of possible arrangement of j_ζ non-overlapping sequences of length ζ is then given by

$$W(n,\{j\}) = (n - \sum \zeta j_\zeta)! / \{\prod j_\zeta !\}(n - \sum \zeta j_\zeta - \sum j_\zeta)\,. \tag{1}$$

The canonical statistical weight of a polymer conformation carrying such helices is then given by

$$e^{-\Delta A(\{j\})/k_B T} = W(n,\{j\}) \prod \eta_\zeta^{j_\zeta}\,, \tag{2}$$

where η_ζ is the weight for a helix of the length ζ. In order to focus on the many-chain problem concerning the connectivity of the solution as a whole (see Figure 1), we here employ for a single chain the simplest Zimm-Bragg model of the coil/helix transition, for which η_ζ is given by $\eta_\zeta = \sigma s(T)^\zeta$, where s is given by $s(T) \equiv \exp(\varepsilon_H / k_B T)$ in terms of the hydrogen-bonding energy ε_H between the neighboring groups on a chain for helix formation.

A characteristic feature of gelation by such sequence selection lies in that, for the pairwise association, the total number of helices on a chain must be equal to or larger than 3 in order for the solution to gel. For triple association, it must be equal to or larger than 2. Therefore, as the sequence ζ becomes longer by lowering the temperature, the effective functionality becomes smaller, and, at a certain critical value of ζ, the solution goes back to a non-gelling system. This simple consideration suggests that gelation becomes easiest in the intermediate temperature region where the coil/helix transition takes place.

In what follows we consider pairwise association of single helices as the simplest model system. We apply our recent theory of network-forming polymer solution[8,9] in which functionalities of polymers change depending upon their conformation. We start from the lattice-theoretical free energy of functional polymers, and minimize the total free energy of the solution to find the most probable molecular-weight distribution functions of clusters formed by the polymers whose conformations are given by the statistical weight eq. (2). This process is most systematically done by imposing multiple equilibrium conditions on the chemical potentials of clusters. The equilibrium conditions between the random-coil state and partial helix state are also imposed. Then, from the obtained distribution function, we calculate the weight-average molecular weight of the clusters, and from its divergence point we identify the sol/gel transition point. (We employ conventional definition of the gel point such that the largest cluster percolates onto the entire solution.) The tendency of helix pairing is described by the association constant $\lambda(T) \equiv \lambda_0 \exp(\varepsilon_A / k_B T)$, where ε_A is the association energy between the helices. The competition between the helix growth and helix association can be described by the ratio $b \equiv \varepsilon_A / \varepsilon_H$, *i.e.* the strength of association energy relative to the hydrogen-bonding energy. If b is large, we expect Type I networks, while, if b is small, we expect Type II networks, or pairing in the extreme case (see Figure 2).

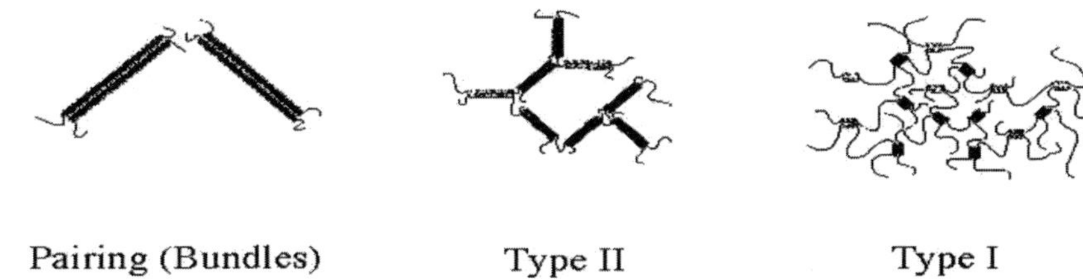

Figure 2. Three fundamental types of association. When the length of a helix reaches the total chain length, pairing (or bundle formation in more general multiple association) of helices occurs. When chains carry small number of long helices, networks of rigid-rod helices connected by short random chains are formed (Type II). The elastic properties of Type II network are highly nonlinear and non-affine. When chains carry many short helices, networks of random coils connected by point-like cross-links are formed (Type I). The elastic properties of Type I networks show those of ordinary affine networks.

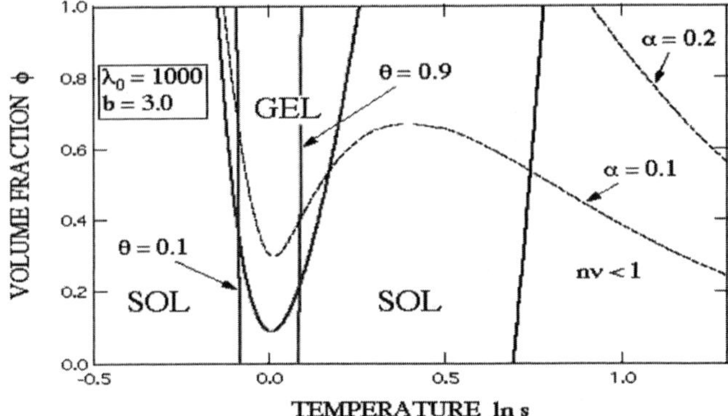

Figure 3. Phase diagram showing sol/gel transition and coil/helix transition in the weak association case (b=3.0). Solid line shows sol/gel transition line. The two lines indicated by θ are the contours at constant helix content (0.1 and 0.9). Broken lines show the contours at constant reactivities. The vertical thick solid line at the right side shows the boundary at which the average number of helices on a chain becomes less than one.

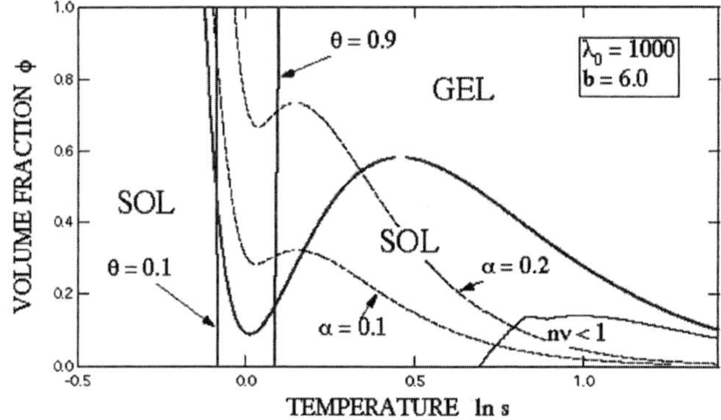

Figure 4. Phase diagram showing sol/gel transition and coil/helix transition in the strong association case (b = 6.0). Lines are the same as in left figure.

Figures 3 and 4 show theoretically calculated typical phase diagrams for weak association and strong association, respectively. The temperature is measured in terms of ln $s = \varepsilon/k_B T$ as in the literature[7] on coil/helix transition, so that it decreases as one moves rightward along the horizontal axis. The vertical axis shows the polymer volume fraction.

In the weak association, the sol/gel transition takes place across the coil-to-helix transition temperature region where the number of helices become large (Fig. 3). The coil-to-helix transition region is shown by the two lines showing the contour with constant helix content θ, *i.e.* $\theta = 0.1$ and $\theta = 0.9$. In the postgel regime, networks of Type I are formed. At high concentration in the postgel regime, transition from random coil to helix takes place inside the network because the transition region is mostly confined in the postgel regime. At low temperatures, a polymer chain carries only one long helix below the line indicated. These rod-like helices may form various anisotropic mesophases either in isolated or paired form, but such phase transitions are not studied in the present theory. The broken lines show the contours at constant reactivity (*i.e.* the number of paired helices divided by the total number of helices) $\alpha = 0.1$ and $\alpha = 0.2$.

In strong association, the gel region is largely expanded. Across the coil/helix transition region, Type I networks are formed, but just below the transition region, helices grow and Type II networks are formed (Fig. 4). At low temperatures, although the helix content becomes nearly full, there are only relatively short helices to reduce the network free energy by increasing the number of helices. Networks of short helices connected by the paired helices are formed. They can be called "reentrant Type I" networks. At low concentration, however, helices can grow because of the small probability of association, and pairing of long helices is possible. Therefore, if we increase the polymer concentration at low temperature, there should be another transition from the pairing state to the network state. This is a new transition characteristic of thermoreversible gelation induced by coil/helix transition

Conclusions

It is shown that two fundamentally different types of polymer networks are formed depending upon the strength of helix-forming tendency and of association among helices, and accordingly, two different phase diagrams are possible. At sufficiently low temperatures, gelation is prevented in both cases because the length of helices reaches the total length of a polymer. There should be a transition from pairing state to network state at such low temperatures with increase in the polymer concentration. Typical phase diagrams are calculated for the simplest case of pairwise association of single helices. Our theory can readily be generalized to multiple association, and also to double helices to compare with the experimental observation of biopolymer gels. Theoretical details will be reported in a forthcoming paper.

[1] Burchard, W., *British Polym. J.* **1985**, *17*, 154.
[2] Clark, A. H.; Ross-Murphy, S. B., *Ad. Polym. Sci.* **1987**, *83*, 57.
[3] Guenet, J. M., *Thermoreversible Gelation of Polymers and Biopolymers*; Academic Press, Harcourt Brace Jovanovich Publishers: **1992**.
[4] Te Nijenhuis, K., *Ad. Polym. Sci.* **1997**, *130*, 1.
[5] Berghmans, M.; Thijs, S.; Cornette, M.; Berghmans, H.; Schryver, F.C.De, *Macromolecules* **1994**, *27*, 7669.
[6] Buyse, K.; Berghmans, H.; Bosco, M; Paoletti, S., *Macromolecules* **1998**, *31*, 9224.
[7] Poland, D..; Scheraga, H.A., *Theory of Helix-Coil Transitions in Biopolymers*; Academic Press: **1970**.
[8] Tanaka, F.; Stockmayer, W.H., *Macromolecules* **1994**, *27*, 3943.
[9] Tanaka, F., *Macromolecules* **2000**, *33*, 4249.

Macromol. Symp. **2004**, *207*, 131-137

Ionic Strength/Temperature-Induced Gelation of Aqueous Poly(N-Isopropylacrylamide-co-Vinylimidazole) Solution

Jin Woo Yi,[1] *Kun Na,*[1] *You Han Bae**[2]

[1]Center for Biomaterial and Biotechnology, Department of Materials Science and Engineering, Kwangju Institute of Science and Technology, 1 Oryong-Dong, Puk-gu, Gwangju 500-712, Korea
[2]Department of Pharmaceutics and Pharmaceutical Chemistry, University of Utah, Slat Lake City, UT 84106, USA

Summary: A high molecular weight copolymer of N-isopropylacrylamide (NiPAAm) and vinyl imidazole (VI) was synthesized and its phase transition behavior in aqueous solutions (5 wt%) by simultaneous changes of ionic strength and temperature was investigated. At low ionic strengths, the copolymer solution showed two phases (clear and opaque solutions), which were freely mobile, as increasing temperatures up to 65 °C due to repulsion interaction of positive charges developed by basic imidazole group on the polymer aggregates. However, at the physiological condition (I=0.15, T=37 °C), four distinctive phases (clear solution, opaque solution, gel and shrunken gel) were observed because of charge shielding effect by added salts. The gel state was stable and maintained from 32 °C to 55°C. In particular, the phase transition from opaque solution to gel rapidly occurred by the change in ionic strength (from ~ 0 to 0.15) at 37 °C. This characteristic can be utilized as a liquid embolic agent.

Introduction

Embolotherapy has become a routine part of malignant tumor surgery and treating arteriovenous malformations due to the pronounced therapeutic effectiveness of endovascular embolization.[1] A battery of synthetic and natural materials in various forms has been investigated as embolic agents during the past several years.[2] Among them, autologous materials, oxidized cellulose, polyvinyl alcohol (PVA), silicon and metallic spheres, liquid agents involving absolute ethanol and cyanoacrylate derivatives have been used as materials for practical embolotherapy. However, for transcatheter embolotherapy in clinical situations, these materials face various technical diffulties.[3-5] For example, in spite of well-established bio- and blood compatibility, autologous blood clot has now been abandoned because of unpredictable duration of occlusion. Oxidized cellulose also failed to produce a permanent occlusion. Although PVA is physiologically inert and causes minimal inflammatory responses, it is different to infect PVA via a small diameter catheter due to its high viscosity. As an alternative, liquid embolic agents composed of responsive-polymers that transform to a gel in response to changes in the

 DOI: 10.1002/masy.200450313

external environmental conditions of temperature, pH and ionic strength, can be considered. Such agents may offer some advantages, over conventional agents, of easy injection (low viscosity in a sol or colloid state), rapid and permanent occlusion (stable gel formation), and the reduction of potential side effects arising from the use of organic solvents.

It has been reported that N-isopropylacrylamide based copolymers showed reversible sol-gel transition by increasing temperature around 32 °C.[6,7] When we tested one of these copolymers for the feasibility for an embolic material, it was found that the temperature-induced gelation blocked the catheter pathway during injection due to thermal equilibration before reaching to a target site. For a more effective embolic material, we synthesized a copolymer of N-isopropylacrylamide (NiPAAm) and vinyl imidazole (VI) (poly(NiPPAm-co-VI), feed mole ratio = 95:5). Vinyl imidazole (VI) is a weak basic monomer and its polymer has been used for a metal chelating agent and for stabilization of colloidal suspension.[8-10] In this paper, we report the gelation behavior of aqueous poly(NiPPAm-co-VI) solution is response to changes in temperature and ionic strength with the emphasis on the role of a weak basic imidazole group in the gelation process.

Synthesis of poly(NiPAAm-co-VI)

A high molecular weight copolymer of NiPAAm (95 mol%) with VI (5 mol%) was synthesized by free radical polymerization in benzene (10 wt% monomer concentration) with AIBN as an initiator (7×10^{3} moles/moles of monomer) (Fig. 1). The polymerization was performed at 60°C for 16 h under dried nitrogen gas. The polymer became phase-separated in the solvent as polymerization proceeded. The precipitated product was dissolved in acetone/methanol mixture (90/10 v/v%) by warming. This polymer solution was precipitated in an excess amount of diethyl ether and then dried in vacuum for 3 days. The product was dialyzed using dialysis membrane (MW cut off; 15,000) against distilled water for at least 7 days to remove unreacted monomers and low molecular weight compounds. Finally, the polymer in a powder form was obtained by freeze-drying with additional drying under P_2O_5. The molecular weight of the polymer determined by light scattering was approximately 1.5×10^{6} Dalton. In order to get the information of copolymer composition, non-aqueous titration using ASTM D 974; solvent (methanol, 30ml), pH indicator (phenophtalate), standard base (0.5 N KOH) and standard acid (0.01 N $C_8H_5O_4K$) was tried. The calculated result indicated that the mole ratio of N-isopropylacrylamide to vinyl imidazole in the copolymer is 94.73 to 5.27.

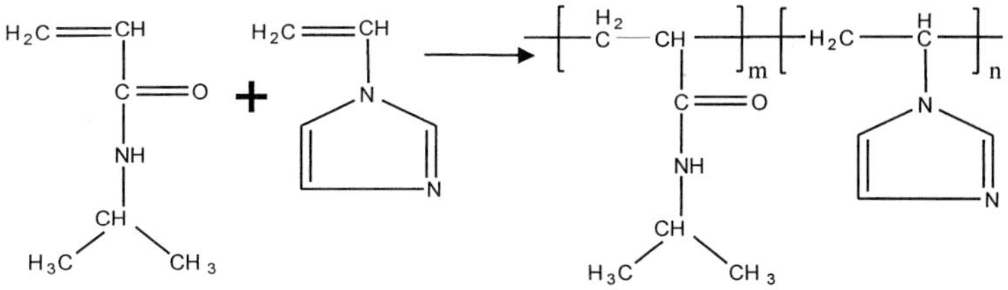

Figure 1. Synthesis scheme of poly (NiPAAm-co-VI) (m:n= 95:5 in mole ratio)

Effect of ionic strength on gelation of poly(NiPPAm-co-VI)

The phase transition behaviors of poly(NiPPAm-co-VI) in an aqueous solution (5 wt%) were investigated by monitoring the change in transmittance and by inversion test at various ionic strengths and temperatures. The temperature of the solutions was raised from 5 to 65 °C at a rate of 0.1 °C/min and the transmittance at 500 nm was measured by a Varian UV-VIS spectrometer (CARY IE). The cloud temperature was defined as the temperature at which the transmittance drops 50 % of the initial value at 5 °C. In the inversion test, the polymer solutions (5 wt%) at various ionic strength in glass vials were kept in a thermostated water bath and observed at 1 °C intervals. The gelation temperature was that at which the polymer solution did not flow by vial inverting. The shrinking temperature was also determined visually at the temperature when the polymer gel began to expel entrapped water and a vial was not fully filled with the polymer gel (syneresis).

Fig. 2 shows the resulting phase diagram as a function of ionic strength adjusted with NaCl. The clear solution (phase A) at low temperature became cloudy as temperature increased (defined as cloud temperature, phase B). The phases A and B were freely mobile. With further increase in the temperature, the opaque polymer solution subsequently became motionless (defined as the gelation temperature, phase C). Upon cooling, the gel turned mobile at the same temperature as the gelation temperature. The gel started to shrink by expelling water at the shrinking temperature (phase D). Once the gel was formed, it did not dissolve nor change its water content in the presence of an excess amount of water. In particular, the time required for gelation was apparently of the same order as the thermal induction time and the gel had a tendency to adhere to the wall of the glass vial. At an ionic strength close to zero, the solution did not form phases

C or D. Increasing the ionic strength produced the gel and shrunken gel phases. This result suggests that it requires a critical ionic strength to form a stable gel phase at the given polymer concentration (5 wt%).

In our previous report, aqueous poly(NiPAAm-co-acrylic acid) solutions above a critical polymer concentration showed a similar phase transition pattern as the increasing temperature. This polymer showed distributions in polymer composition as well as molecular weight, thus forming a mixture of polymer chains having different lower critical solution temperatures. The chain entanglement and the formation of physical associations of collapsed chains at 32 °C seemed to be responsible for the gelation[7)]. In the case of poly(NiPAAm-co-VI) used in this study, the polymer solution at lower ionic strength did not develop physical domains of collapsed polymer chains to act as physical crosslinking junctions. Around the physiological ionic strength (I = 0.1-0.2), the stable gel was maintained in a wide temperature range. It is interesting to note that without further increasing in the ionic strength, the temperature range for the gel became narrower and the upper temperature for the gel region sharply decreased. In the phase B formed at low ionic strengths, the opaque polymer solution was able to flow through a small-diameter (1 mm) tube without noticeable flow resistance. The gelation occured immediately when the polymer solution came in contact with the body fluid where the ionic strength is kept at 0.15. This feature of the polymer solution offers high potential for use as an embolic agent.

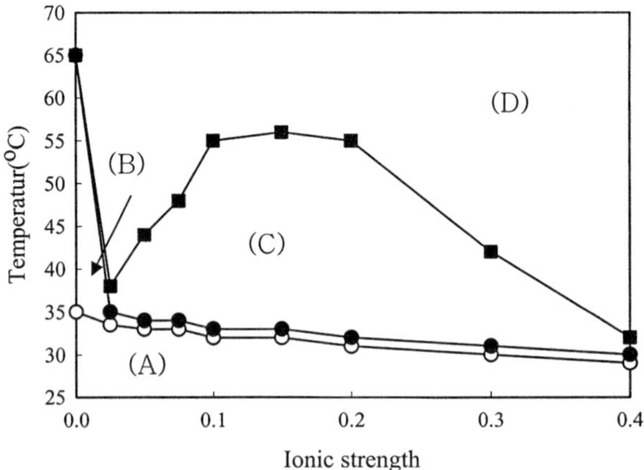

Figure 2. Phase diagram of poly (NiPAAm-co-VI) aqueous solutions (5 wt%) as a function of ionic strength by increasing pH. (A) clear solution; (B) opaque solution; (C) gel phase; and (D) shrunken gel

To partially examine the effect of ionic strength on the solution behavior, the temperature-dependent transmittance of the polymer solution having a low polymer concentration of 0.2 wt% at two different ionic strengths (~ 0 and 0.15) was measured and compared (Fig.3), because the light transmittance reflects the degree of the association or aggregation of the polymer chains. when the temperature was increased from 28 to 50 °C at I = ~ 0, only a slight decrease in transmittance was observed until the temperature reached 35 °C and from 35 °C, which coincided with the clouding temperature, it started to drop much sharply in a 10 °C range and reached slightly below 70 % at 50 °C. However, at I = 0.15, no measurable transmittance changes were observed below 32 °C; but at 32 °C, the transmittance of poly(NiPPAm-co-VI) solution drastically decreased from 100 % at 32 °C to almost 0 % transmitance at 33 °C. This change appeared in a few min at temperature equilibrium state. These results suggests that at I = ~ 0, the polymer could not form high aggregates most likely due to electrostatic repulsive forces between charged imidazole groups which prevent further aggregation of the polymer, while at I = 0.15, the destabilization of polymer dispersion became accelerated, thus forming high aggregates.

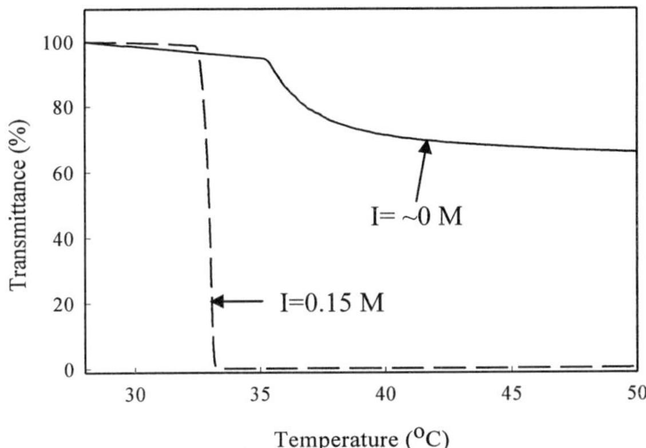

Figure 3. Temperature-dependent transmittance of poly(NiPAAm-co-VI) solutions (0.2 wt%) at different ionic strength (I = ~ 0 M and 0.15 M)

This result indicated that poly(N-isopropylacrylamide-co-vinyl imidazole) is more appropriate compared with poly(N-isopropylacrylamide-co-acrylic acid) for embolization in a deep site in the body by delivering the solution via a long capillary catheter because the temperature-induced gelation of poly(N-isopropylacrylamide-co-acrylic acid) leads to clogging a capillary catheter in the body (37 °C) because the gelation occurs by temperature unless the solution is thermally insulated.

To confirm the charge effect on aggregation, zeta potential measurement was performed at the two ionic strengths. The influence of temperature on the zeta potential of the polymer solution at I = ~ 0 and 0.15 is presented in Fig. 4. The polymer dissolved in deionized water developed a weak positive surface charge (Fig. 4(a)), which was attributed to the ionization of basic imidazole groups. Above the clouding temperature, the surface positive charge was greatly enhanced, most probably due to reorientation and aggregation of polymer molecules above the lower critical solution temperature (LCST), resulting in the redistributing of charged imidazole groups on the surface of the collapsed polymer chains, while some of the polymers remained uncollapsed. This might have led to the two peaks in the zeta potential profile at 37 °C. This positive surface charge seems to be responsible for inhibiting further hydrophobic aggregation between collapsed polymer chains, leading to even lower viscosity. In the case of I = 0.15 solution (Fig. 4(b)), the polymer showed a strong negative surface charge because of adsorbed counter ions. It notes that at 27 °C, only one peak appeare at –270 mV in zeta potential profile, but as the temperature was increased up to 32 °C, a small new peak at -220 mV appeared, just below the LCST.

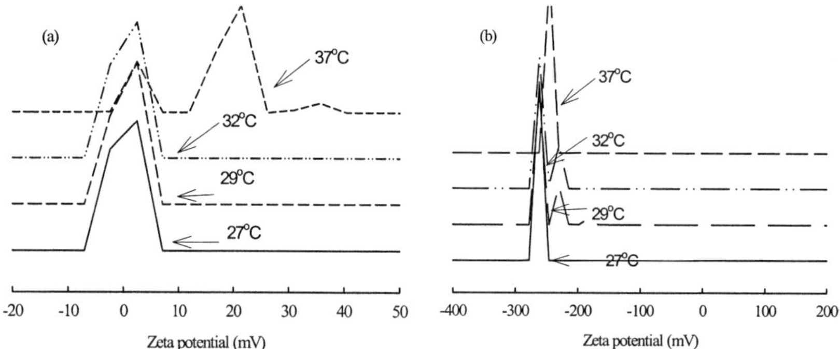

Figure 4. The changes of ζ-potential values as a function of temperature (0.2 wt%, I = ~ 0 (a) and 0.15 (b))

Above LCST, the two peaks combined to form a single peak at –240 mV. This observation is not clearly understood, but it may suggest some salt effects on this particular polymer, possibly charge shielding and specific interactions with imidazole groups. The added salt may contribute its role in the gelation of polymer solution at higher concentration. Further elucidation is needed to understand the salt-temperature induced gelation process of the polymer solution.

Conclusion

A high molecular weight poly(N-isopropylacrylamide-co-vinylimidazole) formed a gel phase at physiological ionic strength and at a temperature range of 34-56 °C. The polymer dissolved in deionized water became opaque at 37 °C but freely flowed through small diameter tubing. When this opaque solution was injected into a phosphate buffered saline solution with an ionic strength of 0.15 adjusted with NaCl and 37 °C, immediate gelation was observed, forming a stable hydrogel. This unique property may offer a new liquid embolic material with injection convenience through a small diameter catheter and the least side effects. The role of ions in the gelation process is not clearly understood and need further investigation. The ions may participate in charge distribution on the collapsed polymer chains at high temperature and polymer chain reorientation possibly via charge shielding effect and/or specific interaction with imidazole rings.

Acknowledgements

This work was financially supported by Grant No HMP-98-G-2-049 from the Ministry of Health and Welfare, Korea.

[1] S. Kondo, N. Hashimoto, H. Kikuchi, F. Hazama, I. Nagata, H. Kataoka, Stroke **29**, 181 (1998)
[2] C. Grandfils, P. Flandroy, N. Nihant, S. Barbette, R. Jérome, P. Teyssié, A. Thibaut, J. Biomed. Mater. Res. **26**, 467 (1992)
[3] K. Kataoka, M. Taneda, T. Asai, A. Kinoshita, M. Ito, R. Kuroda, Stroke **30**, 1396 (1999).
[4] G.J.E. Rinkel, M. Djibuti, A. Algra, J.V. Gijn, Stroke **29**, 251 (1998)
[5] A. Kinoshita, K. Yamada, M. Ito, M. Yamazaki, M. Taneda, T. Hayakawa, Neuroradiology **36**, 65 (1994)
[6] G. Chen, A.S. Hoffman, Macromol. Rapid Commun. **16**, 175 (1995)
[7] C. K. Han, Y.H. Bae, Polymer **39**, 2809 (1998)
[8] B. Mouanda, Polymer **38**, 5301 (1997)
[9] M. Özyalçin, Z. Küçükyavuz, Synthetic Metals **87**, 123 (1997)
[10] A. Kumar, Y. Galaev, B. Mattiasson, Biotechnol. Bioeng. **59**, 695 (1998)

Macromol. Symp. **2004**, *207,* 139-151

Bioconjugates of Smart Polymers and Proteins: Synthesis and Applications

Allan S. Hoffman, Patrick S. Stayton*

Department of Bioengineering, University of Washington, Seattle, WA 98195, USA
E-mail: hoffman@u.washington.edu

Summary: Over the past 20 years we have been deeply involved with the synthesis and applications of stimuli-responsive polymer systems, especially polymer-biomolecule conjugates. The work of Toyoichi Tanaka has been a constant inspiration for our work and this article is dedicated to him. This article summarizes the research that we have carried out along with many collaborators on polymer-protein conjugates. We include conjugates prepared by random polymer conjugation to lysine amino groups, and also those prepared by site-specific conjugation of the polymer to specific amino acid sites that are genetically-engineered into the known amino acid sequence of the protein. We describe the preparation and properties of thermally-sensitive, random conjugates to enzymes and several affinity recognition proteins. We have also prepared site-specific conjugates to streptavidin. with temperature-sensitive polymers, pH-sensitive polymers, and light-sensitive polymers. The preparation of these conjugates and their many fascinating applications are reviewed in this article.

Introduction

Stimuli-responsive, "intelligent" polymers are polymers that respond with large property changes to small physical or chemical stimuli. They are also known as "smart", "stimuli-responsive" or "environmentally-sensitive" polymers. These polymers can take many forms; they may be dissolved in aqueous solution, adsorbed or grafted on aqueous-solid interfaces, or crosslinked in the form of hydrogels (Fig. 1).[1-3] Many

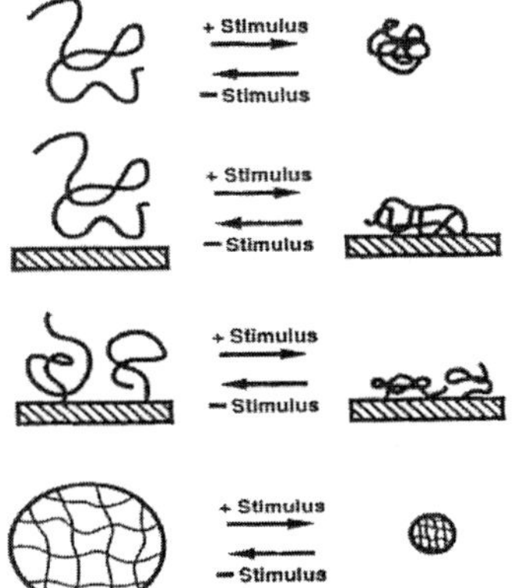

Figure 1. Smart polymers undergo large, abrupt physical and chemical changes in response to a small change in an environmental condition

Table 1: Environmental Stimuli
Physical (T, hν, ionic strength, electric field)
Chemical (pH, specific ions)
Biochemical (metabolite+enzyme)

 DOI: 10.1002/masy.200450314

different stimuli have been investigated, and they are listed in Table 1. Typically, when the polymer's critical response is stimulated, the smart polymer in solution will show a sudden onset of turbidity as it phase separates, the surface-adsorbed or grafted smart polymer will collapse, converting the interface from hydrophilic to hydrophobic, and the smart polymer crosslinked in the form of a hydrogel will exhibit a sharp collapse, and release much of its swelling solution. These phenomena are reversed when the stimulus is reversed, although the rate of reversion is often slower when the polymer has to redissolve or the gel has to reswell in aqueous media.

Smart polymers may be physically mixed with or chemically conjugated to biomolecules to yield a large family of polymer-biomolecule systems that can respond to biological as well as to physical and chemical stimuli. Biomolecules that may be polymer-conjugated include proteins and oligopeptides, sugars and polysaccharides, single and double-stranded oligonucleotides and DNA plasmids, simple lipids and phospholipids, and a wide spectrum of recognition ligands and synthetic drug molecules. In addition, polyethylene glycol (PEG) may also be conjugated to the smart polymer backbone to provide it with "stealth" properties. The conjugate of a synthetic polymer and a biomolecule produces a new, hybrid type of molecule that can synergistically combine the individual properties of the two components to yield new and unusual properties. One could say that these bioconjugates are "doubly smart". Among the most important polymer-biomolecule conjugates are polymer-drug conjugates[4-7] and polymer-protein conjugates.[8] There have been a number of successful applications in medicine and biotechnology for such smart polymer-biomolecule systems, and as such they represent an important extension of polymeric biomaterials beyond their well known uses in implants and medical devices.

We have been combining smart polymers with a wide variety of biomolecules over the last fifteen years.[9-27,32,41] Our major activity in this area has been to conjugate temperature-sensitive polymers to proteins, both randomly[10-20] and at specific sites,[21-27,41] and this work will be the focus of this review article. There are also many others who have randomly conjugated smart polymers to proteins, especially for affinity separations[28-33] and enzyme recovery,[34-40] but to our knowledge, we are the only ones who have synthesized and investigated *site-specific* smart polymer bioconjugates.

There are many polymers that exhibit thermally-induced precipitation (Table 2), and the polymer we have studied most extensively is poly (N-isopropyl acrylamide), or PNIPAAm. This polymer is soluble in water below 32°C, and it precipitates sharply as temperature is raised above 32°C. The precipitation temperature is called the lower critical solution temperature, or LCST. If NIPAAm monomer is copolymerized with more hydrophilic monomers such as acrylamide, the LCST increases and may even disappear. If NIPAAm monomer is copolymerized with more hydrophobic monomers, such as n-butyl acrylamide, the LCST decreases.

Table 2: Some polymers and surfactants that show thermally-induced, reversible phase-separation in aqueous solutions

<u>Polymers with amide groups</u>
Poly (N-substituted acrylamides)
Poly(N-acryloyl pyrrolidine
Poly)N-acryloyl piperidine)
Poly(acryl-L-amino acid amides)
<u>Polymers and surfactants with ether groups</u>
PEO-PPO-PEO triblock surfactants
Alkyl-PEO block surfactants
Random (EO/PO) copolymers
Poly(vinyl methyl ether)
<u>Polymers with alchohol groups</u>
Hydroxypropyl acrylate
Hydroxypropyl methylcellulose
Hydroxypropyl cellulose
Methylcellulose
Poly(vinyl alcohol) derivatives

Synthesis of polymer-biomolecule conjugates

Random conjugation

Proteins may be conjugated randomly to one end of a polymer, or to pendant groups along the polymer backbone (Fig. 2). We have utilized chain transfer free radical

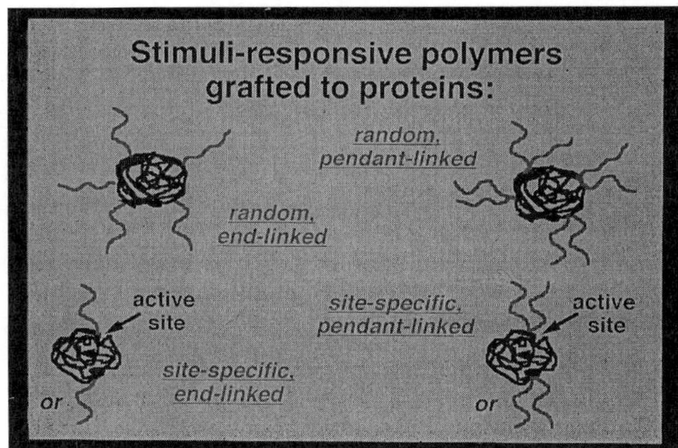

Figure 2. Conjugation of smart (stimuli-responsive) polymers to biomolecules (here proteins)

polymerization to synthesize oligomers with one functional end group, which can then be derivatized to form a reactive group that can be conjugated to the protein. We have also copolymerized NIPAAm with reactive comonomers to yield a random copolymer with pendant reactive groups, which have then been conjugated to the protein.[24]

Normally the lysine amino groups are the reactive sites for random polymer conjugation to proteins, and N-succinimide attachment chemistry is most often utilized. Other possible sites include -COOH groups of aspartic or glutamic acid, -OH groups of serine or tyrosine and -SH groups of cysteine residues. The most likely attachment site will be determined by the reactive group on the polymer and the reaction conditions, especially the pH. Because these conjugations are generally carried out in a non-specific way, they can interfere sterically with the protein's active site or modify its microenvironment, and the bioactivity of the protein is usually reduced. (See ref. 18 for an exception to this).

Site-specific conjugation

We have also been able to conjugate the smart polymer to a specific site on some proteins by inserting a specific reactive amino acid, such as a cysteine with its reactive –SH thiol group, at a selected site (Fig. 3). This is done by genetically-engineering a site-specific mutation into the DNA sequence of the protein, and then cloning the mutant in cell culture. Thus, this method is applicable only to proteins whose complete peptide sequence is known. The preparation of the reactive smart polymer is similar to the

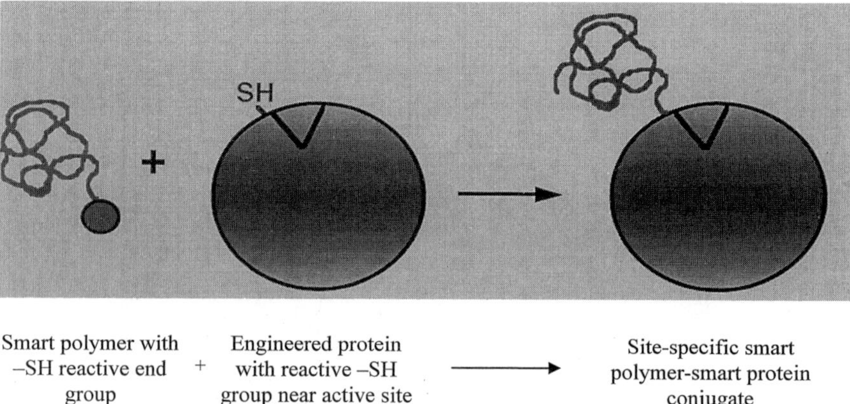

Smart polymer with –SH reactive end group + Engineered protein with reactive –SH group near active site ⟶ Site-specific smart polymer-smart protein conjugate

Figure 3. Site-specific conjugation of environmentally-sensitive polymer to genetically engineered protein

method described above, but now the reactive end or pendant groups and the reaction conditions are specifically designed to favor conjugation to –SH groups rather than to –NH2 groups. Typical mercaptyl-reactive polymer groups include maleimide and vinyl sulfone groups.

For "site-specific" conjugation to –NH2 groups, it should be noted that control of the reaction pH can provide more favorable conjugation to the terminal amine group vs. pendant lysine amine groups. Furthermore, concerning the latter, some lysine amine groups will be more accessible, and therefore more reactive than others.

The specific site for polymer conjugation can be located far away from the active site, in order to avoid interference with the biological functioning of the protein, or nearby or even within the active site, in order to control the ligand-protein recognition process and the biological activity of the protein (Fig. 3).[21,22]

Properties and applications of random, smart polymer-protein bioconjugates

Phase separations for recovery

The thermally-induced precipitation of a PNIPAAm-protein bioconjugate from a complex solution will simultaneously and selectively remove from solution only the protein that is conjugated to the PNIPAAm. We have used this phenomenon for the separation of an enzyme from its reaction solution, in order to enable both recovery of the product from the supernatant and the recycle of the enzyme (Fig. 4).[16-18]

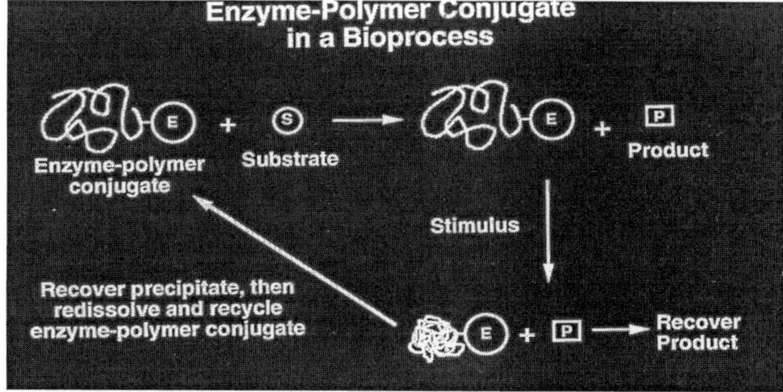

Figure 4. Thermally-induced phase separation of an enzyme-polymer conjugate in a bioprocess

If the conjugated protein forms a complex with another biomolecule, for example by affinity recognition, then the complex will also be selectively precipitated from solution. We have used this phenomenon to selectively remove IgG from solution as a PNIPAAm-protein A••IgG complex, in a fashion similar to affinity chromatography, but in this case it is carried out by reversible phase separation from a solution, instead of flowing through and eluting from a packed column.[15]

This thermally-induced affinity precipitation process may be extended to stimuli-induced phase separation of a biotinylated target molecule which is complexed to avidin or streptavidin. In this case, a biotinylated target molecule is first complexed with an excess of avidin or streptavidin molecules in solution, such that at least one of the four biotin binding sites remains free (on average). Then, an end-linked biotin-smart polymer conjugate is permitted to bind to the free site on the avidin or streptavidin molecules. Following this, the bioconjugate/affinity complex can be phase separated by raising the temperature above the LCST of the smart polymer (or by changing the pH if it is a pH-sensitive smart polymer), which selectively removes the biotinylated target molecule from solution.[32]

Phase separation immunoassays

We have extended the affinity phase separation concept to the selective isolation and assay of an analyte from a complex mixture such as a serum sample, by conjugating a first antibody to the polymer, complexing the analyte by affinity to the first antibody, and then introducing a second, labeled antibody, which then binds to the analyte by affinity to a similar or to a different site (epitope) on the analyte. This yields a temperature-sensitive polymer conjugated to an immune complex sandwich, which can then be selectively removed by thermally-induced precipitation. This is an especially important separation step, because an excess of the labeled, second antibody is usually added to the sample. Washing and resuspension in cold buffer permits easy assay of the analyte.[10-14] This immunoassay resembles ELISA done in solution. This concept has been extended to assay of two different analytes in the same test sample. If NIPAAm is copolymerized with a more hydrophilic or a more hydrophobic comonomer, then copolymers with higher and lower LCSTs can be obtained. If one of each of these two different LCST copolymers is conjugated to a different antibody, then two different analytes may be assayed in the same serum sample by sequentially raising the temperature of the system, to sequentially phase separate the two different polymer-conjugated immune complex sandwiches.[20] One could also carry out such a dual affinity separation or dual immunoassay using combinations of two different pH-sensitive smart polymers, or one temperature-sensitive and one pH-sensitive smart polymer.

Properties and applications of site-specific, smart polymer-protein bioconjugates

Conjugation of a responsive polymer to a specific site near the ligand-binding pocket of a genetically-engineered protein is a powerful new concept. The site-specific placement of a smart polymer near the active site of a protein can permit sensitive environmental control of the ligand•••protein receptor recognition process, which controls all living systems. Then, small changes in environmental conditions can cause large changes in the polymer conformation, leading to reversible "blocking" or "unblocking" of the protein's active site, and such changes also can lead to triggered release of a bound ligand from the protein binding site (Fig. 5).[24,25]

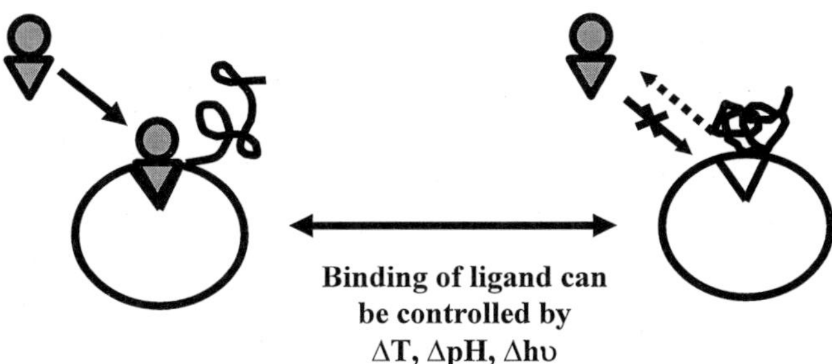

Figure 5. Site-specific conjugates of smart polymers and genetically-engineered (smart) proteins can act as "on-off" molecular switches

Site-specific Cytochrome-b5 conjugates located away from the active site

We first created a mutant cytochrome-b5 with a cysteine located far from the active binding site, and conjugated PNIPAAm to the thiol group using maleimide chemistry.[21] We showed that the cytochrome-b5 was reversibly precipitated and redissolved by thermal cycling through the LCST, and also that the conjugation did not significantly reduce the activity of the protein. This is an extension of the randomly conjugated enzyme phase separation process discussed above, as shown in Fig. 4, and the site-specific conjugation is designed to assure a minimal loss in activity of the protein after conjugation with the smart polymer.

Site-specific Streptavidin (SA) conjugates that control biotin-SA binding

We then selected streptavidin (SA) to study. SA has become one of the most widely used proteins in affinity separations, laboratory assays, and clinical diagnostics because of its very high binding affinity for biotin. There are also several interesting *in vivo* applications under development for use of SA with biotinylated drugs or imaging agents. In order to attempt to control the biotin binding process, we decided to conjugate PNIPAAm near or within the active site of SA. We initially selected the N49 site, which is located on the loop above the biotin-binding pocket of SA, and genetically-engineered a unique cysteine thiol group at that site, called the N49C mutant. (Native SA has no cysteine residues). After that we conjugated a vinyl sulfone-terminated PNIPAAm to the thiol group at that site. We found that the thermally-induced collapse of the polymer acted as a "molecular gate" to control the association of biotin with SA, which had been immobilized on a solid surface to avoid its precipitation from solution (Fig. 95).[22] Using another SA mutant, E116C, we found that the molecular weight of the polymer had an important influence on the gating process, which was only efficient above PNIPAAm MWs of ca. 7,000, supporting the concept of steric interference of the biotin at the binding site.[24,25]

Triggered release of bound biotin from site-specific SA conjugates

When biotin is bound to a PNIPAAm-E116C SA conjugate at temperatures below the LCST, we found that raising the temperature to thermally-induce the collapse of the polymer "triggers" the release of some of the bound biotin molecules (Fig. 5).[24,25] By cycling the temperature through the LCST several times, we were able to release all of the bound biotin.[24]

Such triggered release of bound ligands could be used to release therapeutics, such as for topical drug delivery to the skin or mucosal surfaces of the body, and also for localized delivery of drugs within the body by stimulated release at pre-targeted sites using non-invasive, focused stimuli, or delivery of stimuli from catheters. Triggered release could also be used to release and recover affinity-bound ligands from chromatographic and other supports in eluate-free conditions, including capture and release of specific cell populations to be used in stem cell and bone marrow transplantation. These processes could involve two different stimuli-responsive polymers with sensitivities to the same or different stimuli. For delicate target ligands such as peptides and proteins, recovery could be affected without the need for time-consuming and harsh elution conditions. Triggered release could also be used to remove inhibitors, toxins or fouling agents from the recognition sites of immobilized or free enzymes and affinity molecules, such as those used in biosensors, diagnostic assays, or affinity separations. This could be used to "regenerate" such recognition proteins for extended process use.

Site-specific conjugates of PNIPAAm-SA linked by hybridized DNA sequences to control the distance of the PNIPAAm from the SA binding site

We recently applied an interesting construct to control the distance of the PNIPAAm from the active site. For this purpose, we conjugated one sequence of complementary nucleotides to the E116C site near the binding pocket of SA, and a second sequence to the end of a PNIPAAm chain. Then, by controlling the location and length of the complementary sequence, the self-assembly via hybridization of the two single chain DNA sequences could be used to control the distance of the polymer from the SA binding site.[23,27]

Size-selective blocking of biotinylated proteins by site-specific PNIPAAm-SA conjugates

We have most recently investigated the gating of biotinylated proteins by a site-specific poly (N-N,diethylacrylamide) or (PDEAAm)-SA conjugate, and discovered that the gating process is sensitive to the size of the protein. For this study, we cloned a double lysine SA mutant, E51K/N118K, that was designed to localize two reactive lysine amino groups near the biotin binding pocket and thus to enhance the conjugation efficiency for amino-reactive succinimidyl groups on PDEAAm. Computer modeling has suggested that the two engineered lysine residues are more solvent accessible than the native SA lysines or the N-terminal amine. Previous characterization of the existing amine-reactivity of native SA showed that its few native lysine amino groups are relatively unreactive. Thus, we were able to preferentially conjugate one of the genetically-engineered lysines at the E51K/N118K sites with amine-reactive PDEAAm. (It is unlikely that both lysine groups could be simultaneously conjugated with high efficiency).

We found that the largest biotinylated protein we studied, IgG (MW 150 kD), was unable to bind to the SA conjugate whether the polymer was above or below its LCST, while the smallest biotinylated protein we studied, Protein G (MW 6.2 kD), was able to bind whether the polymer was hydrated or collapsed. An intermediate size biotinylated protein, bovine serum albumin, or BSA (MW 67 kD), exhibited increased binding as temperature was raised through the LCST of the PDEAAm (ca. 23°C in PBS). This surprising result seemed to contradict our observations with biotin binding by itself, which was blocked from binding above the LCST of PNIPAAm (coincidentally, the LCSTs of both polymers in water are ca. 32°C; addition of buffer and salt will lower their LCSTs). We have concluded that the result with biotinylated BSA is related to the tetrameric structure of SA, which has four binding sites, two each on opposite faces. Our data reveal that we have at best only conjugated one polymer molecule on each face, probably due to steric considerations. Thus, the increase in binding of biotinylated BSA as temperature is increased through the LCST reflects the increasing exposure of the *adjacent* SA binding site, which is on the same face of SA as the conjugated site. This result is not related to binding at the conjugated site itself, which probably blocks binding of any size biotinylated protein, whether the polymer is above or below its LCST. Thus, below the LCST the hydrated polymer sterically interferes with the access of the biotinylated BSA at the *adjacent* binding site, while above the LCST, its collapse is enough to sterically expose that adjacent site, permitting the biotinylated BSA to bind.[41]

Site-specific conjugation of a pH- and temperature-sensitive polymer to SA

We have observed similar results to those seen with the PNIPAAm-E116C SA conjugates when we prepared an E116C SA conjugate with a PNIPAAm- acrylic acid (AAc) copolymer that exhibits combined temperature- and pH-sensitivity. By incorporation of as little as 5.5 mol% of AAc in a random copolymer with NIPAAm, we obtained a copolymer that is completely soluble at 37°C and pH 7.4, and insoluble at 37°C and pH 4.0. When this copolymer is conjugated to the E116C site of SA, biotin binding at 37°C is significantly reduced at pH 4.0 compared to pH 7.4. This is most likely due to the more compact copolymer coil at pH 4.0 and 37°C compared to that at pH 7.4 and 37°C. After we allowed biotin to bind at 37°C and pH 7.4, we were able to trigger the release of some of the bound biotin by changing the conditions to pH 4.0.

Site-specific conjugates of temperature- and light-sensitive polymers to SA

In order to create photo-switchable enzymes, we have synthesized two different copolymers that exhibit combined temperature- and photo-sensitivity. The copolymers were prepared by copolymerizing N,N-dimethyl acrylamide (DMA) with two different light-sensitive comonomers, 4-phenylazophenyl acrylate (AZAA) and N-4-phenylazophenyl acrylamide (AZAAm) (Fig. 6). The copolymers were identified as DMAA and DMAAm, respectively. DMAAm and DMAA showed ca. 8-10 °C differences in their LCSTs when irradiated with UV and visible (VIS) light. Surprisingly, they showed the opposite photoresponses from each other using the same photo-stimuli. Under the isothermal conditions at their photo-responsive temperatures, DMAAm precipitated under UV light irradiation (350 nm), while DMAA dissolved under the same UV light irradiation. This difference is related to the difference of the linkage between the azobenzene groups and the polymer backbone, which are amide and ester bonds. We derivatized the copolymers with vinyl sulfone end groups for specific conjugation to the E116C SA thiol group.[41]

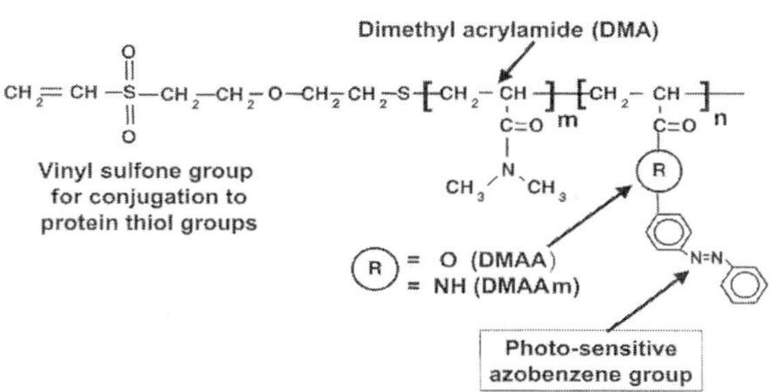

Figure 6. Photoresponsive polymers

One of these copolymer-E116C SA conjugates exhibited blocking of free biotin and triggered release of bound biotin under VIS irradiation, while the other demonstrated the same phenomena under UV irradiation. These opposite photo-responsive biotin-blocking or releasing responses corresponded to the original photo-induced phase transition properties of the copolymers. The vinyl sulfone-terminated DMAAm and DMAA copolymers are also being utilized for site-specific conjugation to a mutant enzyme.[42] That site-specific enzyme conjugate is currently in preparation for use as a photo-switchable enzyme. Such light-controlled binding and release of site-specific protein conjugates may be utilized as a molecular switch for various applications in biotechnology, medicine, and bioelectronics.

Conclusions

We have conjugated stimuli-responsive polymers to random and specific sites on a variety of useful and important proteins, and in the latter case, we demonstrated the ability of the polymer to control the protein-ligand recognition and binding process. Site-specific conjugation combines genetic engineering techniques that allow us to place the mutant residue in one (or more) selected location(s), with polymerization techniques that permit us to vary the molecular weight (thus the polymer coil size) and the composition of the stimuli-responsive polymer, (thus the type and magnitude of the stimulus to be applied). Therefore, each polymer-protein conjugate can be molecularly-engineered for each particular application. It should be emphasized that the site-specific placement of responsive polymers near the binding sites of recognition proteins can provide very sensitive and precise environmental control of the receptor protein-ligand recognition process, which is the process contolling all living systems. Furthermore, triggered-release of a bound ligand may have many interesting applications in drug delivery, delivery of chemical or biochemical reagents in a process reaction, and signal generation in general.

There have been a number of successful uses and there are many more potential applications for these exciting "doubly smart" polymer-protein conjugates in medicine and biotechnology, and as such they represent an important extension of polymeric biomaterials beyond their well-known uses in implants and medical devices.

[1] R.H. Pelton, P. Chibante, *Colloids and Surfaces*, **1986**, *20*, 247.
[2] R. Pelton, X. Wu, W. McPhee, K. C. Tam, in: "Colloidal Polymer Particles," J.W. Goodwin, R. Buscall (Eds.), Academic Press, **1995,** 81.
[3] A.S. Hoffman, *Artif. Org,* **1995,** *19*, 458. Also in *Macromol. Symp.* **1995**, *98*, 645.
[4] F. Fuertges, A. Abuchowski, *J. Control. Rel.* **1990**, *11*, 139.
[5] H. Maeda, *Adv. Drug Deliv. Rev.* **1991**, *6*, 181.
[6] H. Maeda, L.W. Seymour, Y. Miyamoto, *Bioconj. Chem.* **1992**, *3*, 351.
[7] R. Duncan, J. Kopecek, *Adv. Polym. Sci.* **1984**, *57*, 53.
[8] A.S. Hoffman, in: "Biorelated Functional Polymers: Controlled Release and Applications in Biomedical Engineering," T. Okano (Ed.), Academic Press, **1998**, 231.
[9] P.S. Stayton and A.S. Hoffman, in: "Immobilized Biomolecules in Analysis," T. Cass, F.S. Ligler (Eds.), Oxford, **1998**, 135.
[10] N. Monji, A.S. Hoffman, *Appl. Biochem. Biotech.* **1987**, *14*, 107.
[11] C.A. Cole, S.M. Schreiner, J.H. Priest, N. Monji, A.S. Hoffman, in "Reversible Polymeric Gels and Related Systems," P. Russo (Ed.), ACS Symp. Ser. Vol. 350, **1987**, 245.
[12] R.C. Nowinski, A.S. Hoffman, U.S. Patent 4,711,840 (12/8/87).
[13] N. Monji, A.S. Hoffman, J.H. Priest, R.L. Houghton, U.S. Patent 4,780,409 (10/25/88).
[14] E. Kinney-Thomas, R.C. Nowinski, D. Schwartz, J.H. Priest, A.S. Hoffman, U.S. Patent 4,749,647 (6/7/88).
[15] J.P. Chen, A.S. Hoffman, *Biomaterials* **1990**, *11*, 631.
[16] G. Chen, A.S. Hoffman, *Bioconj. Chem.* **1993**, *4*, 509.
[17] G.H. Chen, A.S. Hoffman, *J. Biomater. Sci. Polym. Ed.* **1994**, *5*, 371.
[18] Z. Ding, G. Chen, A.S. Hoffman, *J. Biomed. Mater. Res.* **1998**, *39*, 498.
[19] C.A. Lackey, N. Murthy, O.W. Press, D.A. Tirrell, A.S. Hoffman, P.S. Stayton, *Bioconj. Chem.* **1999**, *10*, 401.
[20] N. Monji, C.A. Cole, A.S. Hoffman, to be submitted to *Bioconj. Chem.*
[21] A Chilkoti, G.H. Chen, P.S. Stayton, A.S. Hoffman, *Bioconj. Chem.* **1994**, 5, 504.

[22] P.S. Stayton, T. Shimoboji, C. Long, A. Chilkoti, G. Chen, J.M. Harris, A.S. Hoffman, *Nature* **1995**, *378*, 472.
[23] P.S. Stayton, A.S. Hoffman, U.S. Patent 5,998,558 (12/7/99).
[24] Z. Ding, C.J. Long, Y. Hayashi, E.V. Elmus, A.S. Hoffman, P.S.Stayton, *Bioconj. Chem.* **1999**, *10*, 395.
[25] Y. Hayashi, P.S. Stayton, A.S. Hoffman, unpublished results.
[26] E.V. Bulmus, Z. Ding, C.J. Long, P.S. Stayton, A.S. Hoffman, *Bioconj. Chem.* **1999**, *11*, 78.
[27] R.B. Fong, Z. Ding, C.J. Long, A.S. Hoffman, P.S. Stayton, *Bioconj. Chem.* **1999**, *10*, 720.
[28] P.O. Larsson, K. Mosbach, *FEBS Lett.* **1979**, *98*, 333.
[29] M. Schneider, C. Guillot, B. Lamy, *Ann. N.Y. Acad. Sci.* **1981**, *369*, 257.
[30] M. Pecs, M. Eggert, K. Schügerl, *J. Biotechnol.* **1991**, *21*, 137.
[31] I.Y. Galaev, B. Mattiasson, *Biotech. Bioeng.* **1993**, *41*, 1101.
[32] J.E. Morris, A.S. Hoffman, R.R. Fisher, *Biotech. Bioeng.* **1993**, *41*, 991.
[33] L.J. Karr, D.L. Donnelly, A. Kozlowski, J.M. Harris, *Methods in Enzymology* **1994**, *228*, 377.
[34] K. Okamura, K. Ikura, M. Yoshikawa, R. Sakaki, H. Chiba, *Agric. Biol. Chem.* **1984**, *48*, 2435.
[35] A.L. Nguyen, J.H.T. Luong, *Biotech. Bioeng.* **1989**, *34*, 1186.
[36] M. Taniguchi, M. Kobayashi, M. Fujii, , *Biotech. Bioeng.* **1989**, *34*, 1092.
[37] M. Taniguchi, K. Hoshino, K. Watanabe, K. Sugai, M. Fujii, , *Biotech. Bioeng.* **1992**, *39*, 287.
[38] Y.G. Takaei, T. Aoki, K. Sanui, N. Ogata, T. Okano, Y. Sakurai, *Bioconj. Chem.* **1993**, *4*, 42.
[39] Y.G. Takaei, T. Aoki, K. Sanui, N. Ogata, T. Okano, Y. Sakurai, *Bioconj. Chem.* **1993**, *4*, 341.
[40] Y.G. Takaei, M. Matsukata, T. Aoki, K. Sanui, N. Ogata, A. Kikuchi, Y. Sakurai, T. Okano, *Bioconj. Chem.* **1994**, *5, 577.*
[41] Z. Ding, R.B. Fong, C.J. Long, P.S. Stayton, A.S. Hoffman, submitted to *Nature*.
[42] T. Shimboji, P.S. Stayton, A.S. Hoffman, unpublished work.

Synthesis, Characterization and Stimuli-Sensitive Properties of Novel Linear and Crosslinked Polybetaines Based on Acrylic Acid and Ethyl 3-Aminocrotonate

Sarkyt E. Kudaibergenov,[1] *Alexander G. Didukh,*[1] *Gulmira T. Zhumadilova,*[1] *Raushan B. Koizhaiganova,*[1] *Larisa A. Bimendina,*[1] *Noh Jung-Gyun,*[2] *Kurt E. Geckeler*[2]

[1] Institute of Polymer Materials and Technology, Satpaev str. 18a, Almaty, 480013, Republic of Kazakhstan
E-mail: ipmt-kau@usa.net
[2] Laboratory of Applied Macromolecular Chemistry, Department of Materials Science and Engineering, Kwangju Institute of Science and Technology, 1 Oryong-dong, Puk-gu, Kwangju 500-712, South Korea
E-mail: keg@kjist.ac.kr

Summary: Novel linear and crosslinked polybetaines based on acrylic acid (AA) and ethyl 3-aminocrotonate (CRO) have been synthesized by a Michael addition reaction followed by radical polymerization. The polymerization of AA and CRO was carried out in bulk, water and organic solvents. The dependence of polymer yield on the molar ratio of monomers and water content was found. Primary attention was paid to linear and crosslinked polybetaines synthesized at equimolar ratio of monomers in the feed. The composition and structure of linear polybetaines was determined by elemental analysis, potentiometric titration, FTIR and NMR spectroscopy. The isoelectric points of linear and crosslinked polybetaines determined by electrophoresis, viscometry and swelling experiments corresponded to pH 2.0-2.2. The stimuli-sensitive properties of amphoteric gels were studied as a function of pH, ionic strength, water-organic solvent mixture, electric, and combined electric and magnetic fields. Appearance of pH gradient within the polyampholyte gel matrix under the externally imposed DC electric field was observed.

Introduction

A number of studies on the volume phase transition of anionic, cationic and nonionic hydrogels has been reviewed in refs.[1-3] Little attention, however, has been paid to amphoteric polyelectrolyte hydrogels.[4,5] In contrast to uniformed polyelectrolytes the behavior of polyampholytes depends on the concentration of the constituent ionic groups and the pH of the external solution.[6-10] Polyampholyte gels also show a reentrant

 DOI: 10.1002/masy.200450315

volume transition in response to pH [11,12] and are useful to design artificial muscles, chemo-mechanical, [13] and biochemo-mechanical[14] systems. Polymeric betaines, whose oppositely charged groups are remote one from another and are displaced on pendant substituents have less been considered in literature,[15,16] in comparison to "annealed" and "quenched" polyampholytes.[17-19]

The synthetic strategy for polyampholytes with betaine structure has been outlined in the pioneering work of Salamone et al.[20-22] Polymeric betaines can be synthesized directly by polymerization of betaine monomers[23] or polymer analogous reaction.[24-27] It was shown that polyampholytes could also be synthesized by hydrogen transfer polyaddition[28] or Michael addition reaction.[29] Previously[30,31] we succeeded in synthesizing novel polybetaines based on ethyl 3-aminocrotonate and acrylic acid using the Michael addition reaction followed by radical polymerization. In the present communication we describe, in more detail, the influence of monomer feeds and solvent nature on the conversion and composition of linear polybetaines derived from the acrylic acid and ethyl 3-aminocrotonate. The acid-base properties, viscosity and electrophoretic mobility of linear polybetaines are considered. The effects of pH, ionic strength, water-organic solvent mixtures, and combined electric and magnetic fields on the behavior of novel polybetaine gels, are also described.

Materials

Ethyl 3-aminocrotonate (CRO, 99% purity), acrylic acid (AA, 99% purity), azoisobutyronitrile (AIBN), N,N'-methylenbisacrylamide (MBAA) and the reagent grade solvents acetone, benzene, ethanol, chloroform, DMF, DMSO, hexane were purchased from Aldrich. AA and CRO were purified by distillation under the low pressure and kept in refrigerator. Reagent-grade NaCl and standard titration solutions of 0.1N HCl and NaOH were used.

Methods

FT-IR spectra were recorded using a Perkin Elmer Spectrum 2000 series as pellets with KBr. ^{1}H NMR spectra of monomers and polymers were obtained in $CDCl_3$ and D_2O at room temperature with the help of JROL JNM-LA 300 WB FT-NMR. Molecular modeling was performed with the help of the Single Point Energy calculation method with Hartree-Fock 3-21G*.

Synthesis of linear and crosslinked polybetaines

Polymerization of AA and CRO was carried out in bulk, water and organic solvents. The mixture of CRO and AA containing 2 mg of AIBN was bubbled by nitrogen gas during 2-3 min. to remove the dissolved oxygen and thermostated at 70 °C during 1 h. The total amount of monomer mixture was always 1.5 mL. Polymerization of equimolar mixtures of monomers in water and organic solvents was also carried out in the presence of AIBN (2 mg) at 70 °C. The volume ratio of solvents and monomer mixture was 1:1 vol/vol. Radical polymerization of equimolar mixtures of monomers was also carried out in the presence of various amounts of water at identical conditions described above. Linear polymers obtained in bulk, water and organic solvents were thoroughly washed out by acetone. Samples were dried *in vacuo* to constant mass. Crosslinked polymers were synthesized as justdescribed, but in the presence of MBAA (2 mg). Gel samples were washed in distilled water for 2-3 weeks to remove the sol fraction.

Potentiometric titration

Potentiometric titration of linear polymers was carried out on a 2-point calibrated "CORNING pH meter 240", equipped with combined electrodes. Direct and back potentiometric titration was carried out as follows. 50 mg of polymer samples were dissolved in either 20 mL of water or 0.1N NaCl and titrated by 0.1N NaOH. For back titration 170 mg of polymer samples were dissolved either in 100 mL of 0.1N NaOH and titrated by 0.1N HCl or dissolved in 100 mL of 0.1N NaOH and titrated by 0.1N HCl. Potentiometric titration curves were plotted in integral and differential forms. The acid-base ratio of functional groups was calculated from the inflection points and differential peaks. The ionization constants of acid and base groups were determined from the Henderson-Hasselbalch equation.

Measurement of the isoelectric point (IEP)

In order to determine the isoelectric point (IEP), the dependence of the reduced viscosity on pH was measured using an Ubbelohde viscometer at 25±0.1 °C. Polymer solutions were filtered through the 0.45μm filter and adjusted to different pH by addition of 0.5N NaOH or HCl, with the concentration of polymer held constant. Then these solutions were stirred and the values of viscosity were measured. Electrophoretic mobility of polybetaine was measured by Doppler electrophoretic light scattering with the help of the Coulter DELSA 440SX. Electrophoretic light scattering was carried out at 25±0.1 °C at four angles (8.9, 17.6, 26.3, 35,2°). The electric field was applied at a constant current of 8-14 mA. The electrophoretic cell had a rectangular cross section connecting the hemispherical cavities in each electrode. The total sample volume was about 1 mL. pH of the polymer solutions was adjusted by 0.5N NaOH or HCl. After adjusting pH and stirring about 30 min the solution was injected into the cell through the 0.45μm filter. The measured electrophoretic mobility, U, was the average value at the upper stationary layer for the four scattering angles. The value of pH at U = 0 was adopted as the IEP.

Equilibrium swelling measurements

Gels, swollen to equilibrium in distilled water, were weighed on an analytical balance and then dried. The swelling degree of the gels, α (in g/g), was calculated according to the formula: $\alpha = (m-m_0)/m_0$, where m and m_0 are the masses of swollen and dried gels respectively. The influence of pH on the swelling behavior of hydrogels was studied by comparing the swelling degree of gel samples at various pH. Gel samples were placed in water solutions, the pH values of which were adjusted by adding 0.1N NaOH or HCl. Since mineral acids and bases have low buffer capacity, the pH of gel-containing aqueous solutions was maintained by adding of appropriate amount of acid or base. After equilibration the preliminary weighed gels were dried up to constant mass and the swelling degree was determined. The swelling degree of hydrogels at different ionic strengths, adjusted by NaCl, and in the mixtures of different solvents (water–ethanol, water-acetone, water-N,N-dimethylformamide), was also determined gravimetrically as described above.

Effect of externally imposed electric and magnetic fields

The equipment and measurement technique concerning the influence of externally imposed DC electric field and combined electric and magnetic fields on the behavior of gels was studied as described in.[32-34]

RESULTS AND DISCUSSION

Polymerization of CRO and AA via Michael addition reaction

It is well known that ethyl 3-aminocrotonate (CRO) exists in the enamine (**1**) and imine (**2**) tautomeric forms and in spite of the presence of C=C and C=N double bonds, it does not polymerize.[35,36] The reason is that the C=C and C=O (or C=N and C=O) bonds are conjugated and additionally stabilized via intramolecular hydrogen bonds. According to literature data [36] at room temperature the enamine form is more stable than the imine form. In particular it was found[35] that in ethanol there are 80% enamine and 20% imine isomers. Our results on molecular modeling also confirm the more energetically stable state of the enamine form. The relative energetic stability of enamine tautomeric form (ΔE = -19.67 kcal·mol^{-1}) is lower than that of the imine (ΔE = 0). Therefore the enamine form is preferentially involved in the Michael reaction with AA, and in the course of enamine consumption the enamine↔imine equilibrium is shifted to left side. Recently [30,31] we have shown for the first time that upon the addition of unsaturated carboxylic acids, the intramolecular hydrogen bonds of CRO are disrupted, the hydrogen atoms of primary amine groups migrate to the double bonds of AA, and via the Michael addition reaction the intermediate product, **3**, the so-called betaine monomer, is produced (Scheme 1). It was demonstrated that the polymerization reaction is spontaneous, exothermic and even can start at room temperature without addition of initiator. The spontaneous polymerization of zwitterionic monomers was also observed by the authors.[37] It was shown[29] that the addition of acrylic acid to chloroform solution of polyiminoethylene or polyiminohexamethylene produces polyampholyte with more than 90% betaine structure. The specific feature of this reaction was that both a protonation and Michael addition reaction take place simultaneously.

The mechanism of spontaneous polymerization of CRO upon addition of AA is complicated and not clear yet. Probably the hydrogen atoms shuttling from the enamine, **1**, to the imine, **2**, tautomeric forms are responsible for initiation and spontaneous polymerization. In the latter case some monomer molecules can capture the migrating hydrogen atoms and produce the free radicals to initiate the polymerization reaction. It is supposed that the high reactivity of the intermediate product **3**, is such that it can be involved into homopolymerization and copolymerization reactions with either AA or CRO, and as a result the polybetaine **4** and betaine containing copolymer **5** can be formed (Scheme 1).

It should be noted that the mixtures of CRO and acetic acid as well as the mixtures of AA and primary (or secondary) amines are not able to polymerize spontaneously. This favors the proceeding of polymerization via Michael reaction.

Fig 1 shows the conversion of CRO, AA and mixture of CRO-AA at identical polymerization conditions, e.g. when the amount of both AA and CRO taken for polymerization is equal to those in CRO-AA mixture. As seen from Fig.1, CRO itself does not polymerize in the presence of initiator, while the yield of PAA is approximately 50-60%, independent of amount. The conversion of CRO-AA increases exponentially with increase of added AA. The data of elemental analysis show that when the CRO is in excess the main product is polybetaine **4** (Table 1), while the excess of AA in the monomer feed leads to formation of copolymer **5** containing betaine **3** and AA. Due to the very low reactivity of CRO, its copolymerization with **3** is less probable.

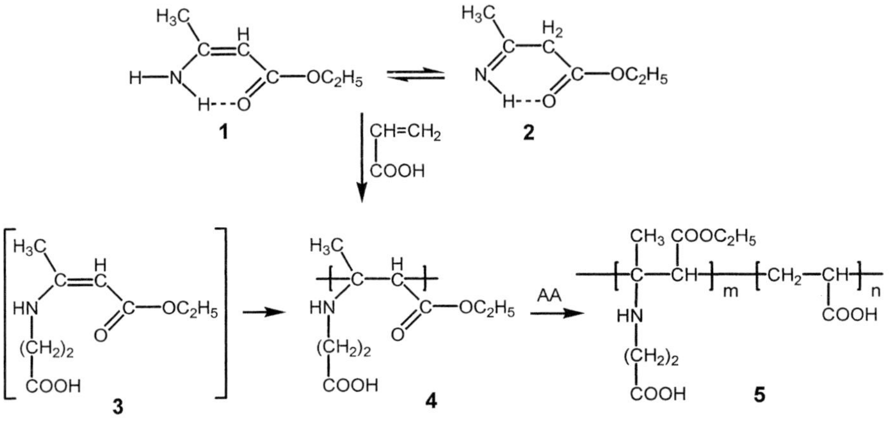

Scheme 1. Polymerization protocol of CRO and AA via Michael addition reaction

This statement is also confirmed by direct and back potentiometric titration of CRO-AA (Table 1, Fig. 2). The content of acidic and basic groups is close to equimolar for polymers synthesized at [CRO]:[AA] = 90:10, 80:20, 70:30, 60:40 and 50:50 mol% in the feed. In the course of back potentiometric titration of [CRO]:[AA] = 50:50 mol%, three well defined peaks are distinguished. The first peak corresponds to titration of the excess of mineral acid or base. The rest two peaks are responsible for successive titration of COOH and NH_2^+ as well as NH and COO^-. The amount of COOH and NH_2^+ calculated from the potentiometric back titration corresponds to 54.5 and 45.5 mol% respectively. The amount of NH and COO^- determined from the potentiometric data is equal to 40 and 60 mol% respectively. Thus the content of acid and base groups calculated from the back potentiometric titration is (45±5 mol.%) and coincides well with direct potentiometric titration data. Ionization constants of carboxylic and secondary amine groups determined from the Henderson-Hasselbalch equation are equal

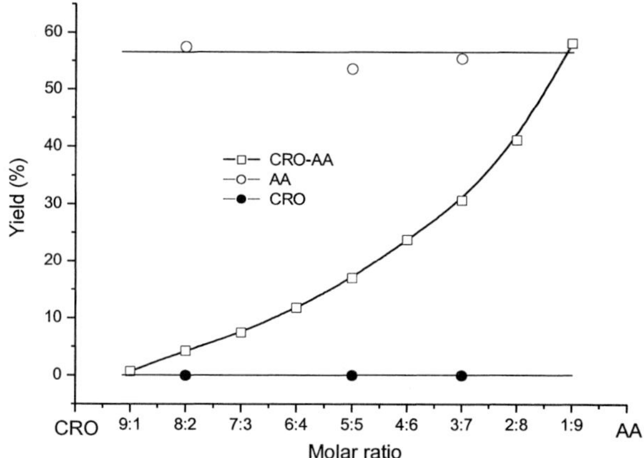

Figure 1. Yield of polymer products on the molar ratio of AA and CRO in bulk. Total volume of monomer mixture is 1.5 mL, [AIBN] = 2 mg. Polymerization time is 1 h. T = 70 °C.

Table 1. Acid-base contents of copolymers found by elemental analysis and potentiometric titration

Monomer ratio in the feed, mol%		Data of elemental analysis, % N		Acid-base contents found from the elemental analysis, mol%		Acid-base contents found from the potentiometric titration curves, mol%	
CRO	AA	Calcd	Found	Carboxylic	Amine	Carboxylic	Amine
10	90					91	9
20	80					77	23
30	70	5.75	5.73	63	37	72	28
40	60					69	31
50	50	6.96	7.18	45	55	59	41
50*	50					54.5	45.5
50**	50					46	54
60	40					46	54
70	30					60	40
80	20					50	50
90	10					43	57

*Found from the back potentiometric titration.
** Found from the direct potentiometric titration in the presence of 0.1N NaCl.

to 6.2 and 9.5 respectively. The higher pK_a of carboxylic groups in comparison with poly(acrylic acid) (pK_a=4.3) may be connected with the nearest-neighbor effect.[38]

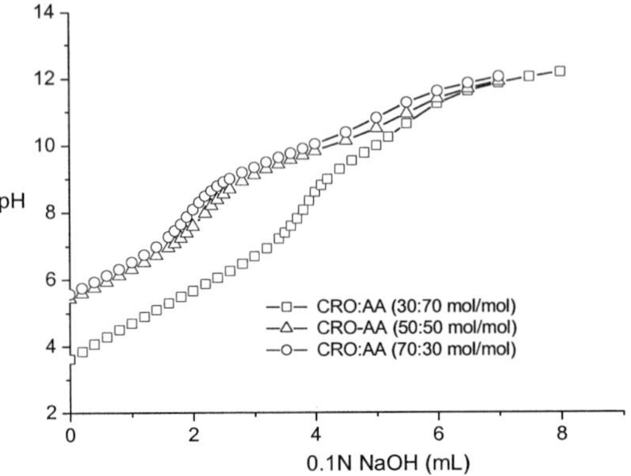

Figure 2. Potentiometric titration curves of CRO-AA synthesized at various molar ratio of monomers in the feed.

Thus, depending on the AA content in the monomer feed, either betaine type polyampholyte **4** or poly(betaine-electrolyte) **5** can be formed. Unique properties of poly(ampholyte-electrolyte) were recently outlined by authors.[39,40]

It is interesting to follow by the influence of solvent nature on the yield and acid/base molar ratio of final products. As seen from Fig. 3, the polymer yield increases according to relative dielectric permittivity of solvents. This is probably explained by active participation of polar solvents in the hydrogen-transfer process. Table 2 presents the content of acid and base groups found from the potentiometric titration data. In water, ethanol, water-ethanol mixture, DMF and benzene the content of carboxylic groups is 50±5 mol% and probably the main product is polybetaine **4,** while in hexane and chloroform the polybetaine **4** is enriched by AA and the structure **5** is formed.

We have found a drastic influence of water content on the conversion of CRO-AA (Fig. 4). Addition of 15-100 vol% of water into the equimolar monomer mixture of CRO and AA increases the yield of polymer 3-4 times in comparison with the bulk (16%). This is probably connected with participation of water molecules in proton transfer process. In particular, it was stated[41] that hydrogen bonds are able to change significantly the reactivity of monomers and radicals if the heteroatom and the reaction center participate in the formation of a unified π-conjugated system.

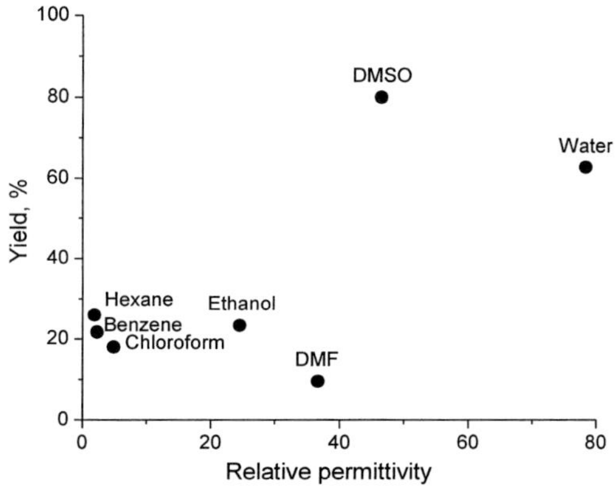

Figure 3. Dependence of the yield of equimolar mixture of CRO-AA on relative permittivity of solvents. The volume ratio of monomer mixture and solvents is 50:50 vol/vol.

Table 2. Acid-base contents of polymer products synthesized in various solvents. The molar ratio of CRO:AA in the monomer feed is 50:50 mol/mol. The volume ratio of solvent and monomer mixture is 50:50 vol/vol.

Solvent	Water	Ethanol	Water/ethanol (50:50 vol/vol)	DMF	Benzene	Hexane	Chloroform
[COOH], mol.%	59	51	48	55	57	69	62
[NH], mol.%	41	49	52	45	43	31	38

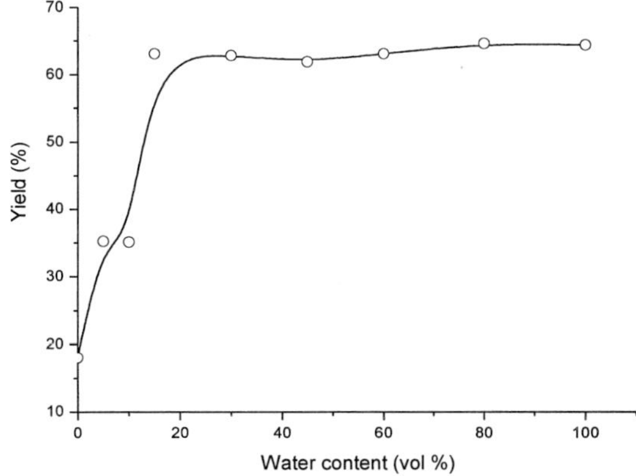

Figure 4. Influence of water content on the yield of the equimolar mixture of CRO-AA.

The structures of formed polymers were identified by FTIR and ^{1}H NMR spectroscopy. FTIR spectra of CRO, AA and CRO-AA (50:50 mol.%) are shown in Fig. 5. Disappearance of intensive peaks of monomers at ν = 1659, 1622 and 1634 cm^{-1} that belong to C=C and C=N of CRO and C=C of AA, and broadening and shifting of intensive lines of primary amine groups of CRO at ν = 3230, 3334 and 3440 cm^{-1} up to 3157 and 3421 cm^{-1} clearly show the formation of polymer product. Stretching vibrations of CH and C=O as well as the symmetric and asymmetric vibrations of carboxylate ions appear respectively at ν = 2958, 1707, 1559 and 1450 cm^{-1}. Besides the intensive rocking vibration of C=C at ν = 789 cm^{-1} from CRO, a diminished but very small shoulder is still seen in the spectrum of polymer. The very small peak at ν = 1632 cm^{-1}, attributed to C=C of AA, indicates the latter's presence in the final product. The admixtures of AA and CRO in the polymer product may be connected with formation of salt bonds between the secondary amine groups of **4** and AA as well as between the carboxylic groups of **4** and amine groups of CRO. Thus analysis of FTIR spectra reveals that the main product of polymerization of CRO and AA is **4,** containing insignificant additives of monomers (AA and CRO). However from the FTIR spectra it is difficult to identify the structure of copolymer **5**.

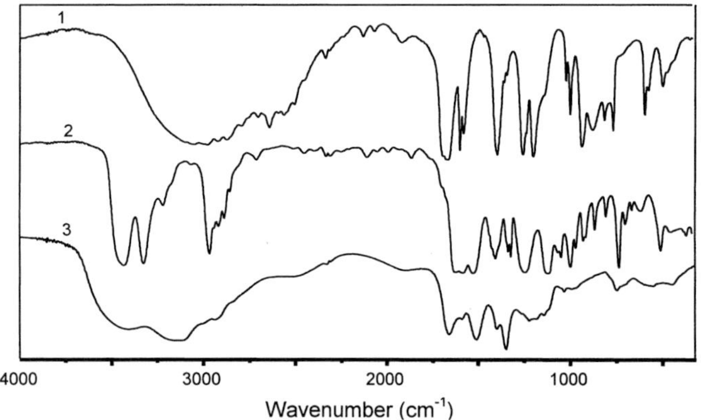

Figure 5. FTIR spectra of AA (1), CRO (2), and CRO-AA (50:50 mol.%) (3).

^{1}H NMR spectra of CRO-AA shown in Fig. 6 coincide well with the results of elemental analysis and potentiometric titration. NMR spectra of CRO-AA (80:20 and 50:50 mol.%) are identical showing that the main polymer is polybetaine **4**. Appearance of

additional proton signals from the AA when the molar ratio of CRO-AA is 20:80 testifies to the formation of polymer structure **5**. Thus NMR spectra are in good agreement with elemental analysis and potentiometric titration and show on the formation of **4** and **5**, depending on polymerization conditions.

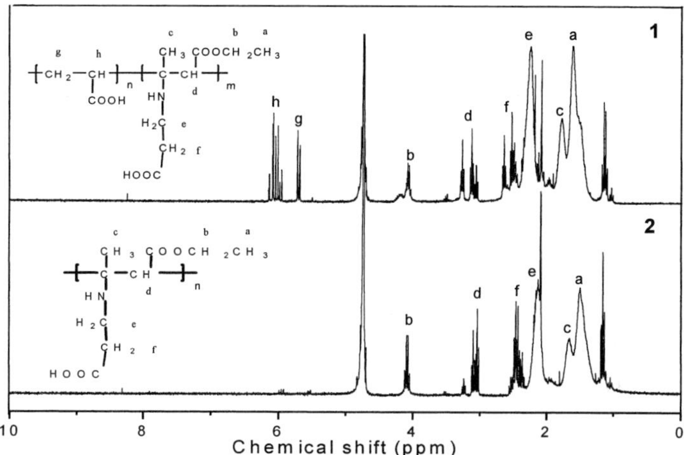

Figure 6. ^{1}H NMR spectra of poly(betaine-acrylic acid) (1) and polybetaine (2).

The properties of linear and crosslinked polybetaines in aqueous and aqueous-salt solutions were studied for samples synthesized at equimolar ratio of AA and CRO in

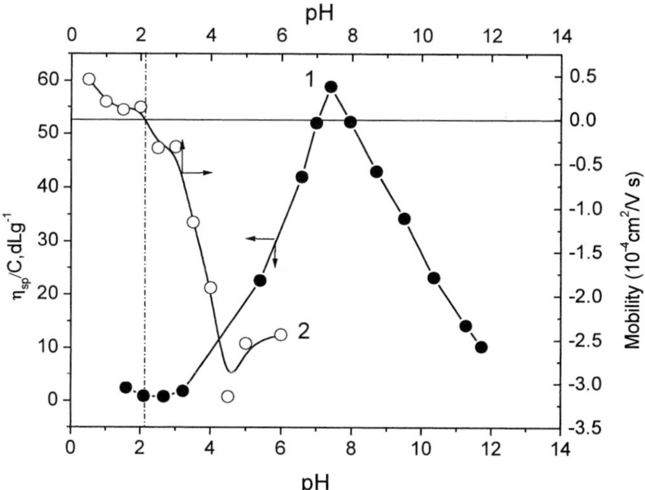

Figure 7. Reduced viscosity (curve 1) and electrophoretic mobility (curve 2) of the polybetaine **4** vs. pH of the solution.

bulk. Fig. 7 shows the influence of pH on the viscosity and electrophoretic mobility of CRO-AA.

The minimal viscosity and the zero mobility are observed near of pH 2.1-2.2 that corresponds to the isoelectric point (IEP). The sharp increase of the reduced viscosity in the pH range 3-7.5 is due to ionization of carboxylic groups. The decrease in viscosity in the basic region is probably accounted for suppression of polyelectrolyte effect by the excess of NaOH that plays the role of low-molecular-weight electrolyte. At pH < 2 a slight increase of the viscosity as a result of protonation of nitrogen atoms is observed. The "asymmetric" character of chain stretching and "anomalous" low magnitude of the IEP ($pH_{IEP} \approx 2.0$-2.1) may be the result of different accessibility of carboxylic and secondary amine groups to ionization. Carboxylic groups replaced far from the main chain can be ionized more easily than the secondary amine groups that are near of the main chain and in a hydrophobic environment. Therefore ionization of carboxylic groups more effectively unfolds the macromolecule than the ionized nitrogen atoms located close to the main chain. Since the requirement of the IEP is the overall electroneutrality of macromolecular chain, the added mineral acid at first suppresses the ionization of carboxylic groups decreasing the negative charges of the whole macromolecules.

At lower pH, however, some parts of the secondary amine groups can also be ionized. These results are in good agreement with the electrophoretic data, e.g. the amount of negative charges on macromolecules is significantly higher at pH>2 than the amount of positive charges at pH<2. Also, the specific binding of Cl^- diminishes the amount of positively charged groups and shifts the IEP to the acidic region. Taking into account that the value of pK_b of secondary amine groups is equal to 4.5 the ionized amount of amine groups at pH = 2.0 was calculated from the Henderson-Hasselbalch equation. It is proved that at the IEP only 3% of secondary amine groups are ionized to make the macromolecules overall electroneutral.

It should be mentioned that the acid/base properties of betaine type polyampholytes are quite different from those of "annealed" polyampholytes.[19] The position of the IEP of "annealed" polyampholytes depends strongly on the molar ratio of acid-base monomers in copolymers. A very low isoelectric state ($pH_{IEP} = 2$-3) is observed for amphoteric copolymers containing up to 90-95 mol.% of acidic monomers. In contrast to "annealed" polyampholytes and polycarboxybetaines, polysulfobetaines at low pH exhibit neither increasing viscosity nor existence of an IEP, due to strong acidity of the sulfonate groups.[42,43]

As seen in Fig. 8, the pH-dependent behavior of polybetaine gel synthesized at the same conditions is also in good agreement with the linear analogue. The swelling degree of polybetaine gel is minimal around the IEP ($pH_{IEP} \approx 2.0$).

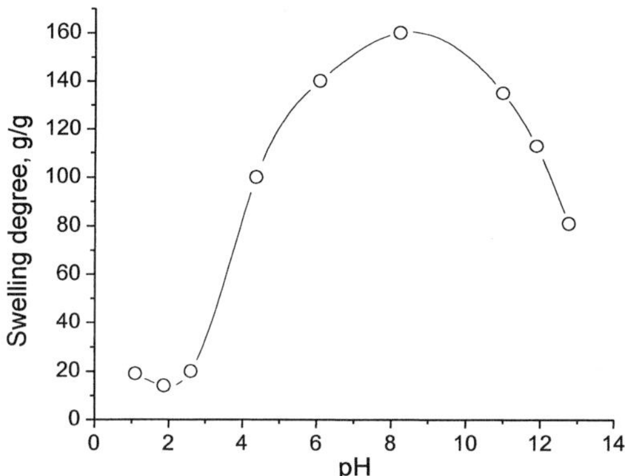

Figure 8. Dependence of the swelling degree of polybetaine gel on pH.

The viscosity of polybetaines is extremely high in aqueous solution. This may be due to intermolecular association of macromolecules via formation of ionic and hydrogen bonds. The value of the intrinsic viscosity determined in $2 \cdot 10^{-3}$ $mol \cdot L^{-1}$ HCl is equal to 26.0 $dL \cdot g^{-1}$.

Effect of the ionic strength

Changes in of the reduced viscosity of linear polybetaines and the swelling degree of crosslinked polybetaines in the presence of salt are shown in Figures 9 and 10. As the ionic strength increases both the reduced viscosity and swelling degree at first sharply decreases, reaches a minimum, and then slightly increases. To interpret these results the ability of polymeric betaines to form intraionic and interionic associates should be taken into account.[44] Probably small amount of added NaCl at first disrupts the interionic aggregates. Further increasing of the ionic strength destroys intraionic contacts allowing minor expansion of the polymer coil.[45] These results are in qualitative agreement with the theoretical[46] and experimental[47] data that show the collapsing of balanced polyampholytes at low salt and swelling at high salt concentrations.

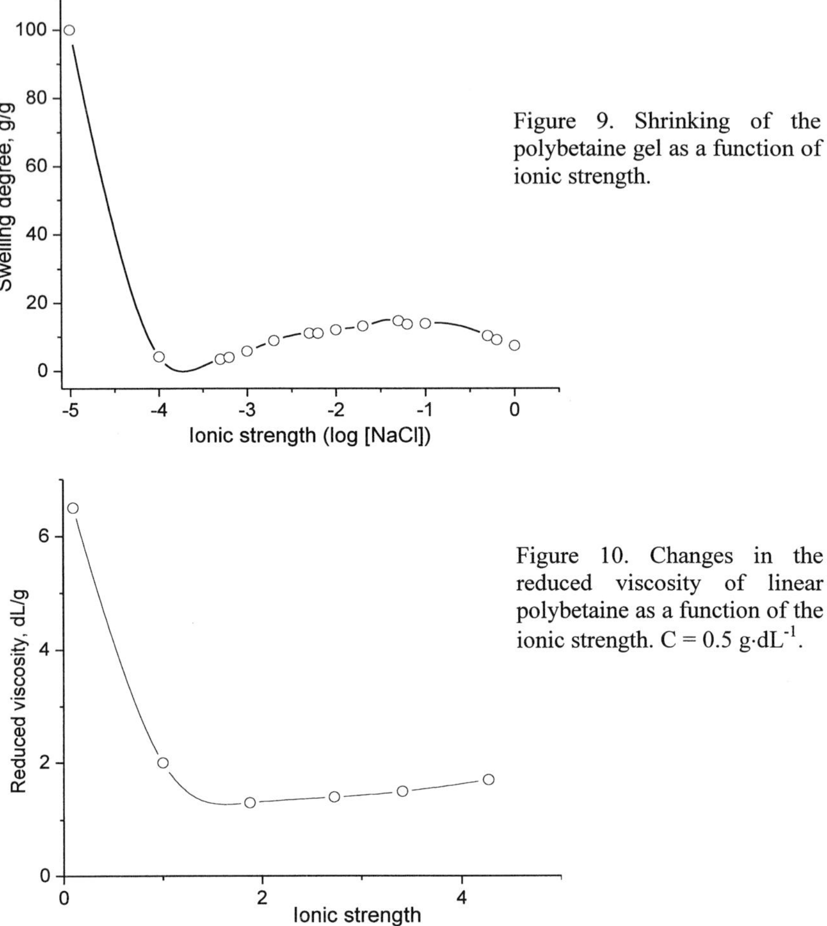

Figure 9. Shrinking of the polybetaine gel as a function of ionic strength.

Figure 10. Changes in the reduced viscosity of linear polybetaine as a function of the ionic strength. C = 0.5 g·dL^{-1}.

Effect of binary solvents

Polybetaine gels shrink in water-acetone, water-alcohol and water-DMF mixtures (Fig. 11). Addition of organic solvents causes gradual deswelling of gel samples, the effectiveness of which changes in the order DMF > acetone > ethanol. This ordering is due to decreasing of dielectric permittivity of the medium, which enhances the condensation of counterions on the macroions. The latter leads to decreasing osmotic pressure within the gel. In organic solvents, macromolecular chains are in the collapsed state due to suppression of ionization. These results confirm the universal behavior of polyelectrolyte gels with respect to the thermodynamic quality of solvents.

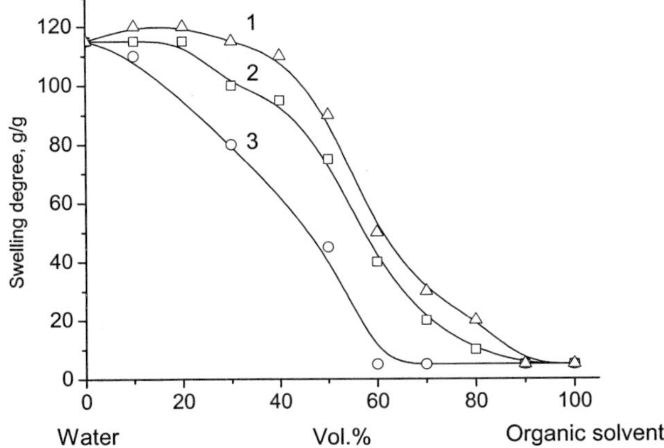

Figure 11. Deswelling of polybetaine gel in water-ethanol (1), water-acetone (2), and water-DMF (3) mixtures.

Effect of combined electric and magnetic fields

Previous work has shown the influence of electric[48] and combined electric and magnetic fields[32, 49] on the behavior of polyampholyte and polyelectrolyte gels. Fig. 12 demonstrates the kinetics of polybetaine gel contraction under the imposed DC electric field and cross-combined electric and magnetic fields. Increasing the electric current leads to enhancement of the gel's contraction rate; the values of $\Delta m/m_0$ are increased linearly, reaching the constant state. Application of a constant magnetic field accelerates gel contraction. Apparently the magnetic field distorts paths of charge carriers and the

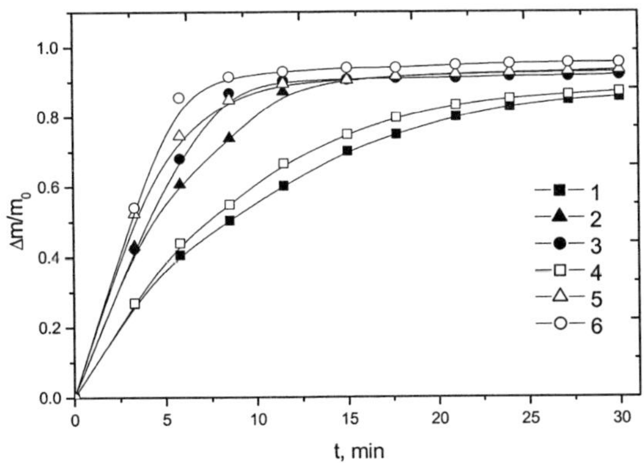

Figure 12. Kinetics of gel contraction under the influence of DC electric field (curves 1-3) and combined electric and magnetic fields (curves 4-6) at E=10 (curves 1,4), 20 (curves 2,5), 30 (curves 3,6) V. Magnetic field is constant and equal to 100 G.

specimen is additionally polarized in the direction perpendicular to the vector of electric current. Consequently the opposite sides of the specimen bear non-zero charge. As a result these charged areas are attracted to oppositely charged electrodes, and the contraction rate increases.

Formation of pH-gradient

Previously[33,34] the formation of a pH-gradient within polyampholyte gel matrix under the externally imposed DC electric field was observed for "annealed" amphoteric gels based on vinyl-2-aminoethyl ether and acrylic acid. Fig. 13 shows the time dependence of pH change within polybetaine gel volume. The applied DC electric field causes a sharp appearance of pH gradient during 1-2 min that is fully stabilized after 5 min. When the electrical stimulus was removed, pH quickly adopted the initial value. The values of $\Delta pH = pH_0 - pH_{t\to 0}$ (where pH_0 is the initial pH, $pH_{t\to 0}$ is the pH value extrapolated to zero time $pH_{t\to 0}$) are a function of DC electric field.

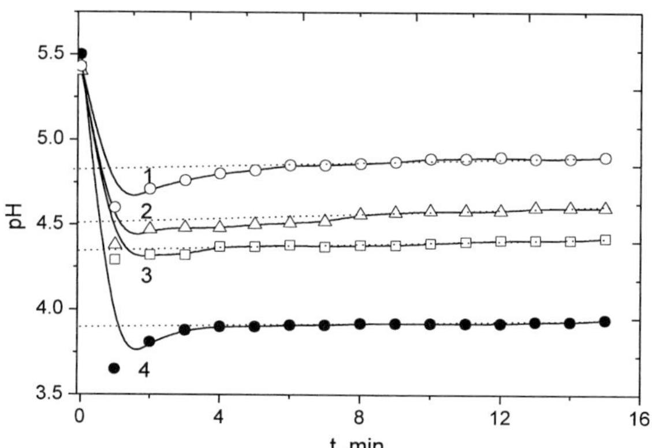

Figure 13. Change in pH with time in the gel near the anode (1), in the center (2,3) and near the cathode (4), with E=10V. The dotted lines indicate on the extrapolation of pH to $pH_{t\to 0}$.

Fig. 14 represents the dependence of pH on distance ± *l* when the glass electrode is placed on cathode or anode sides of gel specimen in comparison with its central section where *l*=0.

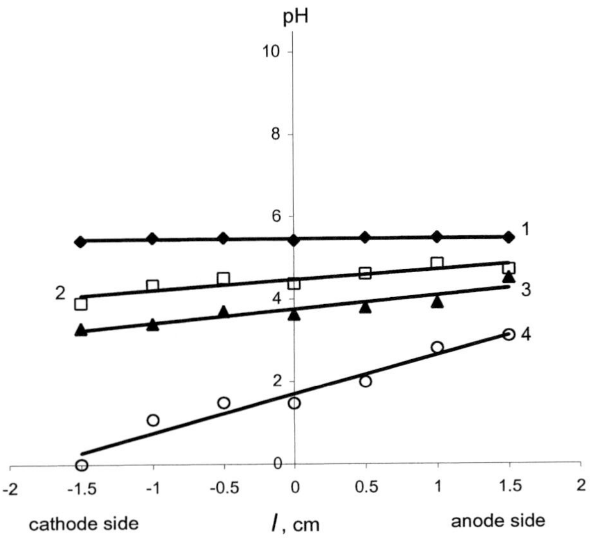

Figure 14. The dependence of pH gradient on distance *l* without the imposed DC electric field (1) and with applied D.C. voltages E=5V (2), 10V (3) and 15V(4).

Without the applied DC electric field, the pH is uniform and equal to 5.46. Increasing of the electric field shifts the pH gradient in the acidic direction. The dependence of pH gradient on *l* has linear character, and the slope of the line is proportional to the strength of applied DC electric field. In order to interpret these data the distribution of fixed network charges and counterions within the gel specimen should be considered. The externally imposed potential across the gel causes the polarization of network charges and counterions. The latter are produced by water electrolysis. Negative fixed charges (COO^-) and mobile ions (OH^-) accumulate at the anodal side (zone **A**) while the accumulation of positive fixed charges (NH_2^+) and mobile ions (H^+) takes place at the cathodal side (zone **C**). The intermediate zone **B** is relatively electroneutral with respect to zones **A** and **C**. Increasing of the DC electric current leads to overall acidification of the gel sample due to enhanced ionization of carboxylic groups (see Fig.14, curve 4). As a result the zones **B** and **C** become narrower, and the zone **A** expands. It should be noted that the existence of a pH gradient is specific only for amphoteric gels and was not detected for uniformly charged polyacid or polybase gels.[19]

Conclusion

Novel linear and crosslinked betaine type polyampholytes were synthesized from the ethyl 3-aminocrotonate and acrylic acid via Michael addition reaction followed by radical polymerization. Polymerization mechanism is probably connected with migration of hydrogen atoms from enamine to imine tautomeric forms. Molecular modeling shows that the enamine tautomeric form is more energetically stable and preferentially involved into Michael addition reaction. The influence of the monomers ratio in the feed and solvent nature on the conversion and composition of polymer products was established. Considerable increasing of the yield of linear polybetaines upon addition of water into the reaction mixture is connected with participation of water molecules in proton transfer process. Elemental analysis together with potentiometric titration, FTIR and NMR data confirm the formation of polybetaine structure and copolymers of betaine with AA, depending on the molar ratio of monomers in the feed. The isoelectric points of linear and crosslinked polybetaines determined by combination of viscometric and electrophoretic methods and from the equilibrium swelling measurements arrange at pH=2.0-2.1. Anomalous low values of the IEP polybetaines are accounted for low ionization degree of secondary amine moieties displaced near of the main chain surrounded by hydrophobic groups. The behavior of linear and crosslinked polyampholytes with respect to ionic strength is explained by destruction of intraionic and interionic associations. Shrinking of polybetaine gel in water-organic solvents mixtures is explained by decreasing of the dielectric permittivity of medium that leads to condensation of counterions to macroions. The combined electric and magnetic fields enhances the gel collapse due to additional polarization of gel specimen. Appearance of pH gradient within the gel volume under imposed externally DC electric field is connected with polarization of network charges and counterions.

Acknowledgement

This work was supported by INTAS-00/57 and INTAS-00/113 Grants. The financial support from the KISTEP, Korea, and the Brain Korea 21 Project is greatly acknowledged.

[1] *Adv.Polym.Sci.* **1992**, *109*, 1.
[2] *Adv.Polym.Sci.* **1993**, *110*, 1.
[3] T. Amiya, T. Tanaka, *Macromolecules* **1987**, *20*, 1162.

[4] S. G. Starodubtzev, V. R. Ryabina, and A. R. Khokhlov, *Vysokomol.Soedin. Ser.A*: **1987**, *29*, 2281.
[5] A. Myoga, S. Katayama, *Kobunshi, High Polym.Jap.* **1988**, *37*, 757.
[6] S. Katayama, A. Myoga, Y. Akahori, *J.Phys.Chem.* **1992**, *96*, 4698.
[7] S. Beltran, J. P. Baker, H. H. Hooper, H. W. Blanch, J. M. Prausnitz, *Macromolecules*, **1991**, *24*, 549.
[8] J. P. Baker, D. R. Stephens, H. W. Blanch, J. M. Prausnitz, *Macromolecules* **1992**, *25*, 1955.
[9] J. P. Baker, H. W. Blanch, and J. M. Prausnitz, *Polymer* **1995**, *36*, 1061.
[10] S. Wen, W. T. K. Stevenson, *Colloid.Polym.Sci.* **1993**, *271*, 38.
[11] M. Annaka, T. Tanaka, *Nature* **1992**, *355*, 430.
[12] M. Shibayama, T. Tanaka, *Adv.Polym.Sci.* **1992**, *109*, 1.
[13] Y. Osada, *Adv.Polym.Sci.*, **1987**, *82*, 2.
[14] Y. Ogawa, K. Ogawa, B. Wang, E.Kokufuta, *Langmuir* **2001**, *17*, 2670.
[15] H.P. Gregor, D.H. Gold, G.K.Hoeschele, *J Amer Chem Soc.* **1955**, *77*, 4743.
[16] T.A. Asonova, E.F. Razvadovskii, A.B. Zezin, V.A.Kargin, *Dokl Akad Nauk SSSR* **1969**, *188*, 583.
[17] F. Candau, J.F. Joanny, *Polymeric Materials Encyclopedia*. Ed. J.F. Salamone, Boca Raton: CRC Press, **1996**, Chapter 7.
[18] S.E.Kudaibergenov, *Adv.Polym.Sci.* **1999**,*144*, 115.
[19] S.E.Kudaibergenov, *Polyampholytes: Synthesis, Characterization, and Application*, New York: Kluwer Academic/Plenum Publications, **2002**, 220 p.
[20] J.C. Salamone, C.C. Tsai, A.C. Watterson, A.P. Olson, *Polymeric Amine and Ammonium Salts*, Oxford, Pergamon Press, **1980**, 105.
[21] J.C. Salamone, A.C. Watterson, T.D. Hsu, C.C. Tsai, M.U. Mahmud, *J Polym Sci Polym Lett Ed.***1977**, *15*, 487.
[22] J.C. Salamone, A.C. Watterson, T.D. Hsu, C.C. Tsai, M.U. Mahmud, A.W. Wisniewski, S.C. Israel, *J Polym Sci Polym Symp.* **1978**, *64*, 229.
[23] D.A. Topchiev, R.A. Mkrtchan, R.A. Simonyan, M.V. Lachinov, V.A. Kabanov, *Vysokomolek Soedin. Ser A*, **1977**, *19*, 506.
[24] H. Ladenheim, H. Morawetz, *J Polym Sci.* **1957**, *26*, 251.
[25] W. Jaeger, U. Wendler, A. Lieske, J.Bohrisch, *Langmuir*, **1999**,*15*, 4026.
[26] A.B. Lowe, N.C. Billingham, S.P.Armes, *J Chem Soc Chem Commun.* **1996**, *13*,1555.
[27] S.A. Ali, A. Rasheed, M.I.Wazeer, *Polymer*, **1999**, *40*, 2439.
[28] M. Casoloro, F. Bignotti, L. Sartore, M. Penco, *Polymer*, **2001**, *42*, 912.
[29] R.C. Schulz, M. Schmidt, E. Schwarzenbach, J.Zoller, *Macromol Chem Macromol Symp.* **1989**, *26*, 221.
[30] R.B. Koizhaiganova, S.E.Kudaibergenov, K.E. Geckeler, *Makromol.Chem.Rapid Commun.* (to be published).
[31] A.G. Didukh, R.B. Koizhaiganova, G. Khamitzhanova, LA.Bimendina, S.E. Kudaibergenov, *Polymer Intern.* (to be published).
[32] I.E. Suleimenov, V.B. Sigitov, S.E. Kudaibergenov, E.A.Bekturov, *Polymer Intern.* **2001**, *50*, 194.
[33] S.E. Kudaibergenov, V.B. Sigitov, A.G. Didukh, S.B. Moldakarimov, *Polym.Prepr.* **2000**, *41*, 724.
[34] Kudaibergenov S.E., Didukh A.G., Moldakarimov S.B. in: "*ACS Symposium Series 833 Polymer Gels. Fundamentals and Applications*", Edited by H.B.Bohidar, P.Dubin and Y.Osada, ACS Washington DC **2002**, Ch.10, pp.149-162.
[35] A.Seher, *Arch. Pharm.* **1951**, *284*, 371.
[36] Y.F.Freimans "*Khimiya Enaminoketonov, Enaminoiminov i Enaminotionov*", English translation is "*Chemistry of Enaminoketones, Enaminoimines and Enaminotiones*", Zinatne, Riga, **1974**, 274p.
[37] D.J.Liaw, K.R.Lee, *Polymer Intern.* **1993**, *30*, 381.
[38] Y. Merle, *J.Phys.Chem.* **1987**, *91*, 3092.
[39] M.M. Ali, H.P. Perzanowski S.A. Ali, *Polymer,* **2000**, *41*, 5591.
[40] R.S. Armentrout, C.L. McCormic, *Macromolecules*, **2000**, *33*, 419.
[41] V.A. Kabanov, V.P. Zubov, Y.D. Semchikov, *Complex-Radical Polymerization* (in Russian), Khimiya, Moscow, **1987**.
[42] E. E. Kathmann, C. L. McCormick, *J.Polym. Sci. Chem.* **1997**, *35*, 231.
[43] E. E. Kathmann, C. L. McCormick, *J. Polym. Sci. Chem.* **1997**, *35*, 243.
[44] A. Niu, D. J. Liaw, H. C. Sang, C. Wu, *Macromolecules*, **2000**, *33*, 3492.
[45] W. F. Lee, C. C. Tsai, *Polymer*, **1995**, *25*, 357.
[46] P.G.Higgs, J.F.Joanny, *J.Chem.Phys.* **1991**, *94*, 1543.
[47] A.E.English, S.Mafe, J.Manzanares, X.Yu, A.Yu.Grosberg, T.Tanaka, *J.Chem.Phys.* **1996**, *104*, 8713.
[48] S.E. Kudaibergenov, V.B. Sigitov, *Langmuir,* **1999**, *15*, 4230.
[49] S.E. Kudaibergenov, V.B. Sigitov, A.G. Didukh, E.A. Bekturov, I.E. Suleimenov, *Polym. Adv. Techn.* **2000**, *11*, 805.

Characterization of Stimuli-Sensitive Polymers for Biomedical Applications

*Mi Kyong Yoo, Won Kyung Seok, Yong Kiel Sung**

Department of Chemistry, Dongguk University, 3Ga 26 Phil-dong, Chung-gu, Seoul 100-715, Korea

Summary: Stimuli-sensitive polymers were synthesized by copolymerizing varying ratios of N-isopropyl acrylamide(NIPAAm) and acrylic acid(AAc). The influence of polyelectrolytes on the lower critical solution temperatures(LCSTs) of these temperature/pH sensitive polymers was investigated in the pH range of 2-12. Polyelectrolyte complexes were prepared by mixing poly(NIPAAm-*co*-AAc) as anionic polyelectrolyte with poly(allyl amine)(PAA) or poly(L-lysine)(PLL) as cationic polyelectrolytes, respectively. Back titration was performed to determine the pK_a values of PAAc in poly(NIPAAm-*co*-AAc) and to study the effect of comonomer ionization on the cloud point temperature. The effect of polyelectrolyte complex formation on the conformation of PLL was studied as a function of temperature by means of circular dichroism(CD). The swelling ratio of poly(NIPAAm-*co*-AAc) hydrogels as a function of pH at various temperature was obtained by measuring the weight of the hydrogels in buffer solutions. The LCSTs of the poly(NIPAAm-*co*-AAc) were strongly affected by pH, polyelectrolyte solutes, AAc content, and charge density. The influence of more hydrophobic PLL as a polyelectrolyte on the cloud point of PNIPAAm/water in the copolymer was stronger than that of poly(allyl amine)(PAA). Indomethacin was loaded into these hydrogels, and controlled release of this molecule from the hydrogel was determined under various temperature and pH conditions using UV/Vis spectrophotometry.

Introduction

Polymers which undergo reversible phase transitions in response to external stimuli such as temperature,[1,2] pH,[3] electric current,[4] ions,[5] and chemical species[6] have been defined as "intelligent materials"[7] having sensor, processor, and effector functions. There are many potential applications of such intelligent polymer system in controlled drug delivery systems(DDS),[8,9] solute separation,[10] and immobilization of enzymes and cells.[11]

 DOI: 10.1002/masy.200450316

Most intelligent polymers known previously are responsive to only one kind of stimulus,[1-6,12,13] and independent response to several factors such as temperature and pH may be required for some applications.[14] These temperature- and pH-sensitive polymers undergo marked solubility changes in response to temperature and pH changes.[15,16]

A typical temperature-sensitive hydrogel such as crosslinked PNIPAAm has a lower critical solution temperature(LCST) near 32 °C.[17] Below that temperature, the gel is in a swollen, hydrated and hydrophilic state. Above the LCST the gel becomes collapsed, dehydrated and hydrophobic. Furthermore, its phase transition behavior can be controlled by incorporating more hydrophilic or hydrophobic monomers in the gel composition.[18] When a small amount of ionic monomer having pH-sensitivity is incorporated into a PNIPAAm gel, the gel is expected to exhibit a higher LCST and a broader phase transition, in addition to the generation of the pH sensitivity. Poly(NIPAAm-*co*-AAc) hydrogels, in which PAAc is introduced as a pH-sensitive component into the temperature-sensitive component PNIPAAm, has both the pH- and temperature-sensitivity.[14.19] However, for a higher pH than pK_a of PAAc and higher AAc content, the swelling transition disappears.[19] In order to utilize the hydrogels for the drug delivery systems, their temperature- and pH-sensitivity in response to small change in physiological conditions should be maintained.

To retain the temperature-induced transition property over a broad and useful pH range, we have used the method of polyelectrolyte complex formation because PNIPAAm incorporated with pH-sensitive component is an ionic polymer. If the pH-sensitive component bearing weakly acidic pendent group is incorporated into the PNIPAAm, the polymer behaves as an anionic polyelectrolyte. When a cationic polyelectrolyte is mixed with an anionic polyelectrolyte solution, a polyelectrolyte complex is formed through electrostatic attraction between two oppositely charged polyelectrolytes.[20,21] It might be expected that the LCST of poly(NIPAAm-*co*-AAc) will be controlled by polyelectrolyte complexation according to the pH value.

The objectives of this work are to study the influence of polyelectrolyte complex formation on the LCST of PNIPAAm in the poly(NIPAAm-*co*-AAc), and to investigate the influence of polyelectrolyte solute on the swelling property of crosslinked poly(NIPAAm-*co*-AAc) hydrogel for the application of drug delivery systems.

Experimental

Materials. N-isopropyl acrylamide(NIPAAm, Polysciences Inc.) was recrystallized in n-hexane. Acrylic acid(AAc, Junsei Chem. Co.) was purified by distillation at 40 °C and 26 mm Hg. N,N'-azobisisobutyronitrile(AIBN, Tokyo Kasei Koygo Co.) was purified in methanol. 1,4-Dioxane(Duksan Co.) was purified by distillation. Poly(allyl amine)(PAA, Aldrich), poly(L-lysine)(PLL, Aldrich), ethylene glycol dimethacrylate(EGDMA, Junsei Chem. Co.), sodium hydroxide standard solution(Katayama Chem.), hydrochloric acid standard solution (Katayama Chem.), poly(acrylic acid)(PAAc Aldrich) and indomethacin(Sigma Chem. Co.) were used as received.

Synthesis of poly(NIPAAm-*co*-AAc). Linear poly(NIPAAm-*co*-AAc) and crosslinked poly(NIPAAm-*co*-AAc) hydrogels, containing 0, 10, 20, 30, 40, and 50 mol% of AAc, were prepared by free radical solution polymerization, as described in previous publications.[22-24]

Titrations. Titrations of the poly(NIPAAm-*co*-AAc) were performed as follows. 100 mg of each linear copolymer was dissolved in 25 ml of 0.1M NaCl. In order to increase the solubility of the polymers during the titrations, back titrations were performed on ionized polymers. The polymers were fully ionized by adding 2-6 ml of 0.1N NaOH. Titrations were performed at 20 °C by adding 0.1N HCl in small quantities and measuring the pH after a stable value has been reached. Titrations of the poly(NIPAAm-*co*-AAc), containing 10, 20, and 30 mol% AAc, were performed at temperatures ranging from 4 °C to 45 °C. 100 mg of each linear copolymer was dissolved in 15 ml of 0.2M NaCl. The solutions were titrated with 0.2N HCl after all carboxyl groups were ionized by adding 0.2N NaOH. After each titration 0.2 N NaOH was added to re-ionized all carboxyl groups allowing subsequent titrations at other temperatures. The temperature was kept constant(±0.3 °C) for 30 min before and during each titration using a waterbath.

Preparation of polyelectrolyte complex. 1.7×10^{-1} monomol/ℓ of poly(NIPAAm-*co*-AAc) solution, containing 0, 10, 20, 30, 40, and 50 mol% of AAc, and 4.2×10^{-1}

monomol/ℓ of PAA solution were prepared by dissolving polymers in distilled water, respectively. Before mixing the two solutions, the pH of each solution was adjusted identically with HCl or NaOH standard solution in the range of 2-12. To each copolymer solution, the PAA solution was added with stirring. Each concentration of poly(NIPAAm-*co*-AAc) and PAA was 4.2×10^{-2} monomol/ℓ and 1.1×10^{-1} monomol/ℓ, respectively. The polyelectrolyte complex solution with PLL was prepared by the above method similarly.

Cloud point temperature determination. Turbidity of the solution was determined by measuring the absorbance at 450 nm of each of the solutions prepared above, using a Hewlett-Packard 8452A Diode Array UV/VIS spectrophotometer with HP89090A Peltier Temperature Control Accessory. The temperature of the solution was raised from 15 °C to 70 °C in 2 °C increments every 10 min. Absorbance was normalized against the absorbance of the polymer solution at temperature where precipitation started to take place. The cloud point temperature was defined as the temperature at the inflection point in the absorbance versus temperature curve. The cloud point of solutions not exhibiting an inflection point was determined at 10 % absorbance.[19]

Swelling measurement. For the studies of pH dependent swelling of the hydrogels, the gels in triplicate were incubated in buffer solutions ranging from pH 2 to 11 at an any particular temperature. 0.2M KCl/0.2M HCl buffer was used for pH 2, 0.2M acetic acid/0.2M sodium acetate buffer for pH 4, 5 and 6, and borate buffer for the remaining pH values. The total ionic strength of each buffer was adjusted to 0.2M with a calculated amount of NaCl. The incubation time was approximately 24 hrs. It was confirmed that 24 hrs was enough time to reach the equilibrium swelling. For the studies of temperature dependent swelling of the hydrogels, the gels were swollen in the particular buffer solution at various temperatures in the expected range of the LCST phase transition. Periodically, the gels were withdrawn from the buffer solution and weighed after removal of excessive surface water by lightly blotting with filter paper. After equilibration at one condition, the gels were reequilibrated at the other temperature. The weight swelling ratio was calculated by Ws/Wp, where Ws and Wp were the fully swollen gel and dried gel weights, respectively. The water content(wt%) was also calculated by the following equation, water content(wt%)=[(Ws-Wp)/Ws]×100. To investigate the effect of polyelectrolyte on the LCST, the gels were immersed in 20

°C buffer solutions containing a calculated amount of PAA, and the swelling ratio was determined by weighing gels as described above.

***In vitro* drug release.** Dried cylindrical gel was soaked in the solution of indomethacin(20g/l) in water at 4 °C for 2 days. After washing the gel surface with ethanol, the swollen gel was dried for 2 days at room temperature, and then dried under vacuum for 2 days at 25 °C. Drug loaded gel was placed in 20 ml of phosphate buffer solution(PBS, pH 7.4) release media, and the media were stirred at 37 °C. At 10 min. time intervals, 3 ml aliquots of the aqueous solution were withdrawn from the release media. After the concentration of indomethacin released was monitored using a UV/Vis spectrophotometer at 319 nm, the solution taken as a sample was replaced in the release medium.

Results and Discussion

Effect of AAc content in poly(NIPAAm-*co*-AAc). Poly(NIPAAm-*co*-AAc) was synthesized from comonomer feed compositions ranging from 0 to 50 mol%. Copolymer compositions and pK_a values determined, by back-titration, are given in Table 1.

Table 1. Monomer ratios in P(NIPAAm-*co*-AAc) gels compared to monomer feed ratios

Comonomer AAc feed molar composition(mol%)	Copolymer AAc molar composition(mol%)	pK_a (at 20 °C)
0	0	0
10	14.0	5.14
20	24.2	4.96
30	36.6	4.86
40	45.2	4.79
50	54.0	4.77

Figure 1 exhibits the cloud point temperatures of PNIPAAm against AAc mol % in the poly(NIPAAm-*co*-AAc) copolymer solutions according to the pH values. It was found that the cloud point temperatures of PNIPAAm in the copolymers were affected by pH and AAc content in the copolymer. The cloud point temperatures of PNIPAAm in the solution decreased with an increase of AAc content in the copolymer around pH value of 2.0-3.0 because more hydrogen bonding between the amide group of PNIPAAm and the carboxylic acid group of PAAc occurs with increasing AAc content in the copolymer. The hydrogen bonding interferes with the access of water molecules to the NIPAAm amide groups.[19] On the other hand, the cloud point temperatures of PNIPAAm in the solution increased with AAc content in the copolymer around pH 4.0 because of increased hydrophilicity resulting from an increase of ionized AAc components.

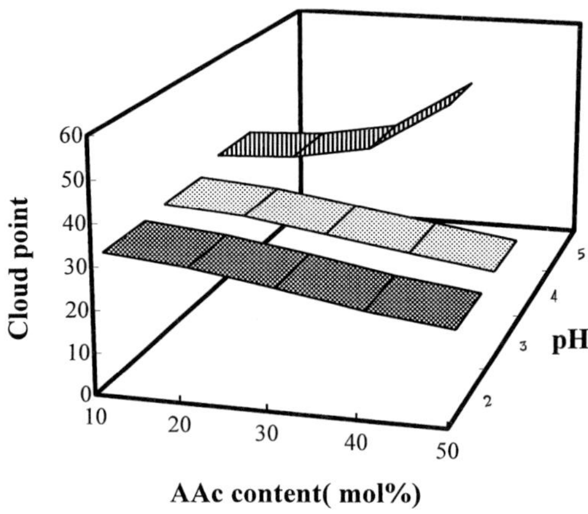

Figure 1. Plot of cloud point of aqueous poly(NIPAAm-*co*-AAc) solution against content of AAc as a function of pH.

In Figure 2, the cloud point temperatures of poly(NIPAAm-*co*-AAc) copolymer solutions PNIPAAm in the presence of PLL at various pH values, are plotted against AAc mol%. The presence or absence of PLL did not have a significant effect on cloud point temperatures for pH values around 2.0, due to hydrogen bonding between PNIPAAm and PAAc. However, at pH 4.0, the cloud point temperature of \poly(NIPAAm-*co*-AAc) with PLL decreased with increasing AAc content in the copolymer, whereas cloud point temperatures of the copolymer solutions without PLL increased with AAc content in the copolymer.

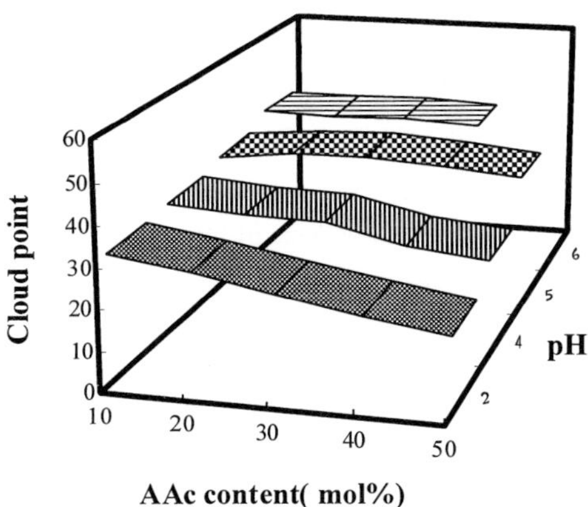

Figure 2. The plot of cloud point of aqueous poly(NIPAAm-*co*-AAc) solutions with PLL against content of AAc, as a function of pH.

When the copolymer of NIPAAm containing partially ionized AAc forms an intermacromolecular complex with oppositely charged PLL, the cloud point temperature of PNIPAAm is affected by the density of complexation.[25] The density of complexing points on poly(NIPAAm-*co*-AAc) increases with AAc content in the copolymer because the complexing points are a carboxylic acid group of AAc in the copolymer and amine group of PLL. An increase of the density of complexing points causes more polyelectrolyte complexes to form between partially ionized PAAc and PLL. This may lead to a decreased cloud point temperature due to the reduced accessibility and mobility of water surrounding the copolymer and dehydration of the copolymer chain by polyelectrolyte complex formation. Therefore, the cloud point temperature of PNIPAAm in the poly(NIPAAm-*co*-AAc) with PLL decreases with increasing AAc content in the copolymer.

On the other hand, the cloud points of the (NIPAAm-*co*-AAc) solutions with PLL did not show an obvious tendency against AAc content around pH 5.0-6.0. With increasing pH or AAc content, the number of $-COO^-$ groups in the copolymer chain increases. The presence of $-COO^-$ groups in the polymer chain increases its rigidity because of electrostatic interchain repulsions and impedes efficient polyelectrolyte complex formation between $-COO^-$ groups of the copolymer and $-NH_3^+$ groups of PLL.[26] However, the formation of polyelectrolyte complex is not hindered completely because the long side chain of PLL facilitates the approach of $-NH_3^+$ to the $-COO^-$ group. Hence, it is considered that the lack of a clear observed trend in cloud point due to AAc content in the copolymer is due to a balance of hydrophilicity of $-COO^-$ groups not participating in polymer complex formation, and the participation of hydrophobic, alkyl side chains of PLL in polymer complex formation.

Effect of comonomer ionization. In order to examine the specific role of charge on phase separation, the amount of charge on poly(NIPAAm-*co*-AAc) copolymers was determined at their respective cloud point temperatures. By determining the pK_a's of AAc in the copolymers as a function of temperature, the amount of charges on the copolymers as a function of temperature can be calculated from the Henderson-Hasselbalch equation,[27] as shown in Figure 3. A decrease in charges on the polymer chain with increasing temperature was observed due to the decreased acidity of AAc. The amount of charge on poly(NIPAAm-*co*-AAc) chains at their respective cloud point temperature as a function of pH is given in Table 2. It was found that percentage of charged AAc on the copolymers was less than 0.06 % at pH 2. It was also found that the amount of charge increased with increasing pH and AAc content. Increasing the amount of charge leads to a large increase in cloud point temperature, with no phase separation taking place, if more than 2-3 mol% of AAc units are charged.

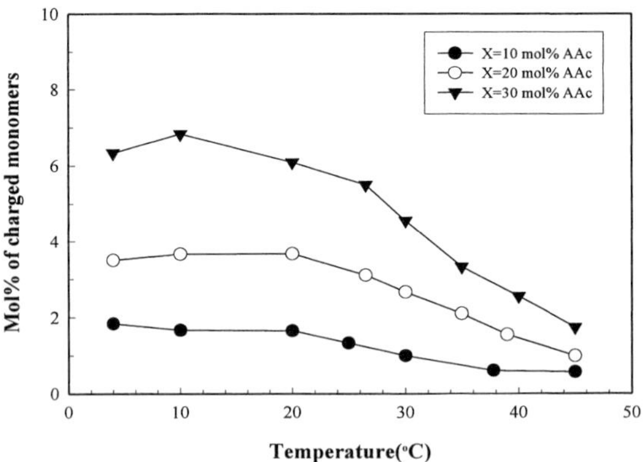

Figure 3. Total amount of charged monomers in poly(NIPAAm-*co*-AAc-X), with X=10, 20, 30 mol%, at pH 4 as a function of temperature.

Table 2. Percentage of charged comonomers in poly(NIPAAm-*co*-AAc) at LCST as a function of pH.

pH	AAc content(mol%)		
	10	20	30
2	0.010%	0.032%	0.065%
3	0.092%	0.298%	0.510%
4	0.768%	2.02%	2.72%
5	3.49%	–	–

Effect of cationic polyelectrolytes and hydrophobicity. Figure 4 shows the effect of polyelectrolytes and the hydrophobic side chain of PLL on the cloud point of PNIPAAm in the poly(NIPAAm-*co*-AAc) as compared with PAA. The cloud point of the copolymer solution in the presence of cationic polyelectrolytes decreased due to polyelectrolyte complex formation between the carboxylic acid group of PAAc and the amine group of PAA or PLL, indicating that dehydration occurred.

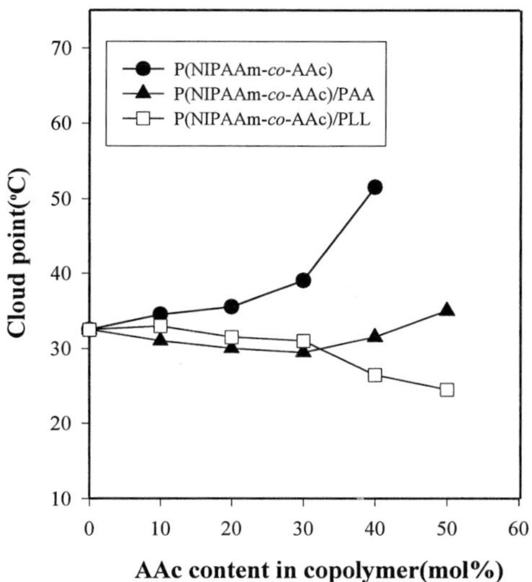

Figure 4. The comparison of cloud point of poly(NIPAAm-*co*-AAc)/PLL with that of poly(NIPAAm-*co*-AAc)/PAA at pH 4.

The influence of PLL on the cloud point of poly(NIPAAm-*co*-AAc) was stronger than that of PAA above 30 mol % of AAc content. It is considered that as the side chain length increases from C_1 for PAA to C_4 for PLL, the hydrophobicity of polyelectrolyte complex increases, indicating that stronger dehydration occurs. That is, the -$(CH_2)_4$ group for PLL is more hydrophobic than -CH_2 for PAA, contributing to an increased hydrophobicity of copolymer chain and subsequently to the stabilization of the formed interpolymer aggregate by synergy with the intrinsic hydrophobicity of NIPAAm.

Effect of polymer complex on conformation of PLL. PLL takes an α-helical conformation at pH above 9.5. Below pH 9.5, PLL exists in an irregular random coil form,[28] in which the flexible backbone undergoes continuous change as a result of thermal motion. This is because at pH below pK_b, all the amine groups of PLL have a positive charge and repel each other so strongly that they overcome the tendency for intrachain hydrogen bonds to form. Ellipticity [θ] at 222 nm is frequently used to estimate the α-helical content, because $[\theta]_{222nm}$ in deg.cm^2·dmol^{-1} is nearly zero for the random coil and $-4\cdot10^4$ for the α-helical conformation, respectively. In a previous paper,[29] we reported that ellipticities at 222 nm in CD spectra of the PLL/PAAc mixture were large negative values at all temperatures, indicating a conformational change of PLL from random-coil to α-helix. It is considered that the positive charges on PLL are blocked by negative charges on PAAc through the polymer-polymer complex, and then the intrachain hydrogen bonds are formed. In the case of PLL/poly(NIPAAm-*co*-AAc-50), the random-coil conformation of PLL did not change, essentially, at all temperatures. It is thought that bulky N-isopropyl groups of PNIPAAm bound in PLL through ionic polymer-polymer interaction inhibit intrachain hydrogen bonding due to steric hinderance.

pH dependent swelling behaviors. Figure 5 shows pH the dependence of swelling ratio for the poly(NIPAAm-*co*-AAc) gels with 10 and 20 mol% of PAAc at 39 °C. These two hydrogels demonstrated a sharp swelling transition according to pH. The equilibrium swelling ratio of poly(NIPAAm-*co*-AAc) gel with 10 mol% AAc greatly increased between pH 5.0 and 6.0, whereas that of poly(NIPAAm-*co*-AAc) gel with 20 mol% of AAc greatly increased between pH 4.0 and 5.0. These swelling transitions are closely related to pK_a values determined by titration curves.

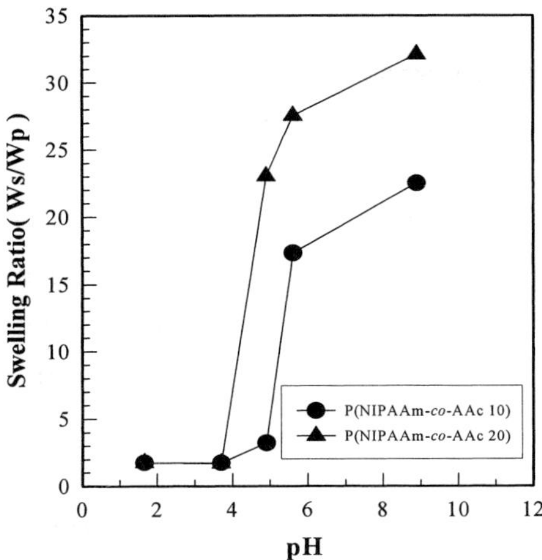

Figure 5. The pH dependent swelling ratio of poly(NIPAAm-*co*-AAc) gels in buffer solution at 39°C.

Figure 6 shows the pH dependence of the swelling ratio of poly(NIPAAm-*co*-AAc) gels with 10 and 20 mol% AAc in the presence of PAA at 39 °C. Comparing Figures 5 and 6, it is seen that the pH sensitivity of these two poly(NIPAAm-*co*-AAc) hydrogels with PAA decreased in comparison with that of the poly(NIPAAm-*co*-AAc) gels in the absence of PAA. Specifically, the swelling pH range widened to 5.6 to 8.9. Also, in the case of poly(NIPAAm-*co*-AAc) gel, there was a large difference between the swelling ratio below pK_a of PAAc (about 4.8)[30] and above swelling ratio above that pK_a, due to the difference in the number of carboxylate side chains resulting from the pH change. In the case of poly(NIPAAm-*co*-AAc) gel with PAA, a polyelectrolyte complex is formed in the pH range from 5.6 to 8.9. This leads to a reduced swelling ratio, since expansion of ionized hydrogel is hindered by the mutual blocking of oppositely charged ionic groups and the accessibility and mobility of water surrounding the hydrogel is decreased. Therefore, the value of Ws/Wp became smaller in spite of the wider pH range.

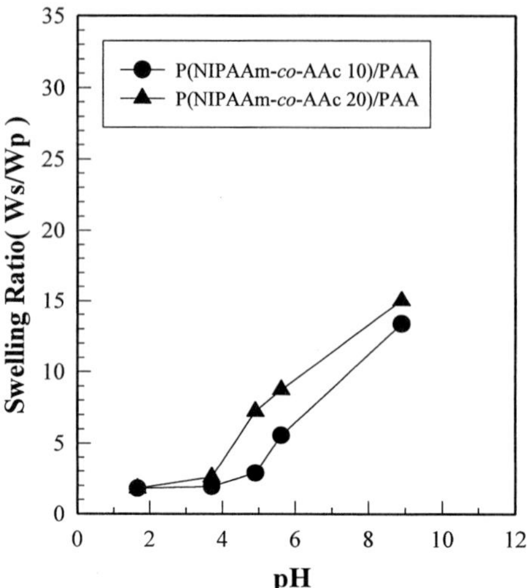

Figure 6. pH dependent swelling ratio of poly(NIPAAm-*co*-AAc) gels with PAA in buffer solution at 39 °C.

Influence of polyelectrolyte on drug release. Figure 7 shows the release profile of indomethacin from poly(NIPAAm-*co*-AAc) hydogel in the absence and presence of PAA, at 37 °C and pH 7.4. In contrast to the rapid release of indomethacin from poly(NIPAAm-*co*-AAc) hydrogel itself, the release of a drug from the hydrogel with PAA was much slower. The swelling ratios increased with pH due to electrostatic repulsion between the carboxylic acid polymer side chain and ions present in the buffer solution. At pH 7.4, the carboxylate side chains are repelled by the anions in the solution, and minimize charge concentration by expanding[31]. Also, the swelling ratios of the hydrogels in the presence of PAA were lower than those of the gels without PAA, since the expansion of ionized hydrogel was hindered by formation of the polyelectrolyte complex, resulting in mutual blocking of oppositely charged ionic groups.

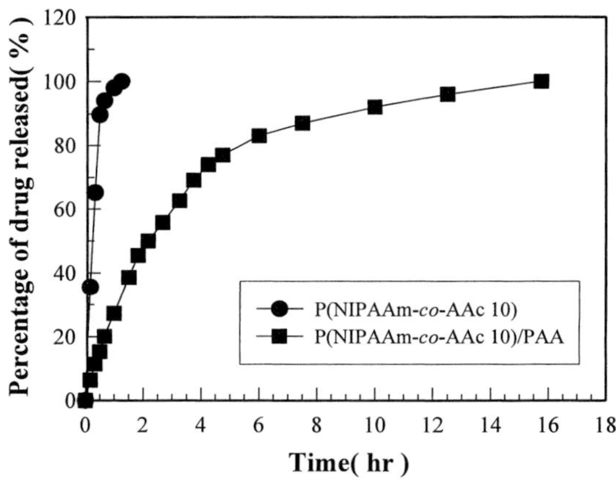

Figure 7. Indomethacin release from poly(NIPAAm-*co*-AAc- 10) and poly(NIPAAm-*co*-AAc 10)/PAA in PBS at 37 °C.

Conclusion

The cloud points of PNIPAAm in aqueous poly(NIPAAm-*co*-AAc) solutions were strongly influenced by pH, AAc content, and the type of added polyelectrolytes such as PAA and PLL. A polyelectrolyte complex in the poly(NIPAAm-*co*-AAc) system with PLL was formed in neutral conditions. The influence of PLL on the LCST of PNIPAAm in the poly(NIPAAm-*co*-AAc) was more stronger than that of PAA above 20 mol% of AAc content due to hydrophobicity of alkyl sidechain of PLL. Although a polymer-polymer complex was formed between poly(NIPAAm-*co*-AAc) and PLL, the conformational change of PLL did not show up clearly due to the steric hinderance of bulky N-isopropyl groups of PNIPAAm.

The LCSTs of the poly(NIPAAm-*co*-AAc) hydrogels increased with pH and disappeared above the pK_a value of AAc, except for the hydrogel containing 10 mol% AAc. The LCSTs of the poly(NIPAAm-*co*-AAc) hydrogels in the presence of PAA were observed even above the pK_a value of AAc. At the same pH, the LCSTs of the poly(NIPAAm-*co*-AAc) hydrogels were lower than those of the same hydrogels in the absence of PAA due to the polyelectrolyte complex formation. Below the pK_a value of AAc, the LCSTs of hydrogels decreased with an increase of AAc content in the hydrogel. On the other hand, the LCSTs of hydrogels increased with AAc content above pK_a value of PAAc. Furthermore, the swelling ratio of poly(NIPAAm-*co*-AAc) gel was decreased by polyelectrolyte complex formation with PAA. The swelling kinetics of poly(NIPAAm-*co*-AAc) hydrogels are strongly dependent on pH value, temperature, and polyelectrolyte solute. In contrast to the rapid release of indomethacin from poly(NIPAAm-*co*-AAc) hydrogel itself, the release of a drug from the hydrogel with PAA was slowed due to polyelectrolyte complex formation between PAA and the copolymer hydrogels.

Acknowledgement: This work has been supported by the BK 21 Dongguk University, Ministry of Education, Seoul, Korea

[1] L. C. Dong, A. S. Hoffman, *J. Controlled Release* **1986**, *4*, 223.
[2] T. Okano, Y. H. Bae, W. S. Kim, *J. Controlled Release* **1990**, *11*, 225.
[3] R. A. Siegel, B. A. Frestone, *Macromolecules* **1988**, *21*, 479.
[4] I. C. Kwon, Y. H. Bae, T. Okano, S. W. Kim, *J. Controlled Release* **1991**, *17*, 149.
[5] J. Ricka, T. Tanaka, *Macromolecules* **1984**, *17*, 2916.
[6] H. Feil, Y. H. Bae, J. Feijen, S. W. Kim, *Makromol. Chem., Rapid Commun.* **1988**, *14*, 243,.
[7] D. Takagi, K. Takahashi, M. Aizawa, S. Miyata, *Proceedings of the 1st International Conference on Intelligent Materials*, Eds.; Technomic: Lancaster, PA, 1992.
[8] L. C., Dong, Q. Y. Yan, A. S. Hoffman, *J. Controlled Release* **1991**, *19*, 171.
[9] A, Gutowaska, Y. H. Bae, S. W. Kim, *J. Controlled Release* **1992**, *22*, 95.
[10] A. S. Hoffman, A. Afrassiabi, L. C. Dong, *J. Controlled Release* **1986**, *4*, 213.
[11] R. F. S. Frietas, E. L. Cussler, *Chem. Eng. Sci.* **1987**, *42*, 97.
[12] Y. H. Bae, T. Okano, S. W. Kim, *J. Polym. Sci, Polym. Phys*. **1990**, *28*, 923.
[13] J. Ricka, T. Tanaka, *Macromolecules* **1984**, *17*, 2916.
[14] L. C. Dong, A. S. Hoffman, *J. Controlled Release* **1991**, *15*, 141.
[15] T. K. Park, A. S. Hoffman, *J. Appl. Polym. Sci.* **1992**, *46*, 659.
[16] C. S. Brazel, N. A. Peppas, *J. Controlled Release* **1996**, *39*, 57.
[17] Y. H. Bae, T. Okano, Y. Sakurai, S. W. Kim, *Pharam. Res.* **1991**, *8*, 624.
[18] T. G. Park, A. S. Hoffman, *J. Biomed. Mater. Res.* **1990**, *24*, 21.
[19] G. Chen, A. S. Hoffman, *Nature* **1995**, *373·5*, 49.
[20] E. A. Bekturov, L. A. Bimendina, *Adv. Polym. Sci.* **1981**, *41*, 99.
[21] E. Tsuchida, K. Abe, *Adv. Polym. Sci.* **1982**, *45*, 1.
[22] M. K. Yoo, C. S. Cho, Y. M. Lee, Y. K. Sung, *Polymer* **1997**, *38*, 2759.
[23] M. K. Yoo, C. S. Cho, Y. M. Lee, Y. K. Sung, *Polymer* **1998**, *39*, 3703.
[24] M. K. Yoo, Y. K. Sung, Y. M. Lee, C. S. Cho, *Polymer* **2000**, *41*, 5713.
[25] H. Feil, Y. H. Bae, J. Feijen, S. W. Kim, *Makromol. Chem., Rapid Commun.* **1993**, *14*, 465.
[26] G. Bokias, G. Stakikos, I. Iliopoulos, R. Audebert, *Macromolecule* **1994**, *27*, 427.
[27] H. Feil, Y. H. Bae, J. Feijen, S. W. Kim, *Macromolecules* **1992**, *25*, 5528.
[28] G. Holzworth, P. Doty, *J. Am. Chem. Soc.* **1965**, 87, 218.
[29] Y. K. Sung, M. K. Yoo, C. S. Cho, *Kor. Polym. J.* **2000**, *8*, 26.
[30] A. B. Zezin, V. A. Kargin, *Vysokomol. Soyed.* **1972**, *A14: No 4*, 772.
[31] C. S. Brazel, N. A. Peppas, *J. Controlled Release* **1996**, *39*, 57.

Copolymers of *N*-Alkylacrylamides as Thermosensitive Hydrogels

X. X. Zhu, D. Avoce, H. Y. Liu, A. Benrebouh*

Département de chimie, Université de Montréal, C. P. 6128, succursale Centre-ville, Montréal, Québec, H3C 3J7, Canada

Summary: Many of the N-alkyl substituted acrylamide polymers can manifest a lower critical solution temperature (LCST) in aqueous solutions. The LCST of such polymers can be easily varied by a free radical copolymerization of the appropriate comonomers. Natural compounds such as bile acids can be introduced into such polymers to modify the LCST and aggregation behavior and to render the material both thermo- and pH-sensitivity.

Introduction

Some polymers can manifest a drastic change in water solubility at a certain temperature, which is known as the lower critical solution temperature (LCST). The behavior is associated with the temperature dependence of hydrogen bonding and hydrophobic interactions. This property of the polymers can be very useful because of the rheological and technological implications and potential biomedical and industrial applications. The thermosensitivity of poly(*N*-isopropylacrylamide) (PNIPAM, which has a LCST at ca. 32 °C) has been extensively studied [1,2]. The variation of the LCST of the thermosensitive polymers can be achieved by copolymerization, chemical modification of the polymer, or the addition of salts or surfactants in the aqueous solutions [2,3]. LCST was also observed for aqueous solutions of other *N*-substituted polyacrylamides [4-7], but these polymers have not drawn as much attention as PNIPAM.

It is interesting and important to tune predictably the LCST of the polymers for the different applications. The thermosensitivity of the polymers can be controlled by the relative hydrophilicity of the copolymers, hence the chemical structure and composition of the constituent monomers. It is also useful to introduce natural compounds for the biological applications of the polymers, which, at the same time, can modify the thermosensitivity and aggregation behavior of such materials. Therefore, different copolymers of N-alkylacrylamides and bile acid derivatives have been prepared to improve and modify the biocompatibility, solution property and aggregation behavior of such materials.

 DOI: 10.1002/masy.200450317

Copolymers of *N*-substituted acrylamide monomers

In the case of *N*-substituted acrylamides, the hydrophilicity of the *N*-alkyl substitution group should be important in the determination of the LCST and the properties of the polymers. We have selected a number of *N*-substituted acrylamide monomers, including acrylamide (AA), *N,N*-dimethylacrylamide (DMA), *N*-ethylacrylamide (EA), *N,N*-diethylacrylamide (DEA), *N*-tert-butylacrylamide (TBA), in the preparation of several series of copolymers to vary the LCST of the polymers in a systematic fashion. We have found that the LCSTs of the copolymers are quite sharp and can be varied between the LCST values of the corresponding homopolymers. The results are shown in Figure 1 for the copolymers made. The LCST was found to vary as a function of the chemical composition of the copolymers according to the following equation:

$$T = \frac{\mu_1 T_1 + K\mu_2 T_2}{\mu_1 + K\mu_2} \qquad (1)$$

where μ is the molar fraction of a given monomer (note that $\mu_1 + \mu_2 = 1$), *T* is the LCST of the corresponding homopolymers 1 and 2, and *K* is a weighting parameter inferior to 1 ranging from 0.51 for poly(TBA-co-EA) to 0.13 for poly(TBA-co-DMA) (Figure 1).

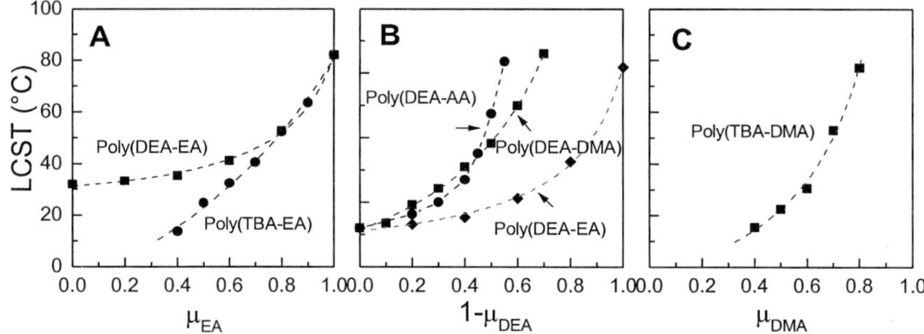

Figure 1. Tuning LCST by copolymerization of N-alkyl substituted acrylamide

Copolymers of *N*-alkylacrylamides with bile acid derivatives

Bile acids are natural amphiphilic compounds with interesting properties. Methacrylic and acrylic derivatives of bile acids have been prepared by attaching a polymerizable double bond to one of the functional groups of the bile acid molecule. We have made copolymers of *N*-substituted acrylamides with the (meth)acrylamide derivatives of cholic acid (Figure 2, top, where the content of cholic acid residues ranged from 0 to 5 mol%) by free radical copolymerization in solution [8] and studied the LCST and aggregation behavior of the copolymers [9]. In general, the LCST of the acrylamide copolymers can be lowered by the presence of the bile acid residues, as observed clearly by turbidimetry shown in Figure 3A. The lowering of the LCST by the the presence of bile acid in the copolymers shows a linear relationship with the molar content of the methacrylamide derivatives of cholic acid (Figure 3B).

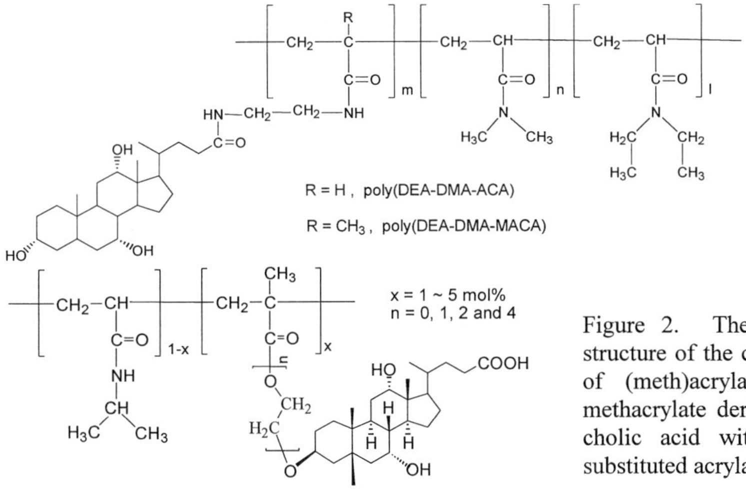

R = H , poly(DEA-DMA-ACA)

R = CH_3 , poly(DEA-DMA-MACA)

x = 1 ~ 5 mol%
n = 0, 1, 2 and 4

Figure 2. The chemical structure of the copolymers of (meth)acrylamide and methacrylate derivatives of cholic acid with N-alkyl substituted acrylamides.

Alternatively, a double bond can also be attached to one of the OH groups of cholic acid through a short poly(ethylene glycol) (PEG) spacer [10] (Figure 2, bottom structure). The PEG spacer makes the copolymers more hydrophilic and the free carboxylic acid group of the bile acid renders the copolymers with N-alkylacrylamide both thermo- and pH-sensitive [11]. As shown in Figure 4, the LCST of the copolymer is much higher in its ionized form (at higher pH values). The decrease in the LCST at lower pH values is certainly related to the protonation of the carboxylic acid group of the cholic acid moiety in the copolymers, which makes the polymer less hydrophilic.

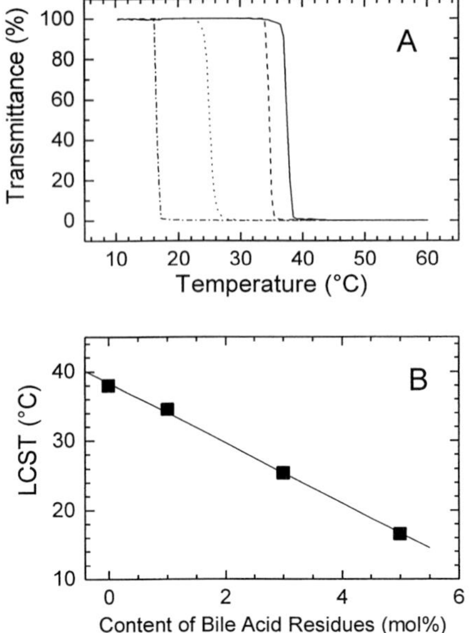

Figure 3. The variation of LCST of poly(DEA-co-DMA-co-MACA) with the molar percentage of MACA. (A) Light transmittance of the 1 wt% polymer solutions at a wavelength of 500 nm with a heating rate of 0.5°C/min. (B) Linear dependence of the LCST of the copolymers on the content of MACA.

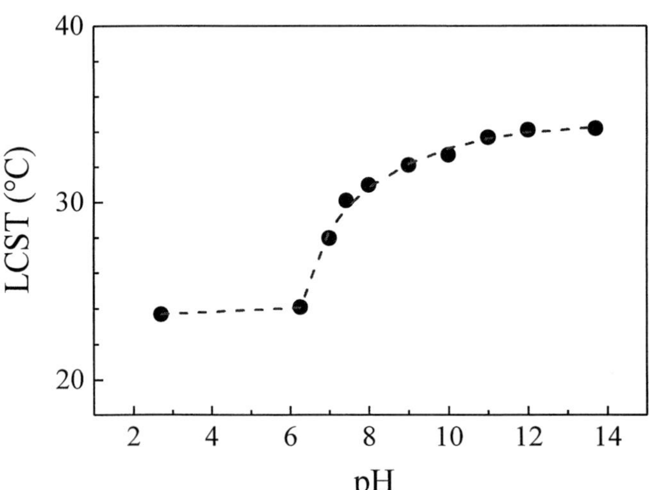

Figure 4. The variation of the LCST as a function of temperature for a copolymer of NIPAM with 3 mol% of methacrylate derivative of cholic acid with a tetra(ethylene glycol) spacer.

Conclusion

The LCST of the N-alkylacrylamide polymers in water can be modified easily and predictably by copolymerization of the appropriate monomers. Natural compounds such as bile acids can be introduced into such copolymers to modify the aggregation behavior of the linear copolymers and to make new hydrogels. The change of the LCST of such copolymers has been shown to be directly related to the content of the bile acid residues in the materials. The free carboxylic acid group of bile acid makes such polymers sensitive to both temperature and pH changes in the aqueous environment. These properties should be interesting for the potential biomedical applications of such materials.

The financial support from NSERC of Canada and Fonds FCAR of Quebec is grafully acknowledged.

[1] A.S. Hoffman, *J. Controlled Release* **1987**, *6*, 297.
[2] H.G. Schild, *Prog. Polym. Sci.* **1992**, *17*, 163.
[3] G. Chen, A.S. Hoffman, *Macromol. Rapid Commun.* **1995**, *15*, 175.
[4] L. D. Taylor, L. D. Cerankowski, *J. Polym. Sci., Polym. Chem. Ed.* **1975**, *13*, 2551.
[5] H.Y. Liu, X.X. Zhu, *Polymer* **1999**, *40*, 6985.
[6] I. Idziak, D. Avoce, D. Lessard, D. Gravel, X.X. Zhu, *Macromolecules* **1999**, *32*, 1260.
[7] I. Idziak, D. Gravel, X.X. Zhu, *Tetrahedron Lett.* **1999**, *40*, 9167.
[8] H.Y. Liu, D. Avoce, Z. Song, X.X. Zhu, *Macromol. Rapid Commun.*, **22**, 675 (2001).
[9] D. Avoce, X.X. Zhu, in preparation, 2002.
[10] A. Benrebouh, Y. Zhang, X.X. Zhu, *Macromol. Rapid Commun*, **21**, 685 (2000).
[11] A. Benrebouh, D. Avoce, X.X. Zhu, *Polymer*, **42**, 4031 (2001).

The Use of Hydrophobic Spacers in the Development of New Temperature- and pH-Sensitive Polymers

*A. Licea-Claverie,**[1] *E. Rogel-Hernández,*[1,2] *R. Salgado-Rodríguez,*[1] *J. A. López-Sánchez,*[1] *L. A. Castillo,*[1,2] *J. M. Cornejo-Bravo,*[2] *K.-F. Arndt*[3]

[1] Instituto Tecnológico de Tijuana, Centro de Graduados e Investigación, A.P. 1166, 22000 Tijuana, B.C., México
[2] Universidad Autónoma de Baja California, Facultad de Ciencias Químicas e Ingeniería, Calzada Universidad S/N, Mesa de Otay, 22300 Tijuana, B.C., México
[3] Technische Universität Dresden, Institut für Physikalische Chemie und Elektrochemie, D-01062 Dresden, Germany

Summary: The preparation and characterization of two series of methacrylic acid derivative polymers is described. One series contains aliphatic spacers with one to ten methylene units, while the other series includes an aromatic ring with changes in the position of the acid, as spacer. Both series of polymers were obtained as methoxy-ester protected acid polymers and as polymers containing free acid groups in different amounts. pH-sensitive gels and temperature-sensitive *N*-isopropylacrylamide (NIPAAm) copolymers were prepared by using some of the monomeric structures described. The pH of the swelling transition of the gels changed from 3.5 up to 9.0 as a function of the spacer length and type. The lower critical solution temperature (LCST) of NIPAAm copolymers in water was lowered from 33.6 °C to 6 °C as a function of the co-monomer content and type. The observed changes in the pH of the swelling transition of gels and in the LCST of NIPAAm copolymers can only be explained if hydrophobic-hydrophobic and hydrogen-bonding interactions are considered in connection with the specific chemical structure of the monomers used.

Introduction

Sensitive polymers and gels change their volume in dependence on the properties of the environment in them they are, mostly through a liquid. This volume change is reversible and can be continuous or discontinuous, depending on the chemical structure of the sensitive polymer or gel. Toyoichi Tanaka brought these materials to the attention of the public worldwide through his pioneering work on temperature sensitive *N*-Isopropylacrylamide (NIPAAm) polymers and copolymers, and on solvent sensitive acrylamide polymers.[1,2] The thermodynamics of the behavior of sensitive gels is qualitatively described by the Gibbs free energy of the system resulting from three main contributions: an elastic contribution, a mixing contribution, and an ionic contribution in the case of gels with ionizable groups.[3]

 DOI: 10.1002/masy.200450318

As pointed out by Tanaka, there are other fundamental interactions that influence greatly the phase transition behavior of these materials, such as hydrophobic and hydrogen-bonding interactions.[4] These specific interactions are not included explicitly in most theories of phase-transitions or the swelling behavior of gels. Exceptions are the work by Saito *et al.*[5] which includes a term dealing with the hydrophobic contribution of isopropyl groups of NIPAAm to the temperature driven phase transition of that polymer gel, and the work by Prausnitz *et al.*[6] dealing with hydrogen-bonding as the main factor to explain the behavior of temperature driven phase-transitions.

For pH-sensitive polymers a similar case is found: no explicit terms are provided for hydrophobic interactions or for hydrogen-bonding interactions affecting the pH-driven phase transition of gels. However, hydrophobic interactions could be implicitly considered through changes in the Flory-Huggins χ-parameter. For example, in the model published by Peppas *et al.*[7] for pH-sensitive gels, it is predicted that through changes in χ from 0.1 to 0.9 the equilibrium swelling degree of the gel is changed in the non-ionized state, while the pH for the swelling transition remains the same. The same model predicts that changes in the pK_a of the acid groups of the gel at constant χ results in changes in the pH of the swelling transition. According to experimental results the former is not correct. For example, Siegel *et al.*[8,9] reported a shift to lower values of the pH of the swelling transition for polybasic gels of *N,N'* dimethylaminoethylmethacrylate (DMAEM) when hydrophobic co-monomers of n-alkylmethacrylates are introduced. Khokhlov *et al*[10] later reported a shift to higher values in the pH of swelling transition in the case of hydrophobically modified polyacrylic acid gels. Although a change of the pK_a as a result of the more hydrophobic environment was considered, from both works it can be concluded that the main factor is additional hydrophobic attractive forces opposing the electrostatic repulsion forces.

The role of hydrogen-bonding interactions is most likely to be observed in amphoteric hydrogels, since they usually combine acid groups with basic groups in their structure. In fact, Tanaka *et al.*[11] reported in their study of amphoteric hydrogels that multiple equilibrium states for the same system in the same pH-value were observed, and they suggested that additional hydrogen-bonding forces can play an important role in that behavior. Since the chemical structure of NIPAAm includes a secondary amine group capable of hydrogen bonding interactions, there is also a possibility that such interactions play an important role in the behavior of NIPAAm-copolymeric gels in addition to the well documented mixing, elastic and ionic contributions. Interestingly, T. Tanaka *et al.*[12] reported hydrogen-bonding as the driving force for phase transition in the case of polyacrylamide-poly(acrylic acid) interpenetrating networks (IPN's) but not in NIPAAm copolymers whose chemical structure is very similar.

The goal of our work is to contribute to the understanding of, and to quantify the role of hydrophobic spacer groups of controlled hydrophobicity to the behavior of both pH-sensitive and temperature sensitive polymers. Furthermore, through protecting-deprotecting synthetic strategies, we study the role of hydrogen-bonding on the behavior of temperature sensitive NIPAAm copolymers.

Experimental

Reagents and Solvents

All chemicals and solvents used in the synthetic procedures were obtained either by Aldrich Chemicals, or by Productos Químicos Monterrey and were used as received. The only exceptions were the following: 2-hydroxybenzoic acid, 3-hydroxybenzoic acid, 4-hydroxybenzoic acid, 4-hydroxyphenylacetic acid and methyl 4-hydroxybenzoate were obtained by Acros Organics and were used as received. 2,2'-Azobisisobutyronitrile (AIBN, Aldrich) was purified by re-crystallization from methanol. *N*-isopropylacrylamide (NIPAAm, Aldrich) was re-crystallized from hexane. Ethyleneglycoldimethacrylate (EGDMA, Aldrich), was purified by passing it through a column of inhibitor-remover for hydroquinones and paramethoxyphenol (Aldrich). Pyridine (Aldrich, anhydrous) and 4-dioxane (Spectrum, HPLC-grade) were further dried by using molecular sieves followed by distillation. THF (Spectrum, HPLC-grade), was dried by using sodium/benzophenone followed by distillation.

Synthesis of monomers

Several methoxy-protected methacrylic acid derivatives with hydrophobic spacer groups (Figure 1) were synthesized through esterification reactions by one of the following two different synthetic routes: reaction of methacrylic anhydride with corresponding hydroxy-esters or reaction of potassium methacrylate with corresponding bromo-esters. Details on the synthetic routes used can be found elsewhere,[13] the schematic description of each route is given in Figure 2.

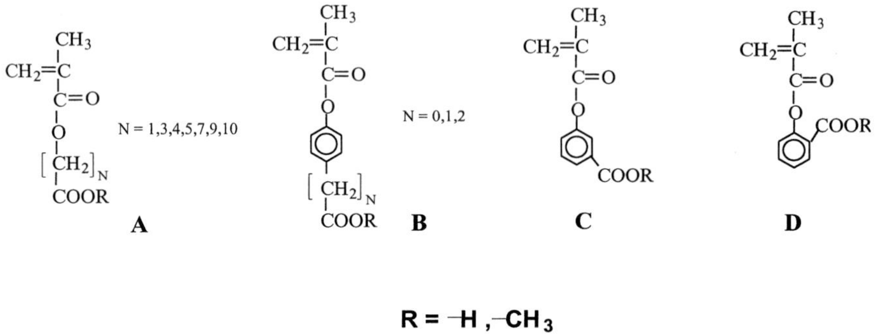

N = 1,3,4,5,7,9,10 (A); N = 0,1,2 (B)

R = –H , –CH_3

Figure 1. Structures of methacrylic acid derivatives with hydrophobic spacers: A-aliphatic, B-aliphatic-aromatic, C-aromatic and D-aromatic.

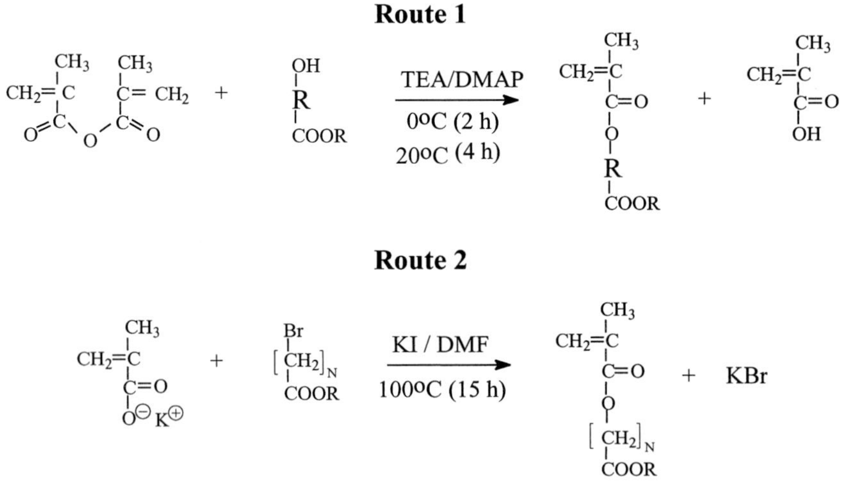

Route 1

TEA/DMAP, 0°C (2 h), 20°C (4 h)

Route 2

KI / DMF, 100°C (15 h)

Figure 2. Synthetic routes used for the preparation of methacrylic acid derivatives with hydrophobic spacer groups.

By a slight modification of the synthesis route #1, methacrylic acid derivatives with aromatic spacers and free acid groups were also prepared (monomers B-0, B-1, C and D). As an example we describe the synthesis of the monomer 2-methacryloyloxy benzoic acid (D):

At 0 °C, 7.5 g (50 mmol) of methacrylic anhydride was added to a solution of 6.91 g (50 mmol) 2-hydroxybenzoic acid and 0.49 g (4 mmol) dimethylaminopyridine in 8.42 mL triethylamine and 20 mL dichloromethane. The solution was stirred for 5h at 0 °C followed by stirring overnight at room temperature. The solution was partitioned between dichloromethane and a HCl solution (3N, 4 x 25 mL), and the organic phase was dried over $MgSO_4$, filtered and concentrated in vacuo. The residue after evaporation was chromatographed on silica gel, 200-400 mesh (eluent: petroleum ether/ether 70:30). The product was a white solid.

Methacrylic acid derivatives with aliphatic spacers and free acid groups (monomers A-4, A-7 and A-10) were prepared by deprotection reaction of the prepared methoxy-protected monomers with aliphatic spacers. As an example we describe the synthesis of 11-methacryloyloxyundecanoic acid (A-10):

To a refluxing solution of 6.16 g (50 mmol) of dry LiI in 50 mL of anhydrous pyridine, 3.31 g (11.65 mmol) of methoxy-11-methacryloyloxyundecanoate (protected A-10) and 0.96 g (11.7 mmol) of sodium acetate were added under nitrogen. Refluxing was continued for 24 h. The solution was then cooled and the solvent removed under vacuum. The residue was dissolved in 20 mL water. Then this solution was acidified to pH=2 by using 6N HCl. The acid solution was extracted with diethylether (3 x 30 mL), and the organic extracts were dried with magnesium sulfate, with the solvent evaporating under reduced pressure. The product was purified by chromatography on silica gel, 70-230 mesh (eluent petroleum ether:diethylether, 70:30).

Synthesis of polymers with aliphatic spacers

Two series of polymers with aliphatic spacers were synthesized, one with methoxy-protected acid groups through solution free radical polymerization of the synthesized A-monomers and the second with free acid groups through selective saponification reactions of the former polymers. Details on the preparation and saponification are reported in our previous paper.[13]

Synthesis of polymers with aromatic and aliphatic-aromatic spacers

Two series of polymers with aromatic spacers were synthesized, one with methoxy-protected acid groups and the second with free acid groups. As an example of the polymerization of monomers with protected acid groups (protected monomers B-0, B-1, B-2, C and D) we describe the synthesis of poly(methyl-methacryloyloxysalicylate) (PolyD):

> 2.86 g (13 mmol) of methylmethacryloyloxysalicylate (protected monomer D) and 21.3 mg (1 mol%) of AIBN were placed in an Schlenk-flask. Dry THF (15 mL) was then added and stirred under cooling until complete dissolution. The Schlenk-flask was degassed at –40 °C and was purged with argon at least five times by conventional freeze-thaw techniques prior to sealing. The polymerization was run at 60 °C for 72 h to allow maximum yield. After polymerization, the polymer was precipitated into a large excess of petroleum ether. The polymer was purified by re-dissolving it in THF and by precipitating it again in petroleum ether. The polymer was then dried in vacuum for 24 h. The polymerization of the monomers with free acid groups (non protected B-0, B-1, C and D) was performed following the same procedure as described above with the following changes: the solvent used was dry 4-dioxane, the polymerization temperature was 70 °C and the polymer products obtained were purified by re-dissolving in 4-dioxane and precipitation in petroleum ether. These polymers were dried in vacuum for only 8 h, but at 50 °C.

Preparation of pH-sensitive Gels

Two series of gels with free acid groups were obtained, one with aliphatic spacers through selective saponification reactions on methoxy-protected acid gels (two steps), and the second series with aromatic spacer groups, directly from the monomers with free acid groups (one step). As an example of the two-step strategy for the preparation of gels with free acid groups of monomers A-1, A-3, A-5, A-9 and A-10, we describe in detail the preparation of the gel from methacryloyloxyacetic acid (A-1):

In a round bottom flask, 2 g (12.66 mmol) of methyl-methacryloyloxyethanoate (protected A-1), 48.72 μL (0.26 mmol) of EGDMA and 21 mg (0.13 mmol, 1 mol%) of AIBN, were dissolved in 3 mL of dry THF. The flask was degassed at –40 °C and was purged with argon at least five times by conventional freeze-thaw techniques prior to sealing with a rubber septum. The solution was transferred using a syringe to a system consisting of two silanized glass plates (10 x 10 cm), separated by a silicone gasket (2 mm in diameter), held together by metal clamps. The assembly was placed in the vertical position in an oven for 24 h at 60 °C, allowing formation of a gel. After cooling at room temperature, the glass plates were separated and the residual solvent was allowed to evaporate. From the gel-sheet obtained, discs were cut (3.75 mm diameter). The gel-discs were washed with THF for 72 h, changing solvent every 4 h. Then the gel-discs were taken out of the solvent, allowed to dry in air for 48 h and finally in vacuum at room temperature for 24 h. For the deprotection reaction 0.6 g of dry gel discs were allowed to swell in dry pyridine, then the gel-discs and 0.3 g (3.8 mmol) of sodium acetate were transferred to a refluxing solution of 2.5 g (18.5 mmol) of LiI in 30 mL of dry pyridine. The mixture was refluxed under argon for 18 h. After cooling, the solution was decanted and the gel discs were washed with THF/water mixtures of the following composition in sequential manner: 100/0, 75/25, 50/50, 25/75, 0/100. Each washing step took 24 h. Afterwards the gel-discs were contracted by THF/water mixtures by using the reverse sequence. The gel discs were finally allowed to dry in air for 48 h and under vacuum for additional 24 h at room temperature.

As an example of the preparation of gels with aromatic spacers and free acid groups in one step (from monomers B-0, C and D) we describe in detail the preparation of gel of 2-methacryloyloxybenzoic acid (non protected monomer D):

In a round bottom flask, 2 g of 2-methacryloyloxybenzoic acid (9.7 mmol), 37.5 μL (0.2 mmol)) of EGDMA and 21 mg (0.13 mmol) of AIBN, were dissolved in a mixture of 4 mL of dry THF and 3 mL of dry dichloromethane. The flask was degassed at –40 °C and was purged with argon at least five times by conventional freeze-thaw techniques prior to sealing with a rubber septum. The solution was transferred using a syringe to the system consisting of two silanized glass plates, separated by a silicone gasket described above. The assembly was placed in vertical position in an oven for 24 h at 60 °C, allowing a gel to form. After cooling at room temperature, the glass plates were separated and the residual solvent was allowed to evaporate. From the gel-sheet obtained, discs were cut (4.45 mm diameter). The gel-discs were washed with THF for 72 h changing solvent every 4 h. Then the gel-discs were taken out of the solvent, allowed to dry in air for 48 h and finally in vacuum at room temperature for 24 h. For the preparation of the gel of 3-methacryloyloxybenzoic acid (unprotected monomer C) and the gel of 4-methacryloyloxybenzoic acid (unprotected monomer B-0), the same procedure was used, except that the solvents used were THF (3 mL) and 4-dioxane (6 mL) respectively. In both cases the discs cut from the gel-sheets obtained were larger (8.53 mm diameter).

Synthesis of NIPAAm Copolymers

Two series of NIPAAm copolymers with aliphatic spacer groups were obtained, one with protected acid groups and the other with free acid groups. For these the monomers A-4, A-7 and A-10 were used in several ratios to NIPAAm (Figure 3). As an example of the procedure we describe in detail the preparation of the copolymer of NIPAAm with 5 mol% of 5-methacryloyloxy pentanoic acid (non protected monomer A-4):

3.33 g (29.45 mmol) of NIPAAm, 0.29 g (1.55 mmol) of 5-methacryloyloxy pentanoic acid (non protected monomer A-4) and 15.3 mg (0.3 mol%) of AIBN were placed in an Schlenk-flask. Dry 4-Dioxane (10.5 mL) was then added and stirred under cooling until complete dissolution. The Schlenk-flask was degassed at –40 °C and was purged with nitrogen at least three times by conventional freeze-thaw techniques prior to sealing. The copolymerization was run at 70 °C for 6 h. After copolymerization, the product was precipitated into a large excess of diethylether. The copolymer was purified by re-dissolving it in THF and precipitating it again in diethylether. The copolymer was finally dried in vacuum at 40 °C for 24 h. Both series of copolymers were prepared by using the same procedure, although in some cases (the copolymers with higher content on hydrophobic co-monomers with free acid groups), hot THF was needed in the purification step.

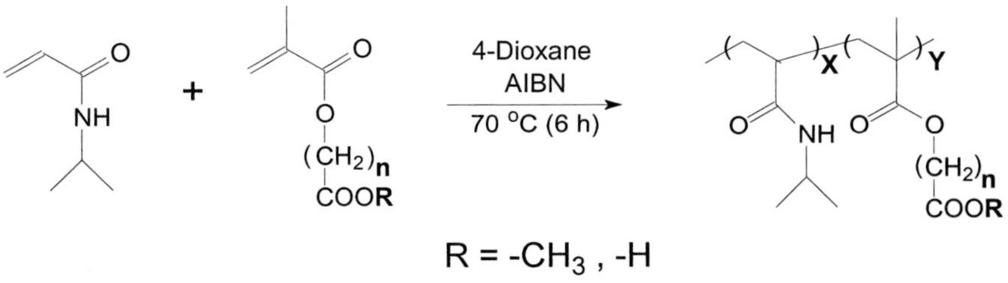

4-Dioxane
AIBN
70 °C (6 h)

$R = -CH_3$, -H

Feed-Composition of Methoxy-protected Copolymers					Feed-Composition of Copolymers with free acid-groups				
X [mol%]	Y [mol%]	n=4	n=7	n=10	X [mol%]	Y [mol%]	n=4	n=7	n=10
95	5	✔	✔	✔	95	5	✔	✔	✔
90	10	✔	✔	✔	90	10	✔	✔	✔
85	15	✔	✔	✔	85	15	✔	✔	✔
80	20	✔	-	-	80	20	✔	-	-

Figure 3. Copolymerization of NIPAAm with methacrylic acid derivatives having aliphatic spacer groups.

Characterization of polymers and copolymers

The methacrylic acid derivative monomers with hydrophobic spacers were characterized by ^{1}H-NMR (200 MHz) spectra recorded in deuterated chloroform ($CDCl_3$) solution on a Varian Gemini 200 spectrometer at room temperature. ^{1}H-NMR-spectra of the protected polymers were obtained on a Varian Gemini 200 spectrometer (200 MHz) (acetone d_6, room temperature) or on a Bruker RDV500 spectrometer (500 MHz) ($CDCl_3$, room temperature). The deprotected polymers (with free acid groups) were characterized by the same NMR-techniques, using either deuterated methanol (methyl-d_3-alcohol-d) or deuterated dimethylsulfoxide (di-methyld_3 sulfoxide) as solvent. The ^{1}H-NMR-spectra of NIPAAm-copolymers with protected acid groups were obtained by the same NMR-techniques using $CDCl_3$ at room temperature and those of NIPAAm-copolymers with free acid groups were obtained using deuterated methanol (methyl-d_3-alcohol-d) at room temperature.

The glass transition temperatures (T_g's) of the polymers and NIPAAm-copolymers were determined using a Mettler DSC-30 calorimeter, at a heating rate of 10 K/min, the only exception being the polymers with aromatic spacers and free acid groups whose T_g's were determined by using a MDSC from TA-Instruments (2920) at a heating rate of 5 K/min using a modulation of 1 K every 60 s. The weight-average molecular weights (Mw) and molecular dimensions ($\langle r^2 \rangle$) were measured by static light scattering (SLS) using an SLS 2-goniometer (SLS-Systemtechnik, Germany). The measurements were performed with a He-Ne laser (λ_o = 632.6 nm) in the angular range 30° - 145° (in 5° steps). The refractive index increment (dn/dc) in the given solvent needed for the evaluation of the light scattering results were measured using a DR-1 differential refractometer (SLS-Systemtechnik, Germany). Measurements were made at 25 °C in acetone (protected polymers with aliphatic spacers), THF (protected polymers with aromatic spacers and NIPAAm-copolymers) and methanol (deprotected polymers with aliphatic spacers and some NIPAAm-copolymers). The deprotected polymers with aromatic spacers were evaluated by solution viscosity in THF at 25 °C by using Ubbelohde-Viscometers. The molecular weights reported were calculated by using the Kuhn-Mark-Houwink constant values reported for polymethylmethacrylate.[14]

Solution behavior of polymers and copolymers

Potentiometric Titrations

Potentiometric titrations were performed on aqueous polymer solutions to study the polyelectrolytic behavior of the methacrylic acid derivative polymers with free acid groups. In order to diminish ionic strength variations during the titration experiments, polymer solutions (0.01 N) were prepared using an aqueous 0.1 N NaCl-solution as solvent. The pH of the polymer solution was adjusted to pH = 11 by adding appropriate amounts of a 0.1 N NaOH-solution. The titration experiments were performed starting with 25 mL of a given polymer solution and adding 0.1 N HCl in 0.1 mL steps under stirring. The measurements were performed at 25°C under argon atmosphere and pH was monitored using a Corning pH meter 430. The pH-value was recorded only after it remains constant (about 10 min after each acid addition). Every polymer solution becomes turbid at different pH-values; nevertheless, the titration was continued until the pH of the solutions reached a fairly acid value (pH≈2.2).

Determination of the lower critical solution temperature (LCST)

Aqueous solutions (pure water and buffer of pH=8) of PolyNIPAAm and NIPAAm-copolymers were prepared at a concentration of 5 mg/mL. For some of the copolymers prepared, cooling at 5 °C was needed to allow dissolution. The solutions were prepared and were stored in a refrigerator at 4 °C until measurement. The LCST was evaluated by modulated differential scanning calorimetry (MDSC) using the endothermic peak observed by heating according to [15]. The equipment used was a MDSC from TA-Instruments (2920) at a heating rate of 2 K/min using a modulation of 1 K every 60 s.

pH-dependent swelling experiments

Eleven phosphate-buffered solutions with adjusted ionic strength (0.15 M) having nominal pH values from 2 to 12 were prepared. The exact pH value was measured using a Corning pH meter 430. Preweighed dry gel-discs were placed in 11 vials, to each of which 15 mL of a particular buffer solution was added, and the gel-discs were allowed to swell for 24 h. Afterwards the buffer-solutions were replaced and the gel-discs allowed to stand in the fresh solutions for an additional 24 h. Finally the swelled gel-discs were taken out from the buffer-solutions, the excess liquid tapped out by using filter paper, and were weighed. The gel swelling, as water sorption capacity was evaluated according to following equation:

$$\%H = (W_s - W_d)/W_d \qquad (1)$$

where W_s is the weight of the swollen gel and W_d is the weight of the dry gel. All swelling experiments were performed in triplicate and the average values (e.g. %H) were taken.

Results and Discussion

pH-sensitive polymers and gels

The preparation of methoxy protected polymers with hydrophobic spacers, aliphatic and aromatic, was straightforward although polymerization times of 72 h were used as needed to obtain reasonable polymerization yields for the monomers with long aliphatic spacer chains (A-5, A-7, A-9 and A-10). Table 1 shows the main characterization parameters for these polymers.

Table 1. Main characterization parameters of methoxy-protected polymers

NAME	Yield [%]	M_w [g mol^{-1}]	$<r^2>$ [nm]	T_g [°C]
Polymers with aliphatic spacers				
PolyA-1	~100	119 000	18	50
PolyA-3	91	68 500	17	-06
PolyA-4	76	74 700	6.5	-24.7
PolyA-5	65	96 000	19	-41
PolyA-7	36	78 100	15	-
PolyA-9	50	99 000	20	-49
PolyA-10	65	84 200	20	-46
Polymers with aromatic and aliphatic spacers				
PolyB-0	88	54 200	7.5	96
PolyB-1	~100	189 000	21	81
PolyB-2	76	90 100	20	68
PolyC	69	67 600	18	72
PolyD	40	70 500	16	64

Most of the polymerization yields and the molecular weights obtained were acceptable. The glass transition temperatures show that with higher number of methylene units in the spacer chain the T_g value is lower. This is not surprising since the methylene chains are flexible and bring increased flexibility to the whole polymer structure. This effect was also observed for the polymers with aromatic spacers when methylene units were included (PolyB-1 and PolyB-2 as compared with PolyB-0). As a consequence these methoxy protected polymers at room temperature range from viscous liquids (PolyA-5 to PolyA-10) to hard powders (PolyA-1 and all polymers including aromatic spacers).

Table 2 summarizes the results obtained in the preparation of polymers with free acid groups. The results depend on the different synthetic strategies used. Polymers with aliphatic spacers and free acid groups were obtained through selective saponification on the preformed methoxy protected polymers, while the polymers including aromatic spacer units were obtained through polymerization of monomers having already free acid groups. By the first described strategy free acid groups in moderate to high yields were obtained (50 to 80%) but complete conversion was not achieved. The so obtained polymers are hydrophobic polyelectrolytes soluble in weak basic aqueous solutions.[13] For the polymers obtained from monomers having already free acid groups, the polymerization route was not very effective as can be seen by the generally lower molecular weights obtained as compared with the polymers with aliphatic spacers (Table 2, lower rows).

Table 2. Main characterization parameters of polymers with free acid groups

NAME	Content on free acid groups [% of units]	M_w [g mol^{-1}]	T_g [°C]
Polymers with aliphatic spacers			
PolyDA-1	81	72 600	116
PolyDA-3	80	43 000	78
PolyDA-5	51	79 300	-18
PolyDA-9	51	90 000	-26
PolyDA-10	54	45 800	-26
Polymers with aromatic and aliphatic spacers			
PolyDB-0	100	19 000*	107
PolyDB-1	100	34 000*	-
PolyDC	100	20 000*	116
PolyDD	100	23 000*	130

* By Solution Viscosity measurements

Nevertheless, the advantage of these polymers is that they have only units with free acid groups and are polyelectrolytes in nature. We are working currently to develop new synthetic strategies that allow the preparation of A-1 to A-10 monomers with free acid groups and high yields. The T_g values obtained for the polymers with aliphatic spacers and free acid groups show the expected tendency: the longer the methylene spacer groups the lower the T_g is. However, in comparison with the starting methoxy protected polymers, the T_g is always higher indicating polymer-polymer hydrogen-bonding in the solid state through the presence of free acid groups, which are absent in the methoxy-protected polymers. The T_g values for the polymers with aromatic spacers and free acid groups are also higher than the values for methoxy protected similar polymers. Interestingly, the position of the acid group (or methoxy ester group) in the aromatic spacer ring results in considerable changes in the T_g values (compare the lower parts of Tables 1 and 2, last row). We will discuss this issue further when we compare gel behaviors.

The swelling behavior of gels, prepared from methacrylic acid derivatives and aliphatic spacers that include free acid groups by using the deprotection strategy, is shown in Figure 4. Error-bars are drawn and are in most cases smaller than the symbols used in the figure. All those gels show a swelling transition at different pH values of the surrounding solution. This transition depends clearly on the spacer groups of the gel: the longer the aliphatic spacer, the higher the pH needed for the swelling transition. The maximum swelling attained is lower for the gels with longer aliphatic spacer chains, with the exception of GelDA-5. Although all these gels have free carboxylic acid groups, the hydrophobic aliphatic spacers influence their behavior so far that the gels with deprotected A-1, A-3 and A-5 units swell in weak acid solutions while the gels with deprotected A-9 and A-10 units swell in weak basic solutions.

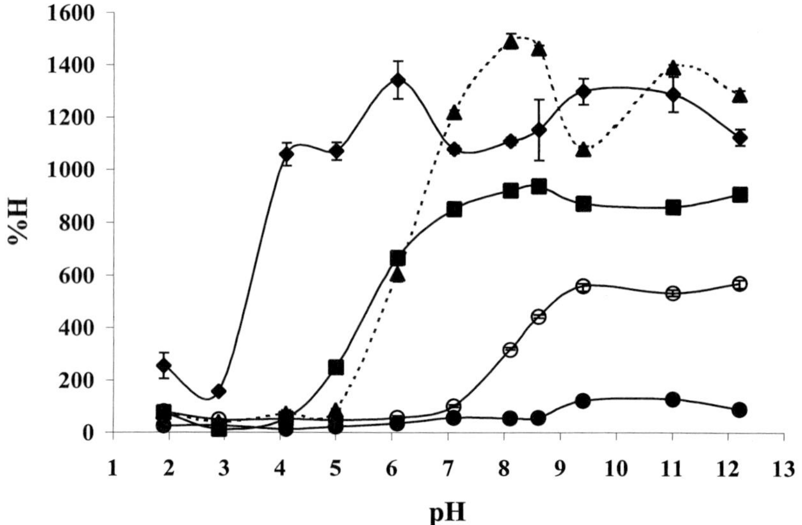

Figure 4. pH-dependent swelling behavior of Gels with aliphatic spacers: ◆ is GelDA-1, ■ is GelDA-3 , ▲ is GelDA-5, ○ is GelDA-9 and ● is GelDA-10.

As pointed out by Khokhlov *et al.*[10] the shift of the pH of the swelling transition to higher values for gels including weak acid groups and some hydrophobic units may have two origins:

1) the decrease of the dissociation constant in the vicinity of hydrophobic units (pK_a shift), or 2) the formation of hydrophobic micro-domains in the collapsed state acting as

additional crosslinks and opposing the gel swelling. Table 3 compares the values of the pH of the swelling transition and the apparent acidity constant at 50% ionization for the parent linear polymers as evaluated by potentiometric titration experiments and calculated using the Henderson-Hasselbach equation [16] at $\alpha = 0.5$.

$$pKa = pH + \log\frac{1-\alpha}{\alpha} \qquad (2)$$

Table 3. Swelling characteristics of partially hydrophobic pH-sensitive gels

NAME	Maximum swelling (%H)	pH of swelling transition	Apparent pK_a (α=0.5) of linear polymers
Gels with aliphatic spacers			
GelDA-1	1325	3.5	3.8*
GelDA-3	939	5.5	5.8*
GelDA-5	1490	6.4	6.3*
GelDA-9	565	8.1	7.1*
GelDA-10	125	9.0	7.3*
Gels with aromatic spacers			
GelDB-0	203.5	6.5	5.3
GelDC	221	7.5	5.5
GelDD	4432	8.3	5.8

* From Reference [13]

For gels with 1, 3 and 5 methylene units as spacers, the pK_a of the linear polymers corresponds very well with the observed pH of the gel-swelling transition. This suggests that indeed the acidity constant of the acid groups is lowered (pK_a is increased) due to the hydrophobicity of the environment provided by the spacers. For gels with 9 and 10 methylene units as spacers the pK_a's of the linear polymers do not correspond exactly to the observed pH's of the swelling transitions, suggesting that the long aliphatic chains not only induce a pK_a shift, but they also create micro-domains with strong hydrophobic-hydrophobic interactions. These interactions would act as additional forces opposing to the gel swelling. The result is that an ionization degree greater than 50% of the carboxylic acid units is required to overcome the additional hydrophobic-hydrophobic interactions.

The behavior of the gels with aromatic spacers, differing only in the position of the free carboxylic acid group (ortho, meta or para) was somewhat surprising. Considering the

pK_a values reported for low molecular weight analogs such as methyl benzoic acid, which show pK_a values of 3.91, 4.27 and 4.38 for ortho, meta and para, respectively,[17] we would expect only slight differences in the pH dependent swelling behavior of this gels. Furthermore, the apparent pK_a values from the potentiometric titration experiments of the linear polymers showed differences in the pK_a of only 0.5 pH-units between them, however in a reverse sequence than the low molecular weight compounds (see Table 3, last row).

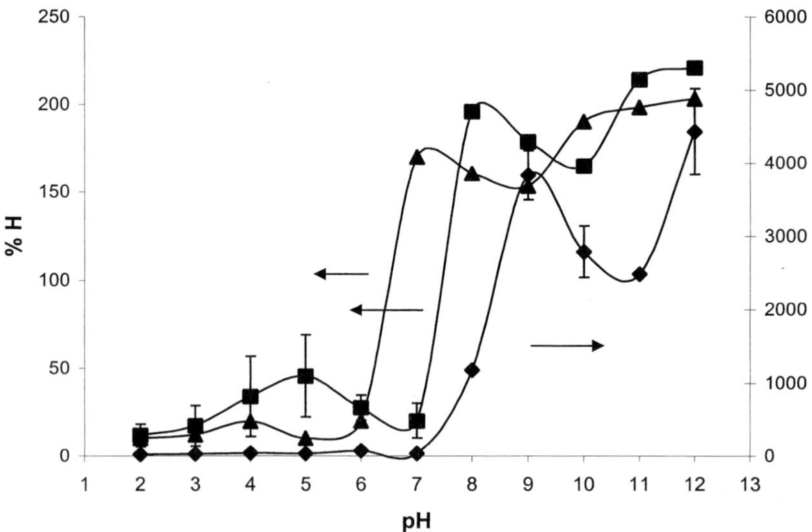

Figure 5. pH-dependent swelling behavior of Gels with aromatic spacers: ▲ is GelDB-0 (para), ■ is GelDC (meta) and ◆ is GelDD (ortho).

Nevertheless, the gel swelling behavior seen in Figure 5 shows that the pH of the swelling transitions varied by almost 2 pH units (error-bars are drawn and are in many cases smaller than the symbols used in the figure). In accordance with our discussion above on gels with aliphatic spacers, and comparing with the values in Table 3, we must conclude that the observed pH values of the swelling transition did not arise from differences in the acidity constant of the carboxylic acid units. Moreover, explaining differences in terms of a more or less "hydrophobic" environment does not seem appropriate. We suggest that the differences observed may arise from the magnitude of hydrogen-bonding interactions resulting from the different positions of the carboxylic acid group in the aromatic ring used as spacer.

Returning to Table 2 and comparing the T_g values for the parent linear polymers, we find the tendency ortho>meta>para. This suggests that polymer-polymer interactions in these otherwise similar polymers, all having free acid groups in 100% of units and molecular weights also similar, are mediated by hydrogen bonds, which strongest when the aromatic sidechains are ortho, followed by meta and para. In the literature we have found one example showing this same trend: the work by J. San Roman *et al.*[18] on methacrylamide polymers with salicylic acids moieties. Those authors reported a T_g value 30 °C higher for the meta (carboxylic acid) substituted polymer compared to the para substituted polymer and discussed their results in terms of stronger hydrogen-bonding interactions in the former polymer. If there are stronger hydrogen-bonding interactions in the ortho structure (deprotected monomer D) than in the meta structure (deprotected monomer C) and these two interactions are stronger than for the para structure (deprotected monomer B-0), then the behavior of the gels prepared from them can be explained as follows: In the collapsed, non-ionized, state the polymer chains have strong hydrogen-bonding interactions in addition to the constraints imposed by the crosslink sites. As soon as the pH of the solution is raised, some ionization of the carboxylic acid groups is achieved and the gels swell accordingly. When 50% of the carboxylic groups are ionized, there is still not enough free energy in the system to overcome the strong hydrogen bonding from the non-ionized groups. The gels do swell but not to a high extent. The pH of the swelling transition therefore does not depend alone on the acidity constant of the polymeric acid groups (apparent pK_a) but also depends on the strong hydrogen-bonding interactions in this kind of polymer chain. Further ionization is needed to overcome these hydrogen-bonding interactions to allow a high swelling degree of these gels.

In summary, in the partial hydrophobic gels prepared we have found three cases:

1. The pH of the swelling transition is shifted to higher values than expected due to a shift in the acidity constant resulting from the more hydrophobic environment imposed by the spacer groups (GelDA-1, GelDA-3 and GelDA-5).

2.- The pH of the swelling transition is shifted to higher values than expected, due partially to a shift in the acidity constant resulting from the more hydrophobic environment imposed by the spacer groups, but also due to hydrophobic-hydrophobic interactions opposing the gel swelling predicted by ionization only (GelDA-9 and GelDA-10).

3. The swelling transition is shifted to higher values than expected due to strong hydrogen-bonding interactions opposing the gel swelling predicted by ionization only (GelDB-0, GelDC and GelDD).

Temperature-sensitive copolymers

As seen in Table 4, the composition of the NIPAAm copolymers prepared is in good agreement with the monomer feed composition. The molecular weights are in all cases fairly high and no special trend was observed during the copolymerization procedure. The glass transition temperature (T_g in Table 4 last row) show the expected trend: since the methacrylic acid derivative co-monomers have flexible methylene chains as spacer groups, the higher the co-monomer content, the lower the T_g . This trend is more pronounced in the copolymers with co-monomer A-10, both protected and deprotected. A content of 11.3% of this co-monomer in NIPPAm copolymer resulted in a drop in T_g of 39 °C. A careful comparison of T_g values for copolymers of similar composition but differing in that one has free acid groups while the other has methoxy protected acid groups, e.g. CoA-10(15) compared with CoDA-10(15), shows that the copolymers with free acid groups always have higher T_g values. This could result from additional hydrogen-bonding interactions in the solid state due to the carboxylic acid groups present in this NIPAAm copolymer series (bottom part of Table 4).

For the copolymer series with methoxy protected acid groups, the solution behavior in pure water corresponds well with the results reported by other researchers[15]: The more hydrophobic the NIPAAm copolymer, e.g. the higher the content of hydrophobic co-monomer, the lower the LCST (Figure 6, solid lines). By deprotecting the acid groups we expected, as generally believed, a less hydrophobic co-monomer structure. Therefore, the hydrophobic effect of the comonomer on the NIPAAm-based coplymer should be weaker. Contrary to the expectation, the LCST of the series of NIPAAm copolymers with comonomers having free acid groups and the same hydrophobic spacers as the protected ones, were found at much lower temperatures (Figure 6, dashed lines). Thinking only in terms of hydrophobicity, this would mean that these copolymers behave more hydrophobically in pure water than do the copolymers with methoxy protected acid groups. As pointed out by H.G. Schild in his review on NIPAAm,[15] there dispute in the literature whether hydrophobic effects or hydrogen-bonding is the main factor responsible for the LCST behavior of polyNIPAAm in particular and of biological polymers in general, in water. In our case, an explanation for this behavior can be found if we consider that carboxylic acid groups could form hydrogen-bonds with the lone nitrogen electronic pairs in the NIPAAm structure (Figure 7).

Table 4. Main characterization parameters of NIPAAm copolymers

NAME	Co-monomer content by NMR [mol%]	M_w [g mol^{-1}]	$<r^2>$ in THF [nm]	T_g [°C]
PNIPAAm	-	237 000	27	138
NIPAAm-Copolymers with methoxy protected acid groups				
CoA-4(05)	6.2	264 000	27	128
CoA-4(10)	9.9	292 000	28	114
CoA-4(15)	16.8	281 000	27	106
CoA-4(20)	20.4	297 000	24	106
CoA-7(05)	7.8	398 000	31	128
CoA-7(10)	11.6	514 000	31	114
CoA-7(15)	15.5	406 000	34	105
CoA-10(05)	4.8	368 000	32	127
CoA-10(10)	8.4	343 000	30	108
CoA-10(15)	11.3	414 000	44	100
NIPAAm-Copolymers with free acid groups				
CoDA-4(05)	7.6	277 000	24	131.6
CoDA-4(10)	12.1	237 000	21	126.0
CoDA-4(15)	18.0	233 000	19	122.2
CoDA-4(20)	21.9	197 000	17	117.2
CoDA-7(05)	7.2	354 000	29	127.9
CoDA-7(10)	12.4	318 000	24	118.6
CoDA-7(15)	17.4	275 000	19*	109.6
CoDA-10(05)	6.6	287 000	24	126.4
CoDA-10(10)	12.8	289 000	22	112.5
CoDA-10(15)	17.3	231 000	23*	99.7

* In Methanol

This would lead to additional polymer-polymer interactions, making phase separation by heating (LCST) a less endothermic process than with pure polyNIPAAm.

If we add a strong base to the solution, the protons of the carboxylic acid groups would be removed and the resulting carboxylates would not be able to bind to the nitrogen electronic pair any more. Moreover, electrostatic charge repulsion would weaken the polymer-polymer interactions, driving the LCST to higher temperatures. This is what we observed indeed by measuring the LCST of the same copolymer in buffer solutions at pH=8 (Figure 8, solid lines). In all cases the LCST is shifted to higher temperatures; moreover, some copolymers that were insoluble in pure water, e.g. copolymers with A-7 and A-10 co-monomers in content higher than 10% or copolymers with A-4 co-monomer in content higher than 15%, are soluble in this buffer-solution. These polymers also show LCST values lower than that of pure polyNIPAAm indicating that the hydrophobicity of the spacer group overcomes the hydrophilicity of the ionized carboxylic acid groups. We believe this is an example of hydrophobicity and hydrogen-bonding working in the same direction: increased polymer-polymer interactions instead of opposing factors for explaining the solubility behavior of partially hydrophobic polymers in water.

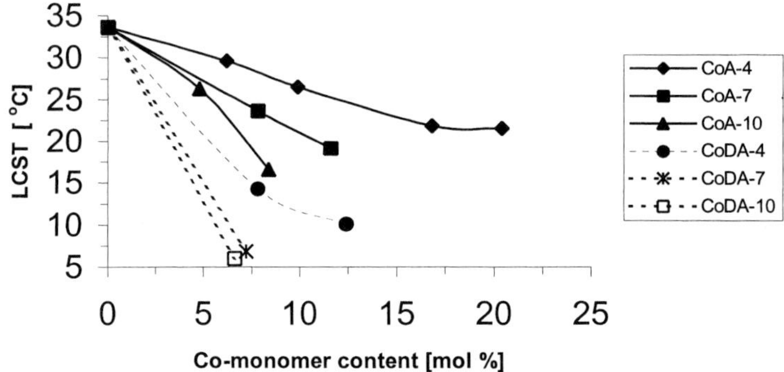

Figure 6. LCST in pure water for NIPAAm copolymers with co-monomers having hydrophobic spacers with protected acid groups (full lines) and free acid groups (dashed lines) as a function of co-monomer content.

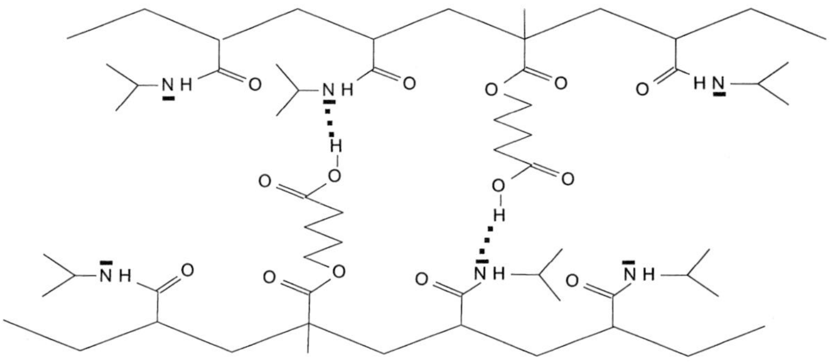

Figure 7. Possible hydrogen-bonding interactions between polymer chains affecting the LCST of NIPAAm-copolymers.

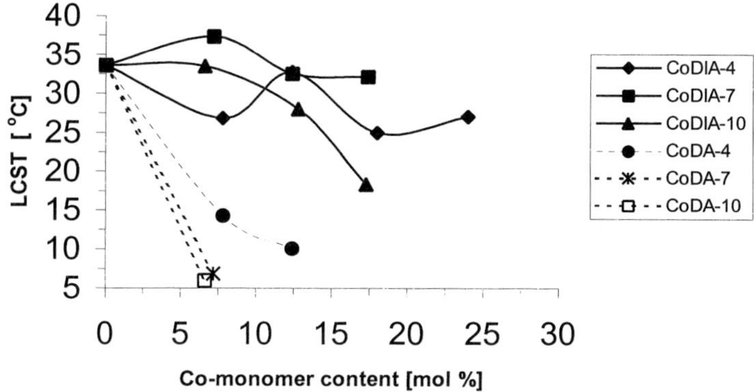

Figure 8. LCST of NIPAAm copolymers with co-monomers having hydrophobic spacers and free acid groups as a function of co-monomer content: In pure water (dashed lines) and in buffer solution of pH=8 (full lines).

Although more experiments are needed to confirm this in similar systems, our results show that both hydrophobic interactions and hydrogen-bonding are very important for the behavior of NIPAAm copolymers depending strongly on the fine chemical structure of the copolymeric units.

Conclusions

The synthesis of methoxy-protected methacrylic acid derivative polymers with aliphatic and with aromatic spacers, was straightforward and yielded polymers of varied physical properties depending on the structure of the spacer used. Either by a deprotection reaction with preformed polymers or by direct polymerization of monomers containing free acid groups, the preparation of polyelectrolytes was achieved. The first route resulted in moderate to high yields (50 – 80%) but with higher molecular weights, and the second one resulted in 100% of free acid units but with lower molecular weights, in comparison.

The pH of the swelling transition of gels prepared varied from 3.5 up to 9.0. This large shift in the pH of the swelling transition results from a combination of three effects:

1) the hydrophobic local environment imposed by the spacer groups altering the acidity constant of the carboxylic acids (pK_a shift).

2) the hydrophobic-hydrophobic interactions in micro-domains acting as a force opposed to the ionization driven swelling.

3) the strong polymer-polymer hydrogen-bonding interactions acting as a force opposed to the ionization driven swelling.

Which effect prevails depends on the fine chemical structure of the monomeric unit.

The LCST of NIPAAm copolymers in water was lowered from 33.6 °C for polyNIPAAm down to 6 °C as a function of the content on hydrophobic co-monomer and also depending on the spacer used. Comonomers with protected acid groups act only in terms of increasing the hydrophobicity of the copolymer chain, which in turns depends on the spacer length used. Comonomers with free acid groups lowered the LCST of NIPAAm copolymers even further through their capability of hydrogen-bonding interactions with the NIPAAm units. This is an example where the water solubility of polymers is decreased by hydrophobic and hydrogen-bonding interactions acting in the same direction.

In general we have demonstrated, with the polymers prepared with hydrophobic spacers, that their solution and swelling behavior cannot be explained only in terms of mixing, elastic and ionic contributions to the free energy of the system. Hydrophobic-hydrophobic and hydrogen-bonding interactions need to be considered in careful connection with their fine chemical structure to understand their behavior.

Acknowledgments

The financial support of the following agencies is gratefully acknowledged: CONACYT-México (Project Nr. 28022-U), COSNET-México (Project Nr. 834.98-P), DFG-Germany (SFB 287), Volkswagen Foundation-Germany (Project Nr. I/76 065). R.S.-R. thanks DAAD-Germany for financing his PHD-Sandwich Program. We thank specially I. Rivero (NMR), D. Scheller (NMR), I. Poitz (DSC) and C. Meissner (SLS) for valuable technical support.

[1] T. Tanaka, S. Ishiwata, C. Ishimoto, *Phys. Rev. Lett.* **1977**, *39*, 474.
[2] T. Tanaka, *Phys. Rev. Lett.* **1978**, *40*, 820.
[3] T. Tanaka, D. Fillmore, S.-T. Sun, I. Nishio, G. Swislow, A. Shah, *Phys. Rev. Lett.* **1980**, *45*, 1636.
[4] M. Shibayama, T. Tanaka, *Adv. Polym. Sci.* **1993**, *109*, 1.
[5] S. Saito, M. Konno, H. Inomata, *Adv. Polym. Sci.* **1993**, *109*, 207.
[6] S. Beltran, H.H. Hooper, H.W. Blanch, J.M. Prausnitz, *J. Chem. Phys.* **1990**, *92*, 2061.
[7] L. Brannon-Peppas, N.A. Peppas, *Chem. Eng. Sci.* **1991**, *46*, 715.
[8] R.A. Siegel, B.A. Firestone, *Macromol.* **1988**, *21*, 3254.
[9] R.A. Siegel, *Adv. Polym. Sci.* **1993**, *109*, 233.
[10] O.E. Philippova, D. Hourdet, R. Audebert, A.R. Khokhlov, *Macromol.* **1997**, *30*, 8278.
[11] M. Annaka, T. Tanaka, *Nature* **1992**, *355*, 430.
[12] F. Ilmain, T. Tanaka, E. Kokufuta, *Nature* **1991**, *349*, 400.
[13] E. Rogel-Hernández, A. Licea-Claveríe, J.M. Cornejo-Bravo, K.-F. Arndt, *Desig. Monom. Polym.* **2001**, *4*, 343.
[14] J. Brandrup, E.H. Immergut, "*Polymer Handbook*", 3rd Ed , J. Wiley & Sons, New York 1989, p. VII/198.
[15] H.G. Schild, *Prog. Polym. Sci.* **1992**, *17*, 163.
[16] G. Conio, E. Patrone, S. Russo, V. Trefileti, *Makromol. Chem.* **1976**, *177*, 49.
[17] R.T. Morrison, R.N. Boyd, "*Organic Chemistry*", Allison and Bacon Inc., New York 1983, p. 796.
[18] C. Elvira, J. San Roman, *Polymer*, **1997**, *38*, 4743.

Temperature-Responsive, Polymer-Modified Surfaces for Green Chromatography

*Akihiko Kikuchi, Teruo Okano**

Institute of Advanced Biomedical Engineering & Science, Tokyo Women's Medical University, 8-1 Kawadacho, Shinjuku, Tokyo 162-8666, Japan

Summary: Thermoresponsive poly(*N*-isopropylacrylamide) (PIPAAm) and its derivatives were utilized as chromatography column matrix modifiers to develop novel supports for thermoresponsive hydrophobic chromatography with aqueous mobile phase. In the column, matrix surfaces show thermoresponsive hydrophilic/hydrophobic property alterations, which alter interaction with and retention of solutes to be separated. We have also demonstrated that the electrostatic interaction of ionic solutes and charged, thermoresponsive polymer-modified surfaces can be modulated temperature changes in the aqueous mobile phase alone.

Introduction

We have been investigating the preparation and characterization of a series of thermoresponsive polymer-modified surfaces. Poly(*N*-isopropylacrylamide) (PIPAAm) shows temperature-responsive soluble-insoluble changes across its lower critical solution temperature (LCST) at 32°C in aqueous solution[1,2] (Fig. 1). The thermoresponsive solubility change can be explained by the hydration/dehydration changes of polymer sidechain isopropyl groups. When PIPAAm is covalently incorporated onto solid surfaces to form densely polymer-grafted gel-like surfaces, these surfaces show temperature responsive hydrophilic/hydrophobic surface property alterations.[3-5] Utilizing these thermoresponsive surface property alterations, PIPAAm-modified surfaces have been utilized to recover cultured cells noninvasively for possible application in tissue engineering[6-11]. Furthermore, PIPAAm-modified surfaces are

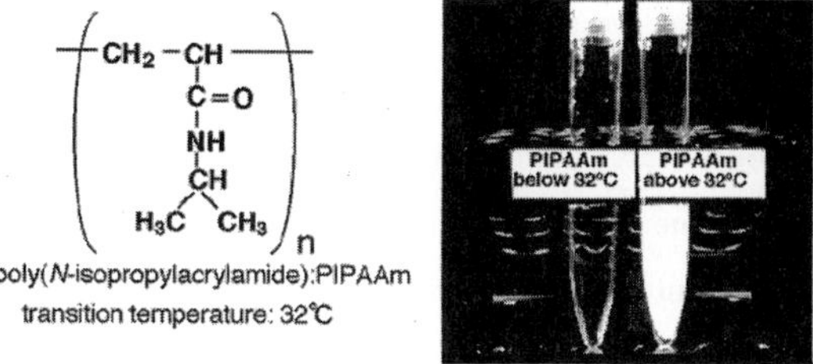

Fig. 1. Structure and solution appearance of PIPAAm.

 DOI: 10.1002/masy.200450319

also applied to develop a series of new column matrixes for chromatography using water as a sole mobile phase.[5,12-15] Several of the thermoresponsive surface types drawn in Fig. 2 are currently being investigated in our research groups. In the present paper, brief discussion will be made on these newly developed aqueous temperature responsive chromatography systems.

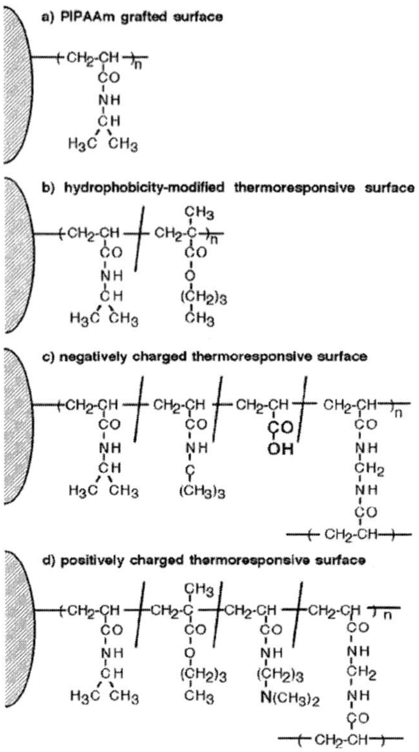

Fig. 2. Several surface types used in our studies.

Preparation and Wettability Changes of PIPAAm-Modified Surfaces

We have prepared four-types of PIPAAm-modified surfaces.[4,5] Those are, A) PIPAAm-grafted surface at one chain end, B) multipoint grafted PIPAAm surfaces, C) combination of the surfaces A) and B), and D) PIPAAm thin hydrogel layer grafted surfaces. PIPAAm-terminally grafted surfaces (type A) were prepared using end-carboxylated PIPAAm to graft onto aminated surfaces. Poly(IPAAm-*co*-*N*-acryloxysuccinimide) was used to prepare the model surface b). In this case, solvent composition was revealed to be an important factor affecting the polymer graft conformation on the surfaces. When 1,4-dioxane, a good solvent of PIPAAm, was used as the reaction solvent, PIPAAm was grafted on the surface with a relatively extended conformation, resulting in the disappearance of the transition in water contact angle changes. By contrast, addition of a small amount of non-solvent (toluene) for PIPAAm in dioxane resulted in a large, discontinuous water contact angle changes at a definite temperature, as described in a latter section. To prepare the surface type C), the latter preparation procedure was followed to obtain a large contact angle change at the polymer transition temperature. Thin PIPAAm hydrogel layer on glass surfaces was prepared using azo initiator-immobilized glass plates. Polymerization was initiated by the surface-modified radical initiators, and a thin hydrogel layer was formed on the surfaces. Those four surface types are illustrated schematically in Fig. 3.

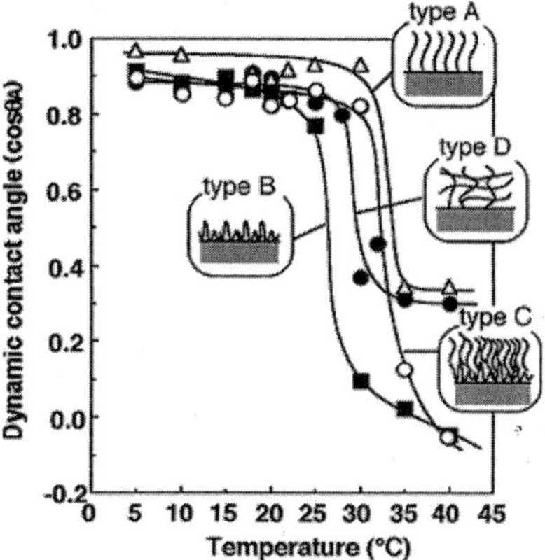

Fig. 3. Thermoresponsive contact angle (cosθ) changes for four types of PIPAAm-modified surfaces.

All four types of PIPAAm-modified surfaces showed thermoresponsive hydrophilic/hydrophobic surface property alterations, being hydrophilic at low temperature below transition temperature, and became hydrophobic above the transition temperature, as shown in Fig. 3. Thermoresponsive wettability changes arose from the characteristic of PIPAAm, which shows hydration changes with temperature. At low temperatures below the LCST, hydrophobic hydration around the side isopropyl groups has considerable influence on solubility of PIPAAm in water. Hydrophobic hydration is destroyed at higher temperature due to the increased entropy of water molecules, and thus, the hydrophobic nature of PIPAAm side chains determines the solubility of PIPAAm molecules in water above the LCST. The transition temperatures were slightly shifted according to the PIPAAm-graft configuration. For surfaces with freely mobile PIPAAm graft chains (surface type A) and C)), the transition temperature was in good accordance with the LCST observed for soluble PIPAAm molecules. On the other hand, transition temperatures observed for surface types B) and D) were slightly lower temperature than for surfaces types A) and C). Surface grafted PIPAAm molecules might have restricted molecular motion arising from the multi-point attachments, which reduces the transition temperature. These differences affected the molecular interaction as shown in the next section.

Hydrophobic Chromatography Using Sole Aqueous Mobile Phases

We then utilized these PIPAAm-modified surfaces for chromatography column matrices. Commercially available aminated silica beads (average diameter 5μm) were modified according to the above described methods to obtain several types of PIPAAm-modified beads with different PIPAAm conformation on the surfaces. These beads were then packed into a stainless steel column (i.d. 4.6mm, 150mm length). The polymer-modified bead-packed column was connected to a high performance liquid chromatography system. Column temperature was controlled within ± 0.1°C with circulating water connected to a thermostated water bath. UV absorbance was used for detection.

We applied thermoresponsive hydrophilic/hydrophobic surface property alteration to the recognition of hydrophobicities of substances by liquid chromatography with an aqueous mobile phase. Fig. 4 shows the effects of temperature on hydrophobic steroid elution from a terminally PIPAAm-modified, silica-packed column.[12]. At lower temperatures where the PIPAAm-modified surfaces are hydrophilic, steroids were eluted in one broad peak near void volume. With increasing temperature each peak can be recognized, but incomplete peak resolution was obtained at 25°C, below the polymer transition temperature. By sharp contrast, steroids were separated nicely above the polymer transition temperature. As a substance's hydrophobicity is usually defined as the logarithm of its octanol/water partition coefficient, log *P*, retention changes of steroids on PIPAAm-modified surfaces at various temperatures were analyzed using the log *P* values. With increasing the log *P* value, retention time became longer regardless of

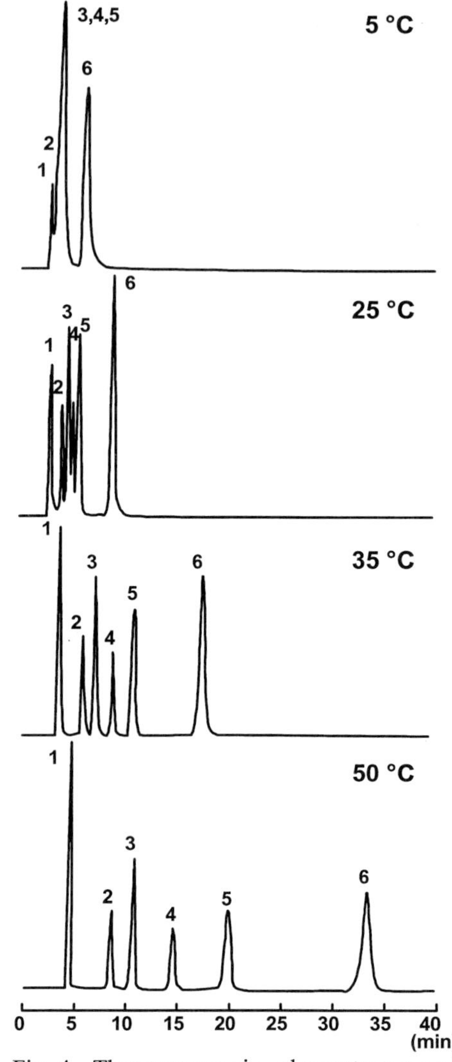

Fig. 4. Thermoresponsive chromatograms of steroids from PIPAAm-end grafted surfaces (type a) with water as a sole mobile phase. 1; benzene, 2; hydrocortisone, 3; prednisolone, 4; dexamethasone, 5; hydrocortisone acetate, 6; testosterone.

temperature, meaning that more hydrophobic substances have stronger interaction with PIPAAm-modified surfaces. Interestingly, when the natural logarithm of the capacity factor, *k*, values (ln *k*) was plotted against inverse temperature (1/T), inflection points were observed in the plots for all steroid types at a definite temperature. The temperature corresponds to the PIPAAm's LCST. This means that steroid interaction toward PIPAAm-modified surfaces is influenced by the polymer property alteration responding to temperature; i.e., the higher the temperature, the stronger the hydrophobic interaction is observed. The elution profiles of steroids from PIPAAm-modified columns are totally different from those observed using conventional reserved-phase chromatography column, octadesylsilane-column, in which elution time was shortened with temperature.

As described in the previous section, graft conformation of PIPAAm on the surface greatly influences the temperature-responsive wettability changes. Thus, it is possible to consider that the graft conformation of the PIPAAm on the column matrix surfaces should affect the elution behavior of steroids. The temperature-dependent elution behavior of steroids was examined on PIPAAm looped chain grafted surfaces, freely PIPAAm grafted onto PIPAAm loops, and PIPAAm thin hydrogel grafted surfaces ([5] for hydrogel surface). On those three PIPAAm-modified surfaces, low separation efficiency was observed in the chromatograms at 5 °C. At 45 °C under surface graft conditions, PIPAAm chains are in dehydrated state, and elution of hydrophobic steroids is retarded. As surface grafted polymer layers have higher grafted polymer densities than the surfaces with freely mobile linear PIPAAm-grafted chains, sample partitioning within densely grafted PIPAAm layers should have an influence on the extension of steroids retention times and peak broadening. When PIPAAm with chains with free ends were grafted onto PIPAAm loop chains, the thickness of the PIPAAm layer increased simultaneously with the graft reaction process. Thus, longer retention was seen on the surface with free end PIPAAm grafted onto PIPAAm loops than with PIPAAm loops alone. Significant peak broadening was seen in the case of PIPAAm thin hydrogel grafted surfaces. Partitioning of steroid molecules should be large for thin PIPAAm hydrogel grafted surfaces, which have a 3-dimensional cross-linked structure. Furthermore, the mobility of the PIPAAm chains is restricted, influencing peak broadening and the retardation of elution of steroids.

Steroid interactions were modulated by incorporating as hydrophobic comonomer, butyl methacrylate (BMA) into PIPAAm sequences, which were utilized as column matrix modifiers (Fig. 2b).[13] Steroids interacted with the matrix, and separated even below the transition temperature of the polymer modifiers, probably due to the hydrophobic interaction with hydrophobic comonomer side chains below the polymer transition temperature. Above the LCST, more pronounced interaction took place, and baseline separation was achieved. Interestingly, steroid elution was easily modulated with stepwise temperature changes, similar to the gradient elution in the reversed-phase chromatography method. These results strongly support the feasibility of temperature-responsive chromatography as an alternative method for separation of bioactive compounds in aqueous systems.

The poly(IPAAm-*co*-BMA) modified column was further utilized to evaluate the interaction with peptides, elution profiles were observed for three peptides; insulin chain A, insulin chain B, and β-endorphin fragment (amino acid residues 1-27) at two different temperatures.[13] All three peptides have approximately the same molecular weight of 3,000 but with different amounts of hydrophobic amino acid residues. At low temperature, where polymer modifiers on the surface of column matrix are in the hydrophilic state, minimum interaction took place with all types of peptide samples, however, these peptides were separated nicely above the polymer transition temperature. Elution became retarded as the amount of hydrophobic amino acid residues in peptide sequences increased. The results indicate that thermoresponsive chromatography using an aqueous mobile phase can be utilized to separate not only low molecular weight substances, but also polypeptides, at least with molecular weights of around 3,000-4,000.

Modulation of Electrostatic Interaction with Ionic, and Thermoresponsive Polymer-modified Columns

Many biologically active substances have both hydrophobic and ionic nature in their molecular structures, such as peptides, DNA, and RNA. As the electrostatic nature can be altered by temperature in many ionic substances as is seen in synthetic polyelectrolytes.[16-18] Therefore, we hypothesized that the electrostatic interaction of ionic biological substances with thermoresponsive IPAAm copolymers bearing charged side groups might be modulated by only changing temperature. Poly(IPAAm-*co*-acrylic acid (AAc),[15] poly(IPAAm-*co*-t-butylacrylamide (tBAAm)-*co*-AAc),[19] or poly(IPAAm-*co*-BMA-*co*-*N,N*-dimethylaminopropylacrylamide (DMAPAA))[20] were

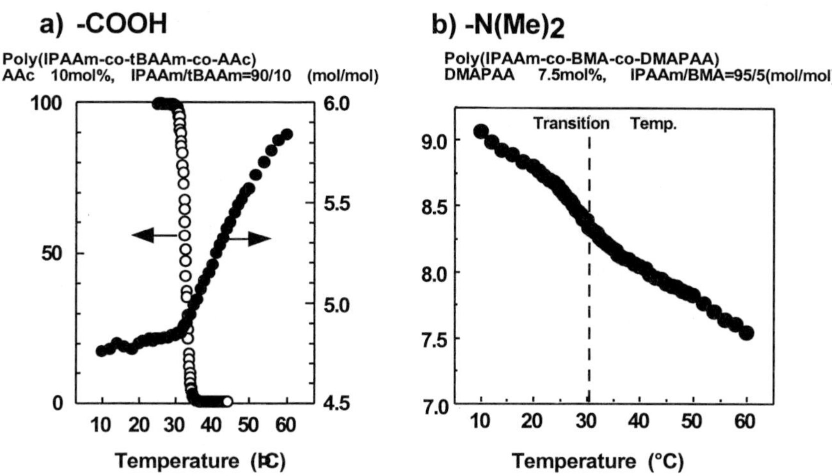

Fig. 5. Temperature dependent pKa' shift for a) carboxyls and b) amines in IPAAm copolymers.

prepared, and p*Ka'* changes were evaluated with temperature change. Both ionic polymer systems showed relatively large p*Ka'* changes around the polymer transition temperature; increasing p*Ka'* values for negatively charged polymer systems, and decreasing p*Ka'* for positively charged polymers (Fig. 5). In the case of positively charged poly(IPAAm-*co*-BMA-*co*-DMAPAA), with increasing surrounding temperature, dehydration of IPAAm isopropyl side chains took place, locally increasing the hydrophobic microenvironment around polymer amino groups. Furthermore, surface potentials of polymer-modified beads decreased and became negative above the polymer transition temperature. Previous reports document the weakening of tertiary amine basicity in IPAAm copolymers at elevated temperature, especially above the LCST, while there is nearly no observable change in amine basicity in water-soluble polyacrylamide derivatives over a wide temperature range.[17] The decrease in amino group basicity in PIPAAm derivatives is due to the decreased dielectric constant around the amino groups at higher temperature where the polymer loses hydration and becomes hydrophobic. Urry[16] also reported a carboxyl p*K*a shift in synthetic polypeptides by modulating peptide hydrophobicity. These observations indicated that the charge density distributions of polymer molecules, or polymer-modified surfaces become lower as polymer is in a hydrophobic, aggregated state above the respective transition temperature even at neutral pH. Therefore, the interaction of charged bioactive compounds with thermoresponsive and ionic polymers can be modulated by temperature alteration alone. Fig. 6 shows the typical result of elution for adenosine nucleotides, adenosine monophosphate (AMP), adenosine diphosphate (ADP), and adenosine triphosphate (ATP), all these share the same adenosine chemistry but different numbers of anionic phosphate units. All of these molecules are important for energy metabolism in living systems. As is shown in this figure, poor separation of adenosine nucleotides was observed in nonionic PIPAAm-modified columns. By sharp contrast, retention time increased with increasing number of anionic phosphate units in the molecules, regardless of temperature, on the amino-containing, terpolymer-modified column. This means that adenosine nucleotides interact with poly(IPAAm-*co*-BMA-*co*-DMAPAA)-modified surfaces through primarily an electrostatic interaction. Interestingly, stronger interaction took place at lower temperatures, and all substrates show longer retention times. By increasing temperature, retention times of adenosine nucleotides dramatically decreased, especially above the polymer transition temperature. These results strongly suggest that polymer matrix charge density alteration with temperature affects to the order of electrostatic interaction with oppositely charged biological molecules. Without charged groups, i.e., PIPAAm-modified surfaces, such retention time changes were not observed. Therefore, the temperature-modulated charge density alteration could be utilized to regulate the interaction with charged, hydrophobic biological compounds. Although the thermoresponsive ionic interaction chromatographic systems need longer analyses times,

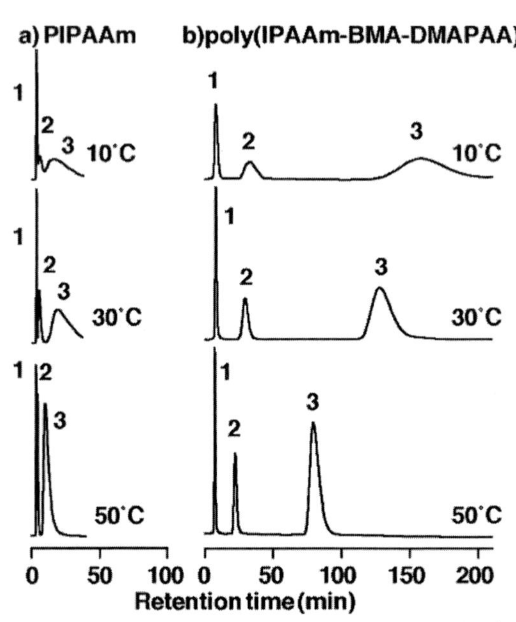

Fig. 6. Temperature dependent adenosine nucleotides elution on a) PIPAAm, and b) cationic thermoresponsive polymer surfaces at pH7.0. 1; AMP, 2; ADP, 3; ATP. flow rate; 0.5mL/min.

column conditions could be optimized through chemical composition of copolymer modifiers, size of column, and silica bead diameter, and so on. Adenosine nucleotides are currently analyzed with reversed-phase columns in the absence of[21,22] or in the presence of additive alkylamine (ion-pair RPC)[23-25] to modulate analyte sample properties, interactions with stationary phase surfaces, and of course, elution time.

To regulate the interaction of positively charged bioactive compounds with negatively charged thermoresponsive polymer modified columns, we introduced the poly(IPAAm-*co*-*N*-*t*-butylacrylamide (*t*BAAm))-*co*-AAc) thin hydrogels onto the silica bead surfaces.[19] A hydrophobic monomer, tBAAm, was incorporated to maintain thermoresponsive property of the polymer molecules, which may be disrupted by the increase in the hydrophilic, negatively charged carboxylate anions. The prepared bead-packed column was then utilized to investigate the elution behavior of positively charged substances, including catecholamines and angiotensin subtypes, as low molecular weight substances and peptides, respectively. Catecholamine interaction with poly(IPAAm-*co*-*t*BAAm-*co*-AAc) with 10mol% of AAc was evaluated. At pH 4.0, where all carboxyl side chains in the surface-grafted polymer were in protonated state (no charge) while amino containing sample analytes were all protonated, negligible interaction took place with catecholamines regardless of temperature change. By sharp contrast, catecholamines, excepting zwitterionic DOPA, are retained on the poly(IPAAm-*co*-*t*BAAm-*co*-AAc) modified column. As catecholamines did not interact with the non-ionic PIPAAm column, these interacted were primarily through electrostatic interaction with poly(IPAAm-*co*-*t*BAAm-*co*-AAc) surfaces at pH 7.0. Elution times were increased in the order DOPA<adrenaline<dopamine<tyramine. This order is in agreement with the molecule hydrophobicity parameter, i.e. the logarithm of the partition coefficient.

The data indicate that the analyte interaction with negatively charged surfaces is affected by the two parameters: electrostatic, or coulombic interaction, and hydrophobic interaction. When positively charged sample molecules interact with negatively charged surfaces, coulombic interaction neutralizes apparent charges. Then, microenvironmental hydrophobicity around the analyte molecules increases due to repulsion of the polarized and hydrated water molecules around charged groups. As analyte hydrophobicity increases, the microenvironmental hydrophobicity also increases. Thus, more hydrophobic substances interact strongly with negatively charged polymer surfaces. Temperature alteration affected peak resolution, and around the surface-modified polymer LCST at pH 7.0 the resolution of tyramine and dopamine was maximal. Above the transition temperature, the resolution become worse due to decreased electrostatic interaction with hydrophobized polymer surfaces.

Three angiotensin subtypes; angiotensin I, II, and III, are peptides differing in sequence and biological affinity. Due to their structural similarity, separation of these subtypes are often difficult with conventional chromatography methods, and combination of reversed-phase chromatography and ion exchange chromatography is applied for separation of these substances. In our newly developed column of poly(IPAAm-*co*-*t*BAAm-*co*-AAc) modified silica, it is possible to separate these three angiotensin subtypes with only one chromatographic run at constant pH and ionic strength, but changing temperature with a sole aqueous mobile phase.

The use of intelligent materials should prove valuable in the design of novel aqueous 'green' chromatography systems.

Conclusions

Poly(*N*-isopropylacrylamide) (PIPAAm) and its derivatives were utilized as chromatographic matrix modifiers to develop novel hydrophobic chromatography using a sole aqueous mobile phase. Thermoresponsive soluble-insoluble changes of the PIPAAm derivatives were used to modulate surface hydrophilic-hydrophobic property alterations. The surface property alteration then influenced the interaction with biological entities, such as peptides, proteins, and also cells. The results briefly demonstrated here indicate that formation of very thin, hydrogel-like, dense thermoresponsive polymer layers on the solid surfaces is a feasible approach for surface property alteration. This technique can be utilized for novel aqueous chromatography matrices and/or cell culture and detachment control substrates without using cell-lytic enzymes. These technologies are promising for future environmentally friendly chromatography developments as well as for tissue engineering.

Acknowledgments

This work was financially supported in part by the Japan Chemical Innovation Institute (JCII) operated under the New Energy and Industrial Technology Development Organization (NEDO), and the Grant-in-Aid for Scientific Research (Grant No. 13680955) by the Japan Society for Promotion of Science. The authors would be grateful to the collaborators, Professor Kiyotaka Sakai of Waseda University, and Professor Hideko Kanazawa of Kyoritsu College of Pharmacy for their collaboration as well as valuable comments, and suggestions throughout the work.

[1] M. Heskins, J. E. Guillet, E. James, *J. Macromol. Sci., Chem. A* **1968**, *2*, 1441.
[2] Y. H. Bae, T. Okano, S. W. Kim, *J. Polym. Sci.: Part B: Polym. Phys.* **1990**, *28*, 923.
[3] Y. G. Takei, T. Aoki, K. Sanui, N. Ogata, Y. Sakurai, T. Okano, *Macromolecules* **1994**, *27*(21), 6163.
[4] T. Yakushiji, K. Sakai, A. Kikuchi, T. Aoyagi, Y. Sakurai, T. Okano, *Langmuir* **1998**, *14*(16), 4657.
[5] T. Yakushiji, K. Sakai, A. Kikuchi, T. Aoyagi, Y. Sakurai, T. Okano, *Anal. Chem.* **1999**, *71*(6), 1125.
[6] T. Okano, N. Yamada, H. Sakai, Y. Sakurai, *J. Biomed. Mater. Res.* **1993**, *27*, 1243.
[7] T. Okano, N. Yamada, M. Okuhara, H. Sakai, Y. Sakurai, *Biomaterials* **1995**, *16*, 297.
[8] A. Kikuchi, M. Okuhara, F. Karikusa, Y. Sakurai, T. Okano, *J. Biomater. Sci., Polym. Ed.* **1998**, *9*(12), 1331.
[9] M. Yamato, A. Kikuchi, S. Kohsaka, T. Terasaki , H. A. von Recum, S. W. Kim, Y. Sakurai, in: „*Tissue Engineering for Therapeutic Use 3*“, Y. Ikada, T. Okano, Eds., Elsevier; Amsterdam 1999, p. 99.
[10] M. Yamato, M. Utsumi, A. Kushida, C. Konno, A. Kikuchi, T. Okano, *Tissue Eng.* **2001**, *7*(4), 473.
[11] T. Shimizu, M. Yamato, A. Kikuchi, T. Okano, *Tissue Eng.* **2001**, *7*(2):141.
[12] H. Kanazawa, K. Yamamoto, Y. Matsushima, N. Takai, A. Kikuchi, Y. Sakurai, T. Okano, *Anal. Chem.* **1996**, *68*(1), 100.
[13] H. Kanazawa, Y. Kashiwase, K. Yamamoto, Y. Matsushima, A. Kikuchi, Y. Sakurai, T. Okano, *Anal. Chem.* **1997**, *69*(5), 823.
[14] H. Kanazawa, T. Sunamoto, Y. Matsushima, A. Kikuchi, T. Okano, Anal. Chem. **2000**, *72*(24), 5961.
[15] J. Kobayashi, K. Sakai, A. Kikuchi, T. Okano, *Anal. Chem.* **2001**, *73*(9), 2027.
[16] D. W. Urry, *Prog. Biophys. Molec. Biol.* **1992**, *57*, 23.
[17] H. Feil, Y. H. Bae, J. Feijen, S. W. Kim, *Macromolecules* **1993**, *26*(10), 2496.
[18] A. Kikuchi, T. Tsuruta, K. Kataoka, J. Biomater. Sci., Polym. Ed. **1994**, *5*(6), 569.
[19] J. Kobayashi, A. Kikuchi, K. Sakai, T. Okano, *J. Chromatogr.* **2002**, *958,* 109.
[20] A. Kikuchi, T. Iwasa, J. Kobayashi, K. Sakai, T. Okano, submitted.
[21] G. Leoncini, E. Buzzi, M. Maresca, M. Mazzei, A. Balbi, *Anal. Biochem.* **1987**, *165*, 379.
[22] T. Teerlink, M. Hennekes, J. Bussemaker, J. Groeneveld, *Anal. Biochem.* **1993**, *214*, 278.
[23] K. K. Tekkanat, I. H. Fox, *Clin. Chem.* **1988**, *34*(5), 925.
[24] B. Schobert, *Anal. Biochem.* **1995**, *226*, 288.
[25] J. Blanco, E. I. Canela, J. Sayos, H. Mallol, C. Lluis, R. Franco, *J. Cell. Physiol.* **1993**, *157*, 77.

Superabsorbent Polymer of Poly[acrylamide-*co*-(acrylic acid)] by Foamed Polymerization. I. Synthesis and Water Swelling Properties

Suda Kiatkamjornwong,[*1] *Rattana Wongwatthanasatien*[2]

[1] Department of Imaging and Printing Technology, Faculty of Science, Chulalongkorn University, Phyathai Road, Bangkok 10330, Thailand
[2] Multidisciplinary Program of Petrochemistry and Polymer Science, Faculty of Science, Chulalongkorn University, Phyathai Road, Bangkok 10330, Thailand

Summary: Superabsorbent polymers of acrylamide and acrylic acid were synthesized by foamed polymerization, initiated by ammonium persulfate and N, N, N', N'-tetramethylenediamine. The crosslinking agent N, N'-methylenebisacrylamide, the foaming agent sodium bicarbonate, and a foam stabilizer, a triblock copolymer of polyoxyethylene/polyoxypropylene/polyoxyethylene, were used in the polymerization. The superabsorbent polymer synthesized with an acrylamide/(acrylic acid) ratio of 1:9 by mole at 250 rpm and 45°C for 30 min of polymerization time produced the highest water absorption of 2300 times its dried weight. This article describes the synthetic method and measurements of water absorption.

Introduction

Superabsorbent polymers containing hydrophilic vinyl monomers have been produced since the 1970s; the polymerization route is mainly radical chain polymerization, by various synthetic techniques. The extent of water absorption depends primarily on the types of hydrophilic monomers, the initiation system, the degree of crosslinking, the method used to precipitate the polymer, and the ionic strength of the test fluid.

Foamed polymerization has been found to be an alternative method for preparing superabsorbent polymers with enhanced absorption rate.[1-2] Ishizaki et al.[3] filed a patent for synthesizing water-absorbent resins by dispersing a solid blowing agent in an aqueous monomer having unsaturated monomers and a vinyl crosslinking agent. The water-absorbent resin was used for various applications in which a rapid absorption rate, high water retention and prevention of water leakage were required.

The present research focuses on the synthetic method and the characterization of

 DOI: 10.1002/masy.200450320

properties affecting water absorption by superabsorbent polymers. High water absorption and a high water absorption rate were achieved.

Procedure

In this research, the foam formation reagents sodium bicarbonate (2g, Merck, Darmstadt, Germany) and 0.05% w v^{-1} Lutrol F®127 (BASF, Ludwigshafen, Germany, analytical grade) were used. Lutrol F®127 is a triblock copolymer of polyoxyethylene/ polyoxypropylene/polyoxyethylene, with molecular weight range 9800–14600 g mol^{-1} [2] which was used as a constant ingredient for the whole experiment. The effects of the crosslinker N,N'-methylenebisacrylamide (N-MBA), the initiator ammonium persulfate (APS)/N,N,N',N'-tetramethylenediamine (TMED) and the ratio of acrylamide (AM)/acrylic acid (AA) on water absorption of the polymer were studied. Swelling kinetics of the resulting superabsorbent polymer were studied.

Basic preparation technique

A mixture of 100-cm^3 aqueous solution containing acrylic acid (AA, 99% pure, Siam Resin & Chemicals, Thailand) and acrylamide (AM, 99% pure, Siam Resin & Chemicals, Thailand), N-MBA, and ammonium persulfate (APS, analytical grade, Merck, Hohenbrunn, Germany) was placed in a 250 cm^3 reaction flask. This round-bottomed flask, equipped with a mechanical stirrer set at 250 rpm, a spiral condenser, a thermometer, and an inlet tube for nitrogen gas, was immersed in a set temperature water bath. The ingredients were stirred by a small-bladed propeller for 5 min while heating at the desired temperature. After the set temperature of 318 K (45°C) was reached, 2 g of sodium bicarbonate powder was added. Lutrol F® 127 of 0.05% w v^{-1}, and TMED (analytical grade, Sigma, Aldrich, St Louis) were then added sequentially. The reaction between acrylic acid and sodium bicarbonate took place immediately to produce carbon dioxide bubbles, which floated upwards to the solution surface. Foam formation and solution polymerization continued for 30 min, after which the bubble production ceased and the reaction mixture became milky white. A fine powder then precipitated. The resulting polymer was filtered, washed several times with distilled water, dewatered with methanol, dried, and cut into small pieces, which were dried again at 323 K (50°C) for 24 h in a vacuum oven to a constant weight and then milled.

Effect of mole fraction of APS/TMED

In this experiment, the mole ratio of AM:AA, percent of neutralization of AA, and the concentration of the crosslinking agent N-MBA concentration were fixed at 0.2, 40% and 6.30 $x10^{-3}$ mol dm^{-3}, respectively. The same experimental procedure was carried out by varying the APS/TMED ratios (on a mol dm^{-3} basis) at: $1.5x10^{-3}/1.0x10^{-2}$, $2.0x10^{3}/1.0x10^{-2}$, $2.3x10^{-3}/1.0x10^{-2}$, $3.0x10^{-3}/1.0x10^{-2}$, $3.5x10^{-3}/1.0x10^{-2}$, $4.4x10^{3}/1.0x10^{-2}$, and $2.3x10^{-3}/2.0x10^{-2}$.

Effect of mole fraction of acrylamide to acrylic acid and neutralization of acrylic acid

The same experimental procedure as described above was carried out, with varying mole fractions of AM to AA (0.1 to 0.9 with an increment of 0.1) and percent neutralization of acrylic acid; 0, 40, 60, 80, or 100%.

Effect of concentration of N, N'-methylenebisacrylamide

The mole fraction of AM : AA was fixed at 0.1 and the percentage neutralization of AA was at 40%. The concentration of crosslinking agent (in mol dm^{-3}) was varied from $1.0x10^{-3}$, $1.5x10^{-3}$, $2.0x10^{-3}$, $2.5x10^{-3}$, $3.0x10^{-3}$, $3.5x10^{-3}$, $4.0x10^{-3}$, $4.5x10^{-3}$ and $5.0x10^{-3}$ and used for the preparation of superabsorbent polymer of poly[acrylamide-*co*-(acrylic acid)] or P(AM-*co*-AA) by following the basic preparation method.

Water absorption in distilled water

Distilled water (400 g) was added to 0.1 g of the dry copolymer in a 600-cm^3 glass beaker with a glass cover. The polymer was allowed to swell for 30 min. The fully swollen gel was then separated by filtering through a 100-mesh sieve aluminum screen for 2 h in a control room. The swollen gel was weighed and the water absorption was calculated by the equation as follows. Water absorption *(Q)* is *(B-A)/A* where *A* and *B* are the weights of the dry polymer (g), and the swollen polymer (g), respectively.

Swelling kinetics in distilled water

Distilled water (400 g) was added to 0.1 g of the dry copolymer in a 600-cm^3 glass beaker with a glass cover. It was allowed to swell for 10 min. The swollen gel was then separated by filtering through a 100-mesh sieve aluminum screen for 2 h in a control room. The swollen gel was weighed. A similar procedure was separately repeated at gel swelling times of 20, 30, 40, 50, and 60 min.

Results and Discussion

As mentioned above, superabsorbent polymers of P(AM-*co*-AA) were prepared by crosslinking polymerization of AM and AA monomers in the presence of gas bubbles. Carbon dioxide gas bubbles were generated by a reaction of sodium bicarbonate with acid solution. The amount of foam was governed by the amount of acid in the reaction mixture. We used an excess amount of sodium bicarbonate so that the foam size and amount were controlled by the amount of added acid, which was AA in this case. AA thus functions as a co-monomer and a foam accelerator. During the crosslinking polymerization, the viscosity of the medium increased gradually. The increased viscosity slowed down the bubble formation.

The foam stabilizer used in the present work is considered as one of the foaming materials with the best properties.[4] Polyoxyethylene/polyoxypropylene/ polyoxyethylene block copolymer (Lutrol® F 127), has the best foaming properties of most hydrophilic monomers; that is, it sustains the foam or bubbles for the longest period of time. In addition, it functions as a surface crosslinking agent, enhancing liquid permeability, improving the water absorption rate, and increasing water absorption capacity.[3]

To produce a highly porous superabsorbent polymer, polymerization and foaming processes had to take place simultaneously. Precise time control of the two processes was critical. Stabilizing a foam longer than a few minutes was rather difficult, because gelation had to start within this period just after the beginning of foaming resulting from the addition of sodium bicarbonate to the monomer mixture containing an acid component. We chose Lutrol® F 127, because it helps gel formation at lower reaction temperatures. The triblock copolymer of polyoxyethylene/polypropylene/polyoxy-ethylene is relatively polar and soluble in water and alcohol, which enhances stabilization of the bubbles and gelling and thickening of the copolymer. Normally, a fast gelation could be achieved by careful choice of monomers, initiators, temperature and solvent. In our case, gel formation was observed when a milky mixture resulted during the first 5 min after the addition of sodium bicarbonate, LF® 127, and TMED.

Crosslinking polymerization takes place in situ during and after formation of carbon dioxide gas bubbles, which acted as porogens. Toward the end of the foamed polymerization, the volume of the mass in the reaction vessel increased by 2 times. The polymerization was completed within 30 min. At the beginning of the crosslinking copolymerization, heat generated from the reaction was dissipated through the carbon dioxide bubbles. At this time, the reaction temperature was relatively constant, but it could increase as the reaction proceeded, which could lead to a decrease in viscosity of the reaction medium. Even though the reaction heat could be dissipated through the reaction solution, an autoacceleration or Trommsdorff effect (Norrish-Smith effect) was observed. The final superabsorbent polymers were thus of high molecular weight.

Effect of mole fraction of APS/TMED

The water absorbency of the crosslinked copolymers synthesized with various initiator (APS/TMED) concentrations for the solution polymerization is shown in Figure 1.

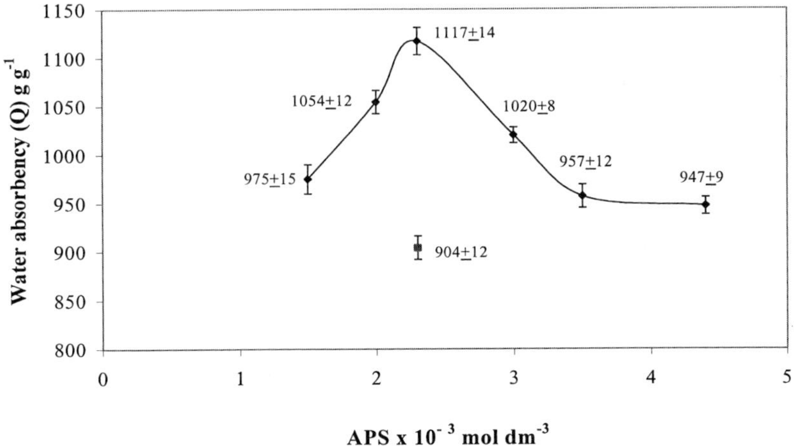

Figure 1. Effect of the initiator concentration on the water absorbency (Q) of the synthesized copolymers (♦ for TMED of 1 x 10^{-2} mol dm^{-3}, ■ for TMED of 2 x10^{-2} mol dm^{-3}).

Copolymers synthesized with APS (2.3×10^{-3} mol dm^{-3})/TMED (1.0×10^{-2} mol dm^{-3}) gave the optimum result, which has the highest water absorbency, 1117±14 g g^{-1}, compared with 904±12 g g^{-1} when twice the amount of TMED was used. Water absorbency increases linearly with an increase in the initiator ratio of APS:TMED at 0.23 and then decreases with increasing initiator ratio. As mentioned above, the rate of polymerization depended on the concentrations of monomers and initiator radicals in a bimolecular termination. When APS to TMED ratio is used as a redox initiator pair for the free radical polymerization, the free radical [$(CH_3)_2NCH_2CH_2(CH_3)N^{\bullet}CH_2)$] is one of the initial free radicals responsible for the initiation of vinyl polymerization in addition to sulfate free radical ($^{\bullet}OSO_3H$).

This redox initiation pair produces two radicals, which may be reasoned to be responsible for the fast polymerization. At the higher APS concentrations ($3x10^{-3}$ and $4.4x10^{-3}$ mol dm^{-3}), too many radicals were produced, resulting in an abundance of short kinetic chains, in which the superabsorbent polymers had lower molecular weights. This inevitably reduced the swelling capacity. Additionally, at constant APS concentration, higher concentrations of TMED ($1x10^{-2}$ and $2.0x10^{-2}$ mol dm^{-3}) hampered the efficiency of the initiators. The TMED accelerated the generation of a higher amount of free radicals to produce the shorter kinetic chains and/or radical recombination. The latter two effects led to the lower swelling superabsorbent polymers.

Effect of concentration of N, N'-methylenebisacrylamide

Water absorbency of the superabsorbent polymers, synthesized by a 0.1 mole fraction of AM with 2.3×10^{-3} mol dm^{-3} APS, 1.0×10^{-2} mol dm^{-3} TMED, 40% neutralization of AA, and various concentrations of the N-MBA crosslinking agent, is shown Figure 2.

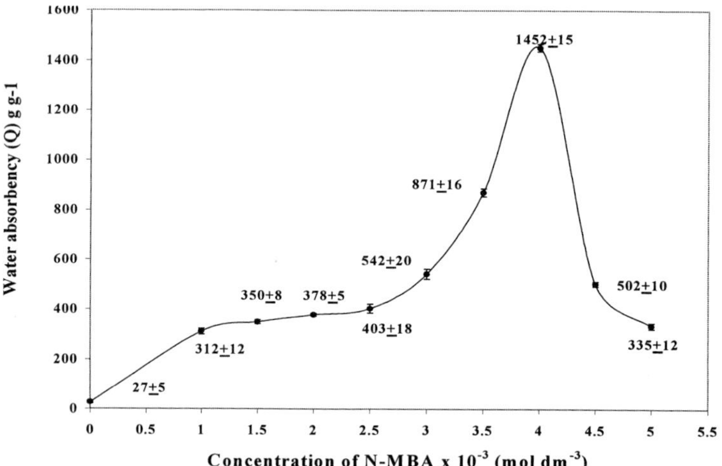

Figure 2. Effect of the concentration of crosslinking agent on the water absorbency (Q) of the synthesized copolymer.

The copolymer crosslinked with N-MBA at a 0.1 mole fraction of AM and 40% degree of neutralization of acrylic acid gave a water absorbency value of 1452±15 g g^{-1}, which is higher than the copolymer synthesized in the absence of the crosslinking agent (27±5 g g^{-1}). In the absence of the crosslinking agent, more soluble copolymers are produced that cannot absorb fluid effectively. Nonetheless, in the presence of high crosslinking agent concentrations, more crosslinks can also be formed to give rigid chains that reduce swelling of the gel as well.

Additionally, we did not extract the soluble part (non-crosslinked moiety) of the polymer before carrying out the water absorption experiment. The soluble portion is anticipated to interfere with the extent of water absorption in a negative way. Buchholz[5] reported that the solubility of a crosslinked agent depended on the extent of neutralization, which could exhibit different effectiveness in the crosslinked products. In general, crosslinking agents have been employed to improve the strength of the swollen gel but they very effectively reduce water absorbency if too high a crosslinking agent concentration is used.

Relatively small changes in crosslinker density can play a major role in modifying the properties of superabsorbent polymers. Figure 2 indicates the dependence of water absorption on the crosslinking agent concentration. It is clear that gel absorbency is diminished at low N-MBA concentration because of an increase in soluble polymers. A maximum water absorbency of 1452±15 g g^{-1} is obtained at the N-MBA concentration of 4.0×10^{-3} mol dm^{-3}. On the other hand, many branching points between two main vinyl backbones induce excessive crosslinking chains with rigidity, drastically decreasing the swelling ability of the superabsorbent polymer.[3,4] Moreover, a high concentration or high reactivity of the crosslinking agent usually leads to early depletion during polymerization. Polymer chains formed in a later stage of the polymerization lack the crosslinking agent, thus resulting in a more soluble polymer.

Effect of mole fraction of acrylamide to acrylic acid

The results on water absorbency of the copolymers synthesized by crosslinking polymerization of various mole fractions of acrylamide (AM) are shown in Figure 3. The synthesized copolymers of acrylamide (AM) and acrylic acid (AA) are swollen in water on account of the hydrophilic pendants (amide and carboxylic groups) in their structure. The synthesized copolymer with the mole fractions of AM and AA at 0.1 and 0.9, respectively, at 40% degree neutralization of acrylic acid and 4.0×10^{-3} mol dm^{-3} N-

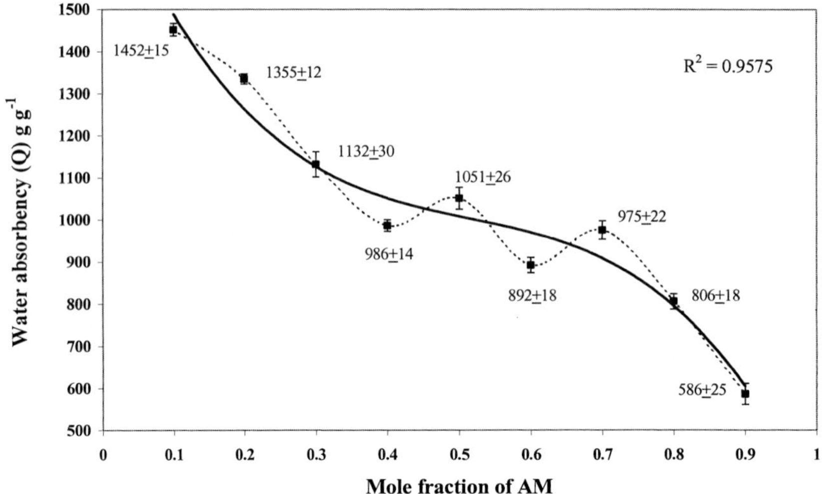

Figure 3. Effect of the mole fraction of AM on the water absorbency (Q) of the synthesized copolymers.

MBA crosslinking agent, produced a high water absorbency value of 1452±15 g g^{-1}. Water absorption capacity decreased with increasing mole fraction of AM, because chain transfer to the polymer increased with monomer concentration. In particular, at mole fraction 0.9 of AM, the water absorbency value was only 586±25 g g^{-1}. The chain transfer reaction, which results in increasing amounts of branching chains, and self-crosslinking reactions of imidization, by the amide moiety, affects the product's properties.[6] The critical mole fraction of AM at 0.1, with no pre-neutralization of acrylic acid, produced the superabsorbent polymer with the highest water absorption; this result indicated that one of the side reactions could be crosslinking of the copolymer with an appropriate crosslinking agent whose concentration is 4.0×10^{-3} mol dm^{-3}. Actually, the structure of the copolymer consists of more acrylamide units than acrylate units owing to the higher hydrophilicity of the carboxylate group, although the reactivity of AM is lower than that of sodium acrylate, because the copolymerization reactivity ratios of acrylamide and sodium acrylate are 0.355 and 2.852, respectively.[5] In contrast, if the flexibility of polymer chain were another important factor, the water absorbency should have been decreased with increasing concentration of acrylate units in the polymer chain, since the glass transition temperature (T_g) of sodium acrylate,

194°C (467 K), is higher than that of acrylamide, 165°C (438 K).[7] The water absorbency of the crosslinked copolymers synthesized with various mole fractions of AM and the degree of neutralization of acrylic acid are shown in Figure 4.

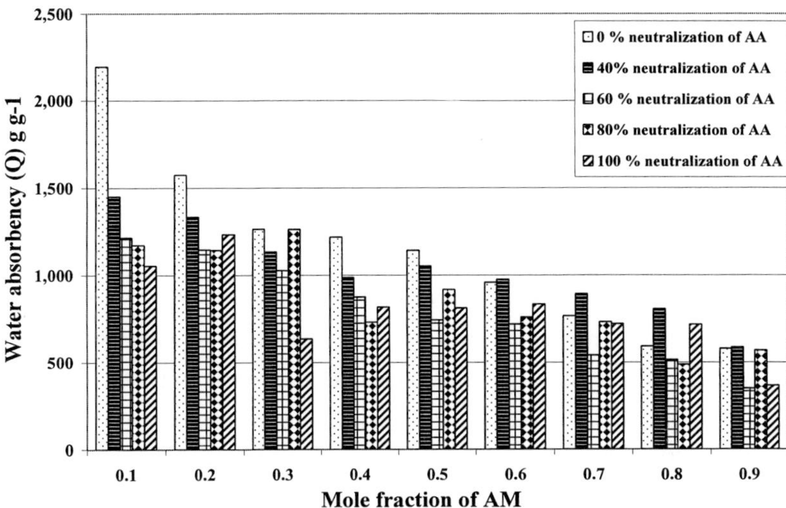

Figure 4. Water absorbency (Q) of the synthesized copolymers vs. mole fraction of AM and the degree of neutralization of acrylic acid.

Water absorbency decreases with increasing degree of neutralization of AA. Partial neutralization of acrylic acid before polymerization did not give the highest water absorption, compared with the non-neutralized AA. Surprisingly, the non-preneutralized AA gives the highest water absorption. However, in the presence of the blowing agent ($NaHCO_3$), the AA portion was eventually neutralized by the concentration of the sodium ion in the solution. Likewise, the preneutralized AA could probably be further neutralized. The presence of 0.1 mole fraction of AM is considered important similarly to the effect of crosslinking agent. Without both of them, the swelling ratio may exceed 2000, possibly compromising gel strength. The gel might then behave just like slightly viscous water, or it may not hold water effectively.

We found that the superabsorbent polymer is a viscous gel. We anticipated that AM could contribute to chain stiffness due to its relatively hydrophobic nature compared to that of sodium acrylate; the association of the amide function groups may also take place. The hydrophobicity of AM contributes to the gel strength of the superabsorbent polymer. Also, N-MBA provides the crosslinking reaction to acrylate portion. We needed to mention that the pre-neutralization of acrylic acid was performed before polymerization, owing to the ease of neutralization of the liquid monomer. As a drawback, sodium acrylate polymerizes less readily than acrylic acid, because ionized monomers react more slowly in the propagation step of polymerization. Higher percentages of soluble polymer and higher residual monomer concentrations in the final product can result. Because an excess amount of sodium bicarbonate was used as a foaming agent, one must consider that there is sodium bicarbonate left in the reaction solution, which can neutralize AA to become sodium acrylate, which is very water soluble.[9]

Flory described the mechanism of swelling of an ionic network.[10] If the polymer chains making up the network contain ionizable groups, the swelling forces may be greatly increased as a result of the localization of charges on the polymer chains. When the crosslinked copolymer is neutralized with sodium hydroxide, the negatively charged carboxyl groups attached to the polymer chains set up an electrostatic repulsion, which tends to expand the network. The concentration of sodium ions also increases with an increase in the degree of neutralization of AA. Sodium ions, which screen the negative charges of carboxyl groups, reduce the electrostatic repulsion tremendously, leading to the decrease in water absorption.

The water absorbency of synthesized copolymer depends not only on the degree of neutralization of AA, but also on the mole fraction of AM, because the electrostatic repulsion may be decreased with an increase of the mole fraction of AM, resulting in decreased water absorbency (Q).

Swelling kinetics in distilled water

Swelling rates were measured for the superabsorbent polymer prepared with AM/AA ratio of 1:9 and without neutralization. This gel absorbed distilled water up to 1628±14 g per gram dry weight of the polymer within 10 min. After 20 min, water absorption rate reached a plateau and remained constant afterwards. The factor determining the swelling time of the dried porous superabsorbent polymer is the time needed for water to enter into the gel matrix and cover most of the surface of the pores. We can determine them by the effectiveness of the capillary action and wettability of the pore surface. The swelling kinetics of superabsorbent polymer can be studied by application of a first-order kinetic expression based upon Fick's second law of diffusion,[9] where C is concentration of the diffusing species, D is the diffusion coefficient of C, and r is the diffusion path length. For a spherical particle, the diffusion equation reduces wherein the rate of swelling is first-order in the amount of swelling capacity remaining unswollen at any time, as given by Equation 1. The values of Q_{max} and Q are the swelling capacities at equilibrium and at the any time t. The first-order rate constant, k, depends on particle radius and on the diffusion coefficient[9] as shown in Figure 5.

$$Q(t) = Q_{\max}(1 - e^{-kt}) \qquad (1)$$

In this model of the dynamic swelling processes, the structural features of the polymer and system affect the maximum swelling capacity Q_{max}. In practical applications, not only a higher swelling capacity, but also a higher swelling rate is required for the absorbent. The synthesized copolymer is also characterized by its rapid swelling kinetics as presented in Figure 5.

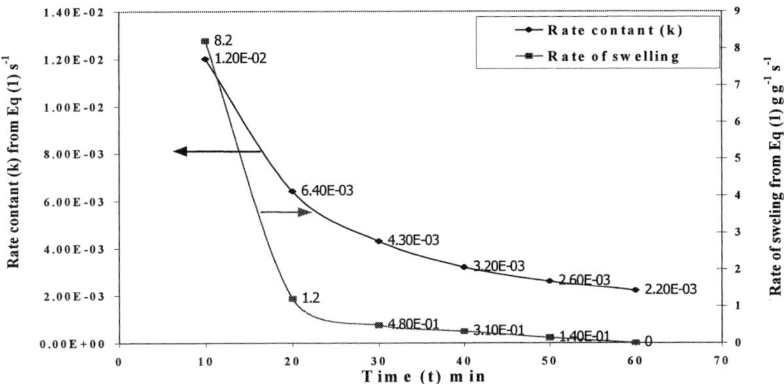

Figure 5. Swelling kinetics of the synthesized copolymer

From the plot of Q vs. time, one gets a slope of the swelling kinetics of 27 g $g^{-1}s^{-1}$. One can also calculate the rate constant of swelling (k) from the first-order kinetics at any time, t, and the k value decreases steadily with time. The k value depends on the diffusion coefficient and the particle radius. We found that the rate of swelling decreased with increasing swelling time. The swelling rate became stable or constant within 20 min, with a rate constant for swelling of 6.4 x 10^{-3} s^{-1}, and a swelling rate of 1.2 g g^{-1} s^{-1}. The faster swelling was then accompanied by a higher degree of swelling.

Conclusions

Superabsorbent polymers of poly[acrylamide-*co*-(acrylic acid)] were synthesized through a crosslinking copolymerization of acrylamide and acrylic acid in the presence of N, N'-methylenebisacrylamide crosslinker under the influence of carbon dioxide gas bubbles, stabilized by a triblock copolymer of polyoxyethylene/polyoxypropylene/ polyoxyethylene stabilizer, from the chemical reaction of acrylic acid and sodium carbonate. The resulting superabsorbent polymers have water absorption values as high as 2300 times their dry weight and much faster volume change kinetics than the non-porous absorbent polymers. The polymer made of the non-preneutralized acrylamide/acrylic acid with a ratio of 1:9 (by mole fraction) gave the highest water absorption with a swelling rate constant and swelling rate of 6.4 x 10^{-3} s^{-1} and 1.2 g g^{-1} s^{-1}, respectively. The non-ionic moiety of acrylamide yielded lower water absorption, but the ionic moiety of acrylic acid has higher water absorption. This research further indicates that the high mole fraction of acrylic acid or ionic part of the poly(acrylic acid) moiety in the copolymer contributes significantly to the high absorption values in distilled water, while the acrylamide portion decreased water absorption due to its less hydrophilicity. The soluble portion of the superabsorbent imposes a negative effect on water absorption, which is caused by both of the amount of sodium bicarbonate and excessive amount of highly reactive crosslinking agent of N, N'-methylene bisacrylamide.

[1] Zhang, L.S.; Wang, Y-Q.; Farrar, D.; *European Patent 0 356 242 A2*, 1990.

[2] Lind, E.J.; *US Patent 5 118 719*, 1992.

[3] Ishizaki, K.; Obara, H.; Harada, N.; Motono, Y.; Miyake, K.; *US Patent 5 985 944,* 1999.

[4] ME 254e Lutrol F127 (Poloxamer 407) Technical Information, **1999**. BASF Aktiengesellschaft, Ludwigshafen, Germany, pp. 3–10.

[5] Buchholz, F. L.; *Preparation Methods of Superabsorbent Polyacrylates* in "*Superabsorbent Polymers, Science and Technology*" (Buchholz, F.L.; Peppas, N.A. Eds). ACS Monograph # 573, American Chemical Society, Washington D.C., **1994**, pp. 32-33.

[6] Zhou, W-J.; Yao, K-J.; Kurt, M. J. **1996**. *J. Appl. Polym. Sci*. 62, 911– 915.

[7] Andrews, R. J.; Grulke, E.A. *Glass Transition Temperature of Polymers*, in "*Polymer Handbook*"; Brandrup, J.; Immergut, E. H. Eds., 4th ed., John Wiley & Sons, New York, **1999**, p. VI/201.

[8] Mizutani, Y. *J. Appl. Polym*. Sci. **1996**, 61, 735-739.

[9] Cutie', S. S.; Smith, P.B.; Reim, R. E.; Graham, A.T. *Analysis and Characterization of Superabsorbent Polymers* in "*Modern Superabsorbent Polymer Technology*" (Buchholz, F.L.; Graham, A.T. Eds.), Wiley-VCH, New York, **1998**, p. 155.

[10] Flory, P.J. Principles of Polymer Chemistry, Cornell University Press, Ithaca, **1953**, p. 589.

Formation of a Charge Distribution within Ionic Gels by Immobilized Enzyme Reaction: Experimental Observations and Mathematical Simulations

*Kazuyoshi Ogawa, Etsuo Kokufuta**

Institute of Applied Biochemistry, University of Tsukuba, Tsukuba, Ibaraki 305-8572, Japan

Summary: Thermally responsive cationic gels with immobilized urease, in the shape of a small cylinder with diameter 290 - 640 μm, were prepared via gelation of an aqueous monomer solution containing the enzyme. We used N-isopropylacrylamide and N-vinylimidazole as a thermo-sensitive and a pH-sensitive monomer, respectively. Diameters at different positions of the cylinder were microscopically measured in a cell through which substrate solution (pH 4; 35 °C) was passed at a constant flow rate; thus, both substrate concentration and pH at the gel surface were maintained at a constant level. It was found that the gel undergoes a shrinking change due to an enzymatically induced increase in pH within the gel phase. There was a marked position dependence of the shrinking degree; the diameter at the center of the cylinder was smaller than those at the top and at the bottom, but the diameters at the top and bottom were identical with each other. This trend was observed at immobilized enzyme concentrations < 1 mg/mL, even after the establishment of swelling equilibrium. By mathematical simulations with a reaction-diffusion model, these results were understood in connection with a charge distribution which is formed as a result of an enzymatically generated pH gradient within the gel phase.

Introduction

The swelling of gels, such as that observed when rubber is placed in benzene, has long been a well-known phenomenon. As early as 1948, Vermaas and Hermans[1] studied the behavior of lightly crosslinked polymer chain networks with ionic charges, i.e., polyelectrolyte gels. Flory[2] and Katchalsky[3], in collaborations with their colleagues, also made great contributions to theories and experiments in the fields of ionic and neutral gels. In early studies by these pioneers, however, little attention was paid to the existence of a critical endpoint in the phase equilibria, although in 1968 Dusek and Patterson[4] published a theoretical paper on this subject. The earliest report of the phase transition in a gel was in 1978 by Tanaka[5], who discovered a discontinuous volume collapse of poly(acrylamide) gels in acetone-water mixtures when varying the temperature or the composition of the mixture. This phenomenon is now called "volume-phase transition" (or simply "phase transition") of gels and is observed in many gel systems consisting of synthetic and natural polymers. The phase transition is

 DOI: 10.1002/masy.200450321

accompanied by reversible, discontinuous (or in some cases, continuous) volume changes, often as large as several hundred times, in response to small variations in the solution conditions surrounding a gel. Variables that trigger the transition include solvent composition, pH, ion concentration, temperature, and small electric fields.

When limiting to the phase transition in a polyelectrolyte gel based on N-isopropylacrylamide (NIPA) which is one of the most extensively studied gels, it is the very well known experimental fact that an increase in the charge density causes a rise in both the transition temperature and the volume at the transition. This was, however, obtained by uses of gel samples composed of crosslinked NIPA polymer networks in which ionic monomers are randomly introduced. In the case of NIPA-based polyelectrolyte gels with inhomogeneous charge distributions,[6-8] we have observed a swelling behavior which is dramatically different from those of ionic NIPA gels with the homogeneous charge distribution. Then, the results were discussed in terms of the role of ionic charges in the swelling of polyelectrolyte gels at the molecular level. Our idea for the formation of a charge inhomogeneity were based on (i) the binding of ionic surfactants to lightly crosslinked NIPA polymer networks,[6],[8] (ii) the physical entrapment of polyions within the NIPA polymer network.[7],[8] and (iii) the use of an immobilized enzyme reaction.[9] In the enzymatic approach, a pH gradient generated as a result of enzyme reaction may establish an inhomogeneous distribution of charges within the NIPA gel with randomly copolymerized ionic monomers. This paper reports an expansion of our previous study[9] on charge distribution formation by the immobilized enzyme reaction.

Morphology

Thermally responsive cationic gels with immobilized urease, in the shape of a small cylinder, were prepared via gelation of an aqueous monomer solution with the following compositions (in mg per mL pure water): NIPA (thermo-sensitive monomer), 71.3; N-vinylimidazole (VI; pH-sensitive ionic monomer), 6.59; N,N'-methylenebisacrylamide (crosslinker), 1.33; and urease (Jack bean source), 0 - 8.0. The gelation was performed in a test tube into which capillaries with different inner diameters (290 - 637 μm) had been inserted. All the samples were taken out of the capillaries, purified by repeated swelling and shrinking procedures, cut into cylinders with a diameter:height ratio = ca. 1:4, and then stored at 3 °C before use.

The overall charge (0.775±0.09 mmol/g dry polymer) of the gel samples was determined by potentiometric titration at ionic strength 0.015 (NaCl) and at 35 °C. The titration for gels was carried out in the same way as described in our previous papers.[10],[11] The gel samples (of a fine grind) with 0 and 8 mg/mL of the immobilized enzyme were taken out of the test tubes from which the capillaries had been drawn out, purified as described above, and converted into the salt form with HCl.

Diameters at different positions on the height of a gel cylinder were microscopically measured at 35 °C and at suitable time intervals in a flow (1 mL/min) of a substrate solution; i.e., maleate buffer (pH 4 and ionic strength 0.015) containing 1 mM urea as the substrate. The details of apparatus and measuring procedures have been described in full in our previous paper.[11]

Results and Discussion

Cationic VI/NIPA gels with immobilized urease undergo a shrinking change at 35 °C in response to the ambient pH change from 4 to 6. Such a shrinking change was also observed at 35 °C in a flow of a substrate solution, even when its pH was maintained at 4.0 throughout the experimental period. Thus, the shrinking change in the substrate solution may be related to an enzymatically induced increase in pH within the gel phase. In the enzymatic gel shrinking, however, there is a marked position dependence of the shrinking degree; the diameter at the center of the cylinder is smaller than those at the top and at the bottom, but the diameters at the top and bottom are identical with each other. This trend was observed during the period of shrinking of gel, as well as after the establishment of swelling equilibrium of gel, at immobilized enzyme concentrations lower than 1 mg/mL. These results can be qualitatively understood in terms of a pH gradient which is enzymatically formed within the gel phase.

To look at an aspect of the enzymatic formation of pH gradients within the gel phase, we have conducted mathematical simulations in the previous study.[9] Then, the central part of our gels was taken as an infinitely long circular cylinder of radius R, in which diffusion proceeds in parallel with reaction. In other words, we neglected the diffusion of substrate molecules through circular area at the top and at the bottom because of consumption of the substrate by enzyme reaction. Thus, we used the following parabolic partial differential equations to simulate the transient change in the concentrations of substrate (S) and product (X) in the gel phase:

$$\frac{\partial[S]}{\partial t} = -v + \frac{D_S}{r}\frac{\partial}{\partial r}\left(r\frac{\partial[S]}{\partial r}\right) \tag{1}$$

$$\frac{\partial[X]}{\partial t} = 2v + \frac{D_X}{r}\frac{\partial}{\partial r}\left(r\frac{\partial[X]}{\partial r}\right) \tag{2}$$

Here D_S and D_X denote diffusion coefficients of S and X, respectively. Moreover, the brackets signify the molar concentration of the corresponding species. For the reaction rate (v), the following equations were employed;

$$v = \frac{k_{cat}[E][S]}{K_m C_1 + [S]C_2} \tag{3}$$

$$C_1 = 1 + 10^a[H^+] + \frac{1}{10^b[H^+]} \tag{4}$$

$$C_2 = 1 + 10^c[H^+] + \frac{1}{10^d[H^+]} \tag{5}$$

where $[E]$ and $[H^+]$ respectively denote molar concentrations of enzymes (urease) and protons, k_{cat} (1.50×10^3 s^{-1}) is reaction rate constant, and K_m (2.82×10^{-3} M) is Michaelis constant. Constants a - d in Eqs (4) and (5) were experimentally determined:[9] a = 5.92, b = 6.86, c = 5.38, and d = 7.12. Also determined by experiments was $[X]$:[9]

$$[X] = 1.05 \times 10^{-7}[H^+]^{-0.731} \tag{6}$$

For Eqs (1) and (2), the accompanying initial and boundary conditions are written as

$$[S] = 0 \text{ and } [X] = 0 \text{ at } t = 0 \tag{7}$$

$$\frac{\partial[S]}{\partial r} = 0 \text{ and } \frac{\partial[X]}{\partial r} = 0 \text{ at } r = 0 \text{ and } t > 0 \tag{8}$$

$$[S] = [S]_o \text{ and } [X] = 0 \text{ at } r = R \text{ and } t > 0 \tag{9}$$

Introducing the dimensionless variables ($U = [S]/[S]_o$, $V = [X]/[S]_o$, $\theta = D_x t / R^2$, and $\eta = r / R$, where $[S]_o$ = [S] at r = R) and the dimensionless parameter ($\phi = k_{cat}R^2 / D_x$) into Eqs (1), (2) and (3), we obtain

$$\frac{\partial U}{\partial \theta} = -v' + \frac{1}{\eta}\frac{D_s}{D_x}\frac{\partial U}{\partial \eta} + \frac{D_s}{D_x}\frac{\partial^2 U}{\partial \eta^2} \tag{10}$$

$$\frac{\partial V}{\partial \theta} = 2v' + \frac{1}{\eta}\frac{\partial V}{\partial \eta} + \frac{\partial^2 V}{\partial \eta^2} \tag{11}$$

$$v' = \frac{\phi[E]U}{K_m C_1 + U[S]_o C_2} \tag{12}$$

and the initial and boundary conditions for Eqs (10) and (11) are written as

$$U = 0 \text{ and } V = 0 \text{ at } \theta = 0 \tag{13}$$

$$\frac{\partial U}{\partial \eta} = 0 \text{ and } \frac{\partial V}{\partial \eta} = 0 \text{ at } \eta = 0 \text{ and } \theta > 0 \tag{14}$$

$$U = 1 \text{ and } V = 0 \text{ at } \eta = 1 \text{ and } \theta > 0 \tag{15}$$

We may solve Eqs (10) - (15) numerically with the aid of a computer using a finite difference method. Then, we approximated that $D_S \sim D_X = 2.80 \times 10^{-6} (cm^2 \cdot s^{-1})$, the value of which is identical with diffusion coefficient of ammonium ion.[9] This approximation would be enough under an assumption that we may neglect effect of the network charges on diffusion of ionic solutes such as NH_4^+ and CO_3^{2-}. As a result, a set of distribution curves of [S] and pH against r/R was obtained as a function of enzyme concentration under conditions where the immobilized enzyme reaction is in a steady state ($\partial U / \partial \theta = \partial V / \partial \theta \sim 0$) and the gel is in an equilibrium state (see Figure 1). It is found that an increase in ϕ[E] (therefore, increase in the immobilized enzyme concentration) enhances the sharpness of a fall in [S] as well as a rise in pH at the range from the surface to the center; therefore, we may point out the following significant prospects: (i) The pH gradient within the gel disappears as the immobilized enzyme concentration approaches infinity, because under such a condition the enzyme reaction takes place only at the liquid-gel interface. (ii) The shrinking due to a broad distribution of pH, as can be seen at $\phi[E] < 0.1$, becomes more remarkable in a gel with a large radius rather than that with a small radius; in other words, a local pH-dependent change in swelling ratio amplifies with increasing R.

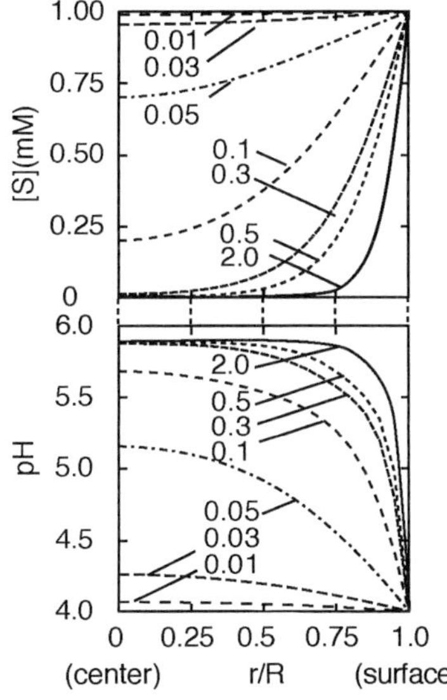

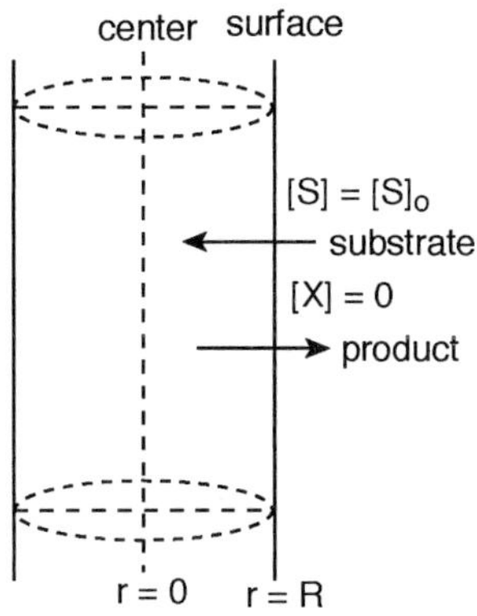

Figure 1. Simulated distribution curves of substrate concentration ([S]) and pH within the gel phase as a function of ϕ[E] at $\theta = 10$. The gel was taken as an infinitely long circular cylinder of radius R. The values of ϕ[E] were given in each figure.

It would be significant to estimate a charge distribution within the gel, which should be formed as a result of an enzymatically generated pH gradient. For this purpose, we must study the pH dependence of the degree of ionization (protonation degree in this case) for the gel. At the present stage, unfortunately, we don't have a good theoretical approach for determining the ionization degree (α) of our polyelectrolyte gels as a function of pH. Thus, we attempted to experimentally estimate α by potentiometric titration of enzyme-free VI/NIPA gel (salt form) with NaOH. Figure 2 shows the pH dependence of α for the gel at ionic strength 0.015 (NaCl) and at 35 °C. It is found that the titration curve can be given by the Henderson-Hasselbalch equation;

$$\mathrm{pH} = \mathrm{pK} + n_i \log\{(1-\alpha)/\alpha\}, \tag{16}$$

where pK (~5.57) represents the ionization constant, and n_i (~1.13) is the empirical constant which denotes the magnitude of the interaction between protons and polyion as the deviation of n_i from unity. Consequently, we may convert the pH distribution curves in Figure 1 into the charge distribution curves for the gel in an equilibrium state (see Figure 3). The charge distribution curves show the same prospects (i) and (ii) as mentioned in the above paragraph.

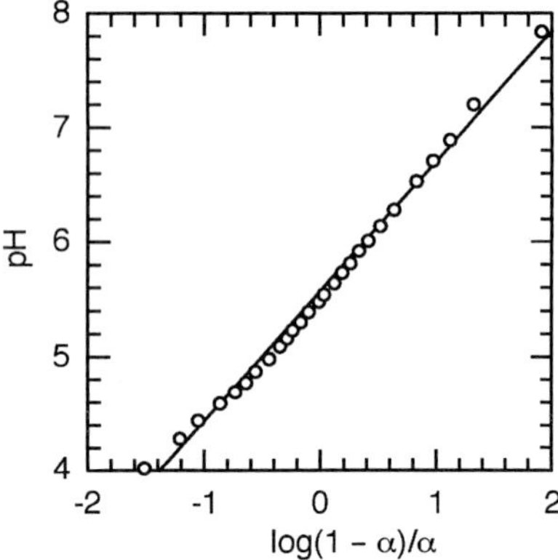

Figure 2. Potentiometric titration curve of enzyme-free VI/NIPA gel in 0.015 M NaCl at 35 °C.

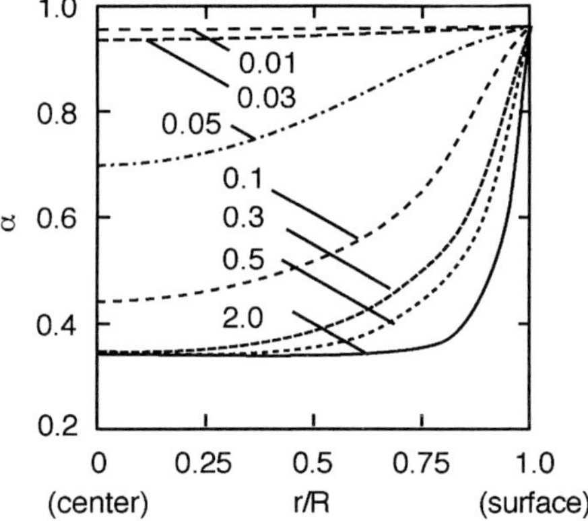

Figure 3. Simulated charge distribution curves within the gel phase as a function of $\phi[E]$ at θ = 10. The values of $\phi[E]$ were shown in the figure.

Now, we attempt to compare the results by simulations and by swelling measurements. For this, we adopted the following empirical equation which was obtained in our previous study[9] to express equilibrium swelling ratio (R_s^e) as a function of pH;

$$R_s^e = 0.068(\mathrm{pH})^3 - 1.11(\mathrm{pH})^2 + 5.55(\mathrm{pH}) - 6.93 \qquad (17)$$

where $R_s^e \sim d_e / d_o$ (d_e and d_o are gel diameters in an equilibrium state and in the preparation, respectively). By combination of Eqs (16) and (17), we may obtain:

$$R_s^e = 0.099f(\alpha)^3 + 0.033f(\alpha)^2 - 0.56f(\alpha) + 1.29 \qquad (18)$$

where $f(\alpha) = \log\{(1-\alpha)/\alpha\}$. Then, we assumed that the overall swelling ratio ($\langle R_s^e \rangle$) of an ionic gel with a charge distribution may be calculated by;

$$\langle R_s^e \rangle = \frac{1}{n}\sum_{i=1}^{n} R_{s,i}^e \qquad (19)$$

where $R_{s,i}^e$ denotes a local swelling ratio at an arbitrary radial division i (i = 1, 2⋯n) and may be determined from the α vs r/R curve using Eq (18), and n is a total of radial divisions. In our simulations we divided dimensionless radius r/R into 20 because at n > 15 there was little difference in the calculated $\langle R_s^e \rangle$ values.

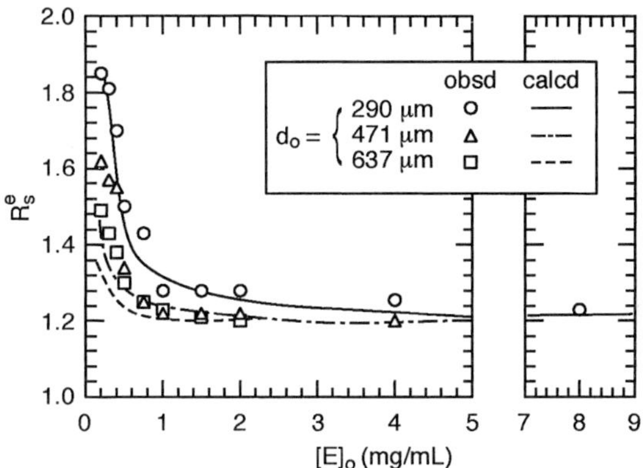

Figure 4. Comparison of equilibrium swelling ratios (R_s^e) obtained from experiments and simulations as a function of immobilized enzyme concentration ($[E]_o$). In simulations, [E] was calculated by $[E] = [E]_o/\{M_w(d_e/d_o)^3\}$, where M_w is molecular weight (480,000) of urease.

In Figure 4 , the equilibrium swelling ratios obtained by experiments using gels with different diameters (d_o) as well as with different amounts ($[E]_o$) of immobilized urease in their preparations were compared with the simulated results ($\langle R_s^e \rangle$) at θ = 10 (i.e., in a steady state). There is a good agreement between the experimental and simulated results. In particular, the following important features are observed in both simulations and experiments. (i) The swelling ratios of gels with different d_0 values decrease and approach a fixed value (R_s^e~1.2) with increasing $[E]_0$, because the sharpness of the r/R vs [S] and r/R vs pH curves from the surface to the center is enhanced. (ii) At a constant $[E]_0$ and at $[E]_0 < 1$ mg/mL, the swelling ratio decreases with an increase in d_0, due to an amplification of the charge distribution. As a result, it has become apparent that the charge distribution formed due to an enzymatically generated pH gradient within the gel phase is the chief factor from which the gel volume is uniquely determined.

There is a lot of studies reporting the generation of a concentration gradient of substrates and of products within the gel phase during the course of immobilized enzyme reactions (e.g., see review articles[12],[13]). Our experiments and mathematical simulations have demonstrated that a charge distribution is formed within the ionic gel due to the concentration gradient of ammonium carbonate as the product. This charge distribution should be a cause of the observed position dependence of the shrinking degree. Therefore, it is reasonable to consider that in our gel system the swelling force results from "electrostatic repulsion" among the fixed cationic charges, the amount of which should vary along the pH gradient formed between the surface and the center of the gel.

Acknowledgments: This work was supported in part by a Grant-in-Aid for Scientific Research to E.K. from the Ministry of Education, Japan (# 08558092).

[1] D. Vermaas, J. J. Hermans, *Recueil*, **1948**; *67*, 983
[2] P. J. Flory, "*Principle of Polymer Chemistry*"; Cornell University Press, New York 1953, pp. 576-593.
[3] A. Katchalsky, I. Michaeli, *J. Polym. Sci.* **1955**, *15*, 69.
[4] K. Dusek , D. Patterson, *J. Polym. Sci., Part A-2*, **1968**, *6*, 1209.
[5] T. Tanaka, *Phys. Rev. Lett.* **1978**, *12*, 820.
[6] E. Kokufuta, H. Suzuki, D. Sakamoto, *Langmuir* **1997**, *13,* 2627.
[7] E. Kokufuta, B. Wang, R. Yoshida, A. R. Khokhlov, M. Hirata, *Macromolecules* **1998**, *31*, 6878.
[8] K. Ogawa, Y. Ogawa, E. Kokufuta, *Colloids Surf. A* **2002**, *209*, 267.
[9] K. Ogawa, E. Kokufuta, *Langmuir* **2002**, *18*, 5661.
[10] H. Suzuki, B. Wang, R. Yoshida, E. Kokufuta, *Langmuir* **1999**, *15*, 4283.
[11] E. Kokufuta, H. Suzuki, R. Yoshida, K. Yamada, M. Hirata, F. Kaneko, *Langmuir* **1998**, *14*, 795.
[12] E. Kokufuta, *Prog. Polym. Sci.*, **1992**, *17*, 647.
[13] E. Kokufuta, in: *"The Polymeric Materials Encyclopedia - Synthesis, Properties and Applications"*, J.C. Salamone, Ed., CRC Press, New York 1996, Vol. 4, F-G, pp. 2615-2621.

Novel Swelling/Shrinking Behaviors of Glucose-Binding Hydrogels and Their Potential Use in a Microfluidic Insulin Delivery System

Ronald A. Siegel,[*1,2] Yuandong Gu,[1] Antonio Baldi,[3] Babak Ziaie*[3]

Departments of [1]Pharmaceutics, [2]Biomedical Engineering, and [3]Electrical and Computer Engineering, University of Minnesota, Minneapolis, MN 55455, USA
E-mail: siege017@umn.edu

Summary: Hydrogels based on lightly crosslinked poly(methacrylamidophenylboronic acid-co-acylamide) swell monototically at pH 7.4 with exposure to increasing concentrations of glucose. At pH 10 these hydrogels first shrink and then reswell with increasing glucose concentration. A mechanism for these disctinct behaviors is proposed and preliminary results validating this mechanism are presented. A simple microfabricated glucose-sensitive valve based on this hydrogel is described.

Introduction

In recent decades, the swelling/shrinking response of hydrogels in response to physical or chemical stimuli has caught the notice of researchers in drug delivery, because swelling changes can be converted to changes in either permeation rate of drugs, or changes in rate of delivery resulting from mechanical forces generated by the hydrogel. Hydrogels that swell or shrink in response to changes in temperature,[1,2] pH,[3,4] glucose concentration,[5-7] or in the presence of an electrical current,[8] have been considered for modulation of drug release rates.

Glucose-responsive hydrogel-based drug delivery systems have received considerable attention due to their potential application in the treatment of diabetes. Two fundamental mechanisms have been investigated. Early work focused on a mechanism in which acidic hydrogen ions (H^+), enzymatically extracted from glucose, binds to sidechains of a hydrogel.[9,10] For polybase gels, such binding increases the hydrogel's charge, causing the gel to swell. For polyacid gels, binding of H^+ has the opposite effects. Usually, a polybase gel is used since an increase in swelling, due ultimately to an increase in glucose concentration, leads to increased insulin release, which is the desired response.

 DOI: 10.1002/masy.200450322

In the second, more recently studied mechanism, glucose binds reversibly to the hydrogel, causing changes in the hydrogel's permeability. For example, sol-gel transitions mediated by glucose have been used to switch permeation of insulin on and off.[5,6] Also, hydrogels containing derivatives of phenylboronic acid (PBA) have been shown to bind glucose and alter their charge, leading to changes in gel swelling and permeability to insulin.[7]

In this contribution, we report studies of a class of PBA-containing hydrogels which can either swell or shrink with increasing glucose level, and suggest a means by which such hydrogels can be incorporated into microfabricated valves that can rapidly gate the flow of insulin solutions in response to changes in blood glucose level.

Swelling and Shrinking of PBA-Containing Hydrogels

Background. As illustrated in Figure 1, PBA moieties form tetrahedral boronate$^-$ ions in the presence of free OH^- ions, and behave as Lewis acids. Molecules containing planar diols, such as glucose, reversibly condense with two of the hydroxyls on the boronate$^-$ and stabilize the charged form of the Lewis acid. This leads to an apparent shift in the Lewis acid's pKa in the acidic direction, according to

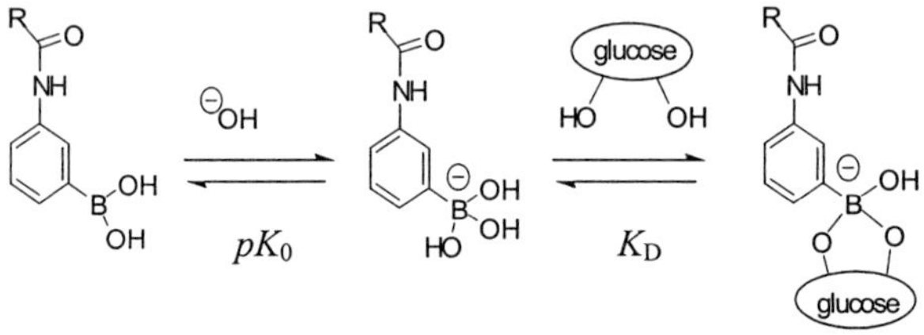

Figure 1. Scheme by which binding of glucose (or other planar diol) stabilizes the charged form of the MBPA moiety. After Aoki et al.[11]

$pK_{app} = pK_0 - \log(1+[S]/K_D)$, where pK_0 is the Lewis acid constant of the PBA moiety, $[S]$ is the concentration of diol (glucose in Fig. 1), K_D is the dissociation constant of the diol-PBA complex (actually the equilibrium constant for the condensation of diol and tetrahedral boronate ion$^-$), and pK_{app} is the apparent acid constant of PBA moiety in the presence of diol.

Methods. We prepared hydrogels consisting of 20 mol% 3-methacrylamidophenylboronic acid (MPBA: see Fig. 1), 80 mol% acrylamide (AAm), with methylene bisacrylamide as crosslinker, mixed together to a total monomer concentration of 200 mg/ml in 1N NaOH solution. Polymerization was initiated by the ammonium persulfate/TEMED couple and carried out at room temperature.

Equilibrium swelling measurements were performed at room temperature in degassed 0.15 M PBS set to particular pH values, and in the presence of glucose or fructose. Uniaxial static compression measurements were carried out on equilibrated hydrogels between the parallel plates of a rheometer.

Results. Figure 2 shows the joint dependence of swelling on pH and fructose concentration. With increasing fructose concentration, the swelling curve shifts in the acid direction. Comparing the midpoints of the swelling curves with the above equation relating apparent pK_A to concentration of diol, the observed shifts are well described by assuming that pK_0=8.86 and K_D=230 μM, which are the literature values for the acidity constant of free MPBA and for the binding constant of fructose to free benzene boronic acid (a close relative of MPBA), respectively.[11]

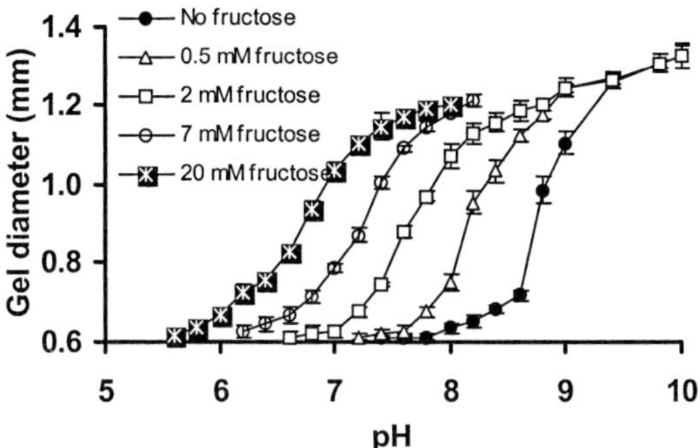

Figure 2. Swelling of p(MPBA-co-AAM) hydrogels as a joint function of pH and fructose concentration.

Figure 3 compares swelling of the poly(MPBA-co-AAm) hydrogels in glucose solutions at pH 7.4 and pH 10, and fructose solutions at pH 10. At pH 7.4, swelling increases monotonically with glucose concentration, as expected from the mechanism in Figure 1 in which glucose stabilizes the charge on the hydrogel. At pH 10, however, the hydrogel initial shrinks as glucose is added, and then reswells. The gel remains swollen

Figure 3. Swelling of p(MPBA-co-AAm) hydrogels in glucose and fructose solutions at pH 7.4 and 10.

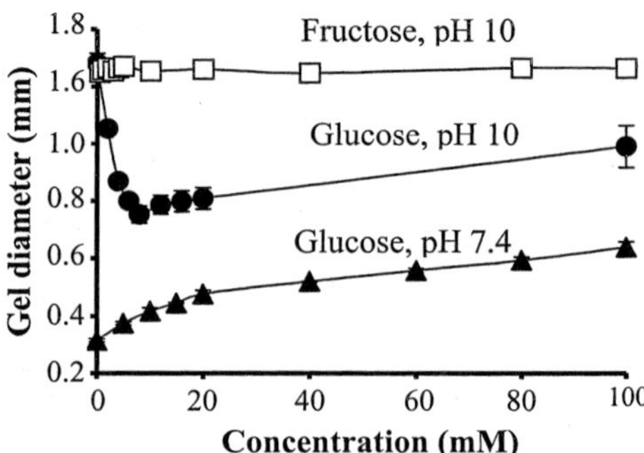

at all tested concentrations of fructose, however.

At pH 10 the hydrogel is almost completely charged, so addition of glucose cannot increase charge and hence swelling. The initial shrinkage at low glucose concentrations at pH 10 is explained by noticing that the glucose molecule contains at least two pairs of planar diols. It therefore can bind simultaneously with two MPBA moieties attached to separate chains, thus forming new, albeit reversible crosslinks in the hydrogel. As glucose concentration increases further, however, free glucose molecules compete with the crosslinking glucose molecules for MPBA sites, causing the crosslinks to be broken. This mechanism is illustrated in Figure 4.

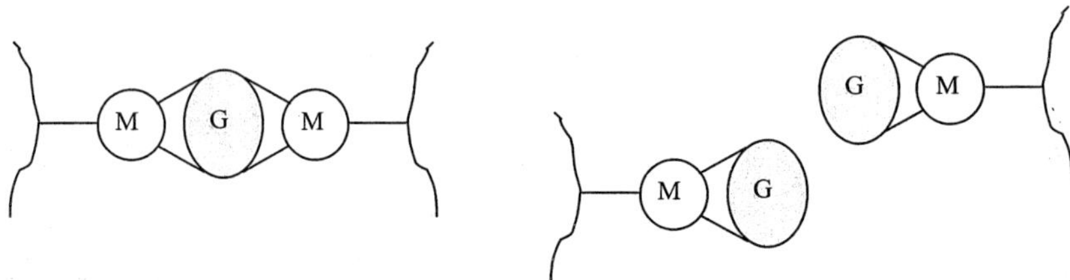

Figure 4. Left: formation of crosslinks between MPBA moieties (M) by glucose (G) at low concentrations. Right: breaking of these crosslinks by competition at higher glucose concentrations.

Confirming evidence for this mechanism is obtained from shear moduli measured by compression of the hydrogels after exposure to varying glucose concentrations at pH 10. As shown in Figure 5, the modulus first rises sharply, and then falls with increasing glucose concentration. The peak modulus is observed at the same concentration as the

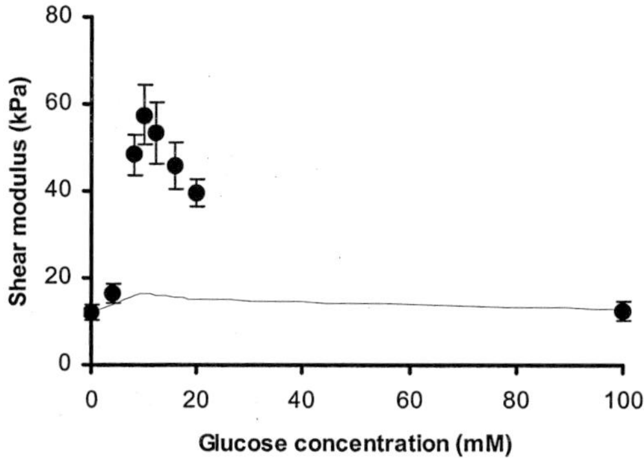

Figure 5. Symbols: Shear moduli of hydrogels determined by uniaxial compression measurements. Solid line: Values of shear moduli predicted based on change in hydrogel volume alone.

minimum swelling observed in Fig. 3. The solid line in Fig. 5 shows the trend in the modulus that would be expected from the change in volume alone, based on the $\phi^{1/3}$ scaling law,[12] where ϕ is the volume fraction of polymer. Since this trend greatly underpredicts the data, extra crosslinks must be formed as the gel contracts.

Fructose, unlike glucose, does not cause the gel to shrink. This is consistent with literature in which glucose is shown to form 1:2 complexes with free PBA molecules, while fructose only forms 1:1 complexes.[13]

Microfluidic Valve for Glucose-Sensitive Insulin Gating

Swelling and shrinking of the poly(MPBA-co-AAm) hydrogel, when the gel is inside a valve, can be used to switch on and off the flow of insulin solution when glucose

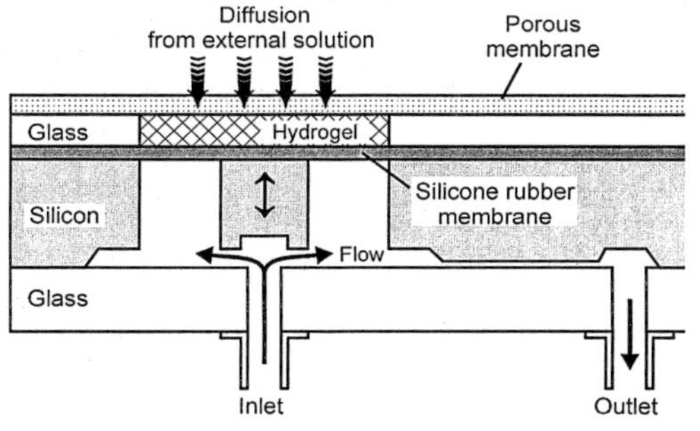

Figure 6. Schematic of hydrogel-based microvalve.[14] (© 2003 IEEE).

concentration changes. A schematic for a microfabricated valve is shown in Figure 6. The hydrogel is sandwiched between a rigid, porous membrane and a flexible, silcone rubber diaphragm, and the latter is connected to an embossment, which can open and shut a microchannel. Glucose diffuses through the porous membrane and affects hydrogel swelling as described above. When the hydrogel swells, it pushes down on the membrane, causing the embossment to close the inlet to a microchannel and stopping fluid flow through the channel. Flow is re-established when the gel shrinks due to elastic retraction of the diaphragm. Figure 7a shows the inside of the microvalve looking from the bottom, while Figure 7b is the bottom view of the microvalve with the glass plate and flow channel inlet and outlet attached.

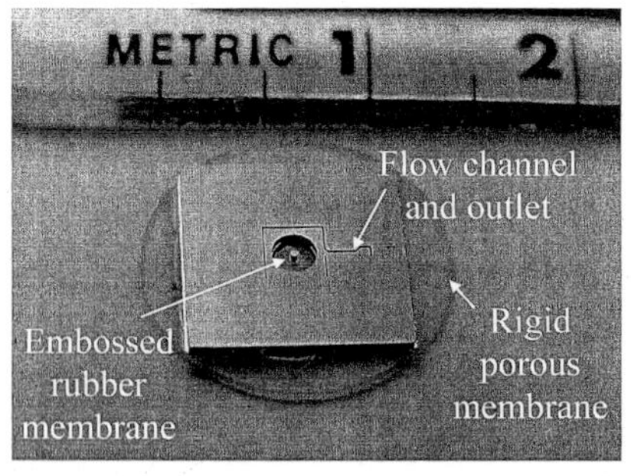

a)

b)

Figure 7. a) Microvalve as seen from below, with interior chamber, flexible rubber membrane and embossment exposed. b) Microvalve as seen from below with glass underplate, inlet and outlet attached.[14] (© 2003 IEEE).

In order to test the function of the microvalve, the assembly was placed in alternating baths containing PBS or PBS with 20 mM glucose. In both cases pH was set to 7.4 and the ionic strength was 0.155 mM, and the test was performed at room temperature. The inlet to the microflow channel was connected to a gravity-fed fluid line with pressure 6 kPa, and the outlet was connected to a fluid line. Fluid velocity was monitored by observing the movement of bubbles injected into the outflow line.

The results of this test are shown in Figure 8. In the absence of glucose, the hydrogel is relatively unswollen, and the microflow channel is open and fluid flow is seen. When glucose is added, however, the hydrogel expands, causing the embossment to close off the inlet to the channel, and shutting off fluid flow. The response time is close to 20 min.[14]

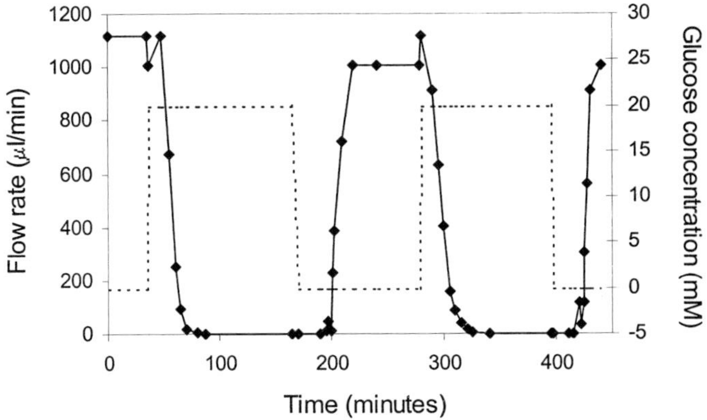

Figure 8. Response of microvalve to changes in glucose concentration between 0 mM and 20 mM PBS solutions at pH 7.4, ionic strength 0.155 mM, and at room temperature. Fluid pressure at inlet is 6 kPa relative to outlet.[14] (© 2003 IEEE).

Discussion

The microvalve prototype just discussed is suggestive of a system for automatic control of insulin delivery based on changes in blood glucose concentration. Such a valve might be placed at the tip of a catheter from a pressurized insulin reservoir, for example. The reservoir could be a small capsule worn outside the body, and the catheter would then cross the skin, leading to the microvalve, which sits in interstitial fluids whose glucose concentrations reflect, to an adequate approximation, the concentration of glucose in blood. With proper design, the reservoir could be replaced every few days, while the microvalve would be implanted indefinitely. Obviously, several improvements are needed for this scenario to be realized.

While the response time of the microvalve (20 min) is much faster than those seen for other polymer-based glucose-responsive systems, this time needs to be shortened to about 5 minutes in order for the microvalve to be therapeutically useful. Generally , the simplest means to speed up response is to miniaturize even further, for example by making the hydrogel, the rigid porous membrane, and the gap between the boss and the microflow channel inlet thinner. Thus far, the limits of miniaturization have not been reached, and work is being pursued in that direction.

A more fundamental problem with the microvalve lies in its polarity. Notice that the present microvalve design permits flow at low glucose concentrations, but shuts off flow at high glucose concentrations. When insulin is involved, the opposite response is desired. One might change the mechanical design of the microvalve. However, polarity can also be reversed using a hydrogel that shrinks with increasing glucose concentration. As shown in Figure 3, when pH is above the pK_0 of the pendant PBA moieties, such shrinkage is observed due to crosslinking of the hydrogel by glucose. While pK_0 for MPBA is 8.86, this characteristic can be lowered by incorporating electron withdrawing groups into the phenyl ring[15] or by placing positively charged groups proximal to the boronate ion.[16,17] Both the mechanical and chemical approaches for reversing polarity are presently under investigation.

Acknowledgments

This work was funded by grants from the US Army Medical Research and Materiel Command and the Drug Delivery Center at the University of Minnesota.

[1] L. C. Dong and A. S. Hoffman, *J. Controlled Release* **1990**, *13*, 21.
[2] T. Okano, Y. H. Bae, H. Jacobs, and S. W. Kim, *J. Controlled. Rel.* **1990**, *11*, 255.
[3] R. A. Siegel, M. Falamarzian, B. A. Firestone, and B. C. Moxley, *J. Controlled Release* **1988**, *8*, 179.
[4] A. Khare and N. A. Peppas, *J. Biomater. Sci. Polym. Edn.* **1993**, *4*, 275.
[5] M. J. Taylor, S. Tanna, P. M. Taylor, and G. Adams, *J. Drug Targeting* **1995**, *3*, 209.
[6] A. A. Obaidat and K. Park, *Biomaterials* **1997**, *18*, 801.
[7] K. Kataoka, H. Miyazaki, M. Bunya, T. Okano, and Y. Sakurai, *J. Am. Chem. Soc.* **1998**, *120*, 12694.
[8] S. Murdan, *J. Controlled Release* **2003**, *92*, 1.
[9] J. Kost, T. A. Horbett, B. D. Ratner, and M. Singh, *J. Biomed. Mater. Res.* **1985**, *19*, 1117.
[10] K. Ishihara and K. Matsui, *J. Polym. Sci. Polym. Lett. Ed.* **1986**, *24*, 413.
[11] J. P. Lorand and J. O. Edwards, *J. Org. Chem.* **1959**, *24*, 709.
[12] F. Horkay and M. Zrinyi, *Macromolecules* **1982**, *15*, 1306.
[13] S. A. Barker, A. K. Chopra, B. W. Hatt, and P. J. Somers, *Carbohydrate Res.* **1973**, *26*, 33.
[14] A. Baldi, Y. Gu, P. Loftness, R. A. Siegel, and B. Ziaie, *J. Microelectromech. Sys.* **2003**, *12*, 613.
[15] B. B. J. Johnson, *Biochemistry* **1981**, *20*, 6103.
[16] I. Hisamitsu, K. Kataoka, T. Okano, and Y. Sakurai, *Pharmaceut. Res.* **1997**, *14*, 289.
[17] S. L. Wiskur, J. J. Lavigne, H. Ait-Haddou, V. Lynch, Y. H. Chiu, J. W. Canaray, and E. V. Anslyn, *Organ. Lett.* **2001**, *3*, 1311.

High Response Smart Gels: Synthesis and Application

Karl-Fr. Arndt,[*1] Thomas Schmidt,[1] Andreas Richter,[1] Dirk Kuckling*[2]

[1] Institut für Physikalische Chemie und Elektrochemie,
[2] Institut für Makromolekulare Chemie und Textilchemie
Technische Universität Dresden, 01062 Dresden, Germany
E-mail: karl-friedrich.arndt@chemie.tu-dresden.de

Summary: Stimuli-responsive hydrogels change their volume and elasticity in dependence on the properties of the liquid phase. The amount of solvent uptake is dependent upon the chemical nature of the gel and the nature of its environment, e.g. solvent composition, temperature, pH, etc. The swelling/deswelling process is time consuming. For many applications the response time must be short. According to Tanaka´s equation the rate of equilibration is inversely proportional to the square of gel size and proportional to the mutual (or cooperative) diffusion of the network and the solvent. D_{coop} is determined by the properties of the cross-linked polymer and can not be changed for a given chemical system. By decreasing the characteristic dimension of the gel, it is possible to reduce the time for equilibration. The synthesis, characterization of thin layers and small particles of T-sensitive polymers and their application in a micro-valve is described.

Introduction

Aqueous solutions of temperature-sensitive polymers like poly(*N*-isopropylacrylamide), and poly(vinyl methyl ether), show a lower critical solution temperature (LCST) at about 34 °C (PNIPAAm), or at 37 °C (PVME), respectively. Both polymers form hydrogels, which were thermo-sensitive similar to the polymer solution. They swell at temperatures below the LCST and shrink above this temperature.

Stimuli-responsive hydrogels that undergo abrupt changes in volume in response to external stimuli such as temperature, pH, concentration of organic solvent in an aqueous phase, have potential application in the creation of "smart" material systems.

Necessarily for the most conceivable applications, the response time must be short. To understand the factors, which determine the response time is a pre-condition for developing strategies for synthesizing gels with fast response. The volume phase transition is describable as a two step process. In a first step the stimuli, which induce the swelling/deswelling must penetrate the gel. We investigate the diffusion process of methanol in a gel by n.m.r.-imaging technique. From the measured time lag at a defined distance to sample surface the diffusion coefficient could be calculated. For methanol in the swollen gel (D_2O at 294 K) we get values from 1.8 10^{-5} cm²/s up to 3.9 10^{-5} cm²/s in dependence of methanol concentration (for D_2O in

 DOI: 10.1002/masy.200450323

water about 2.3 10^{-5} cm²/s). The influence of heating and cooling on the properties of the gel was detected by measurements of the mobility of the network chains (gel swollen in D_2O). From these experimental results, it was clarified, that the molecular motion of the network chains change at the volume phase transition. The rate of the first step seems to be simply dependent on the rate of heat transfer (thermal diffusivity 10^{-3} cm²/s), or on the diffusion of a low molecular weight liquid phase, respectively, which occur much faster than mass transfer. The equilibrium is reached well before any significant change in the swelling degree.[1] If conditions for the volume phase transition are meet, the mechanical properties of the gel change. The reason for this is the formation of a macro-network consisting of bundles of molecules (polymer rich phase) and large pores (polymer poor phase). After formation of the bundles of molecules (isochore process) the process of deswelling starts. The mechanical properties of gels undergoing a volume phase transition were studied by S. Hirotsu[2] (isobar gel) and M. Shibayama et al. (isochore gel).[3]

The kinetic of volume change (second step) is determined by diffusion processes, at least by the cooperative diffusion coefficient of the polymer chains (about 10^{-7} cm²/s). Since the diffusion time scales with the square of dimension, decreasing the characteristic dimension of the network will increase the swelling/deswelling rates dramatically.

The rate of response to the environmental changes may directly influence the system performance. With other words, the swelling/deswelling processes determine the characteristic time for a designed sensor-actuator system.

Different synthetic routes are described to improve the swelling rate or shrinking rate of sensitive hydrogels, i.e.

- formation of additional junction points in dependence of temperature

The formation of hydrophobic cluster in a comb grafted polymer gel results in additional junction points and an increase of cross-linking density.[4]

- porous gels by synthesis at temperatures higher than lcst

A gel body with porous microstructure is expected to give rise to faster volume change, since the effective diffusion distance can be controlled by the average distance between neighboring pores. Porous gels swell or shrink much faster (in order of magnitude) than any other responsive gels of comparable dimension. M. Suzuki and O. Hirasa synthesized PVME gels (cross-linking by γ-irradiation)[5] with a fine porous gel structure to obtain quick response gels. Typical dimensions of the pores are several μm. PNIPAAm was cross-linked in the phase separated state at T>LCST.[6]

- small particles (micro-spheres or nanogels)

Several papers dealing with the synthesis and potential applications of thermo-sensitive *N*-isopropylacrylamide-based micro-gels were published recently (see e.g. ref. [7]). Spherical hydrogel particles of NIPAAm can be formed by inverse suspension polymerization.[8] Another method for the preparation of hydrogel beads in the submicron region (0.2 – 1 μm) is the inverse emulsion polymerization.[9]

The goals of our work are

- Synthesis of gels with fast response in a wide range of dimensions.
  For different application we need large sheets, in mm-range, of gels with fast response.
  We deal with gels in μm-range, for example with thin layers of different pattern of sensitive gels. Recently we start with the synthesis of thermo-sensitive gel-particles in 100 nm-range.
- Application of these gels in chemo-mechanical valves.

Experimental

Synthesis

- Thin layers

Thin layers of sensitive polymers were prepared by photo-crosslinking of linear temperature- and pH-sensitive pre-polymers on a Silicon-wafer as supporting material.[10] By UV-irradiation through a mask it is possible to design different structures on the wafer, and patterned networks in μm-scales can be obtained. The resulting networks show temperature dependent swelling properties. In our experience gel dots of 250 x 250 μm^2 and smaller (thickness 7 - 15 μm) can be formed. If the cross-linking density is high enough, these dots are mechanical stable and can undergo a lot of swelling and shrinking processes without any destruction.

- Porous structure by irradiation cross-linking near or above the phase transition temperature

Irradiation with high-energy radiation (e-beam, γ-radiation) of a concentrated solution of the thermo-sensitive polymer poly(vinyl methyl ether) (PVME) is an interesting method to synthesize porous materials in the mm-range.[11] For irradiation, solutions of different concentrations were prepared by dilution of the high concentrated PVME solution with bi-distilled water. Oxygen can interfere with a free-radical polymerization. Therefore it was necessary to remove all of it by degassing the homogenized polymer solution before starting the irradiation. After degassing, the prepared solutions were poured into petri dishes (e-beam) or poly(ethylene) flasks (γ-rays) and irradiated. Purging the solution with argon and sealing with PARAFILM® helps to reduce the rate of oxygen uptake in the non-crosslinked polymer solution.

During e-beam irradiation the temperature of the solution increases due to the absorption of energy. The temperature of volume-phase-separation is crossed and the phase separated structure is fixed by cross-linking.

- Spherical particles by irradiation of a diluted polymer solution

The non-crosslinked PVME tends to form aggregates in water even at low temperatures and low concentrations. Heating of the diluted polymer solution induces the phase separation. The polymer particles formed by the phase transition can be cross-linked by e-beam irradiation. [12] It is important to mentioned, that surfactants or any other reagents are not necessary for the synthesis.

Dynamic light scattering measurements were performed with a DLS 700 (Otsuka, Japan) at $\theta = 90°$ and a concentration $c_B = 1.0$ g/l. An apparent (no extrapolation to $\theta \rightarrow 0$ and $c_B \rightarrow 0$) value of the hydrodynamic radius $R_{h,app}$ was calculated by using the Stokes-equation.

Design of valves

- Particle based actuator

Particles of "smart polymers" can be used as an actuator for autonomous flow control. Smart gels connect an actuator, which can generate mechanical work with a function of sensing.

In the tested particle-based valve the "gel-actuator" is placed directly in the liquid flow. Spherically particles with a typical dimension of (0.40 ± 0.05) mm were put into a cylinder, which consists of an O-ring (stainless steel, diameter: 5 mm; thickness: 2 mm), and on both ends of a steel net (mesh-size 60 μm). The liquid flows through the space between the polymer particles. A decrease in a diameter leads to an increase in the flow rate at constant pressure, or requires a greater applied pressure to achieve the same flow rate. For a faster change of the properties of the liquid in the valve-body, a circulation circuit (bypass at the inlet) is used (Figure 1). In our first experiments we use these gels for a temperature sensitive chemo-mechanical valve. The temperature alters from 24.5 °C to 35°C and vice versa with a non-linear temperature gradient (Figure 2). The fastest times were determined for the open process at about 25 s and for the close process at about 40 s.

The gel actuator can be filled with gel particles of different sensitivity. The regulation of the flow is possible even in dependence on pH-value and concentration of organic component in water.

For further details see ref. [13].

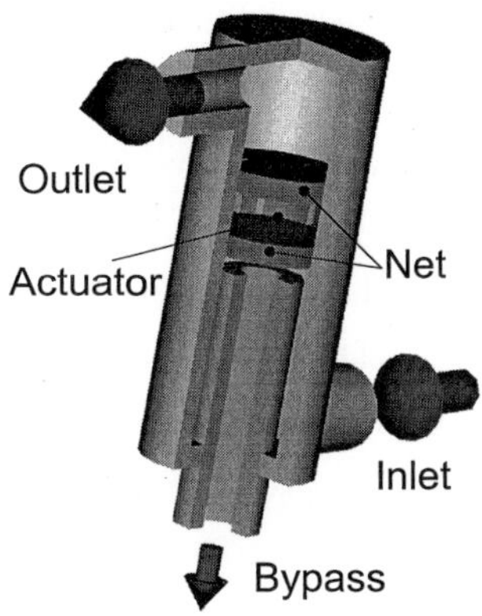

Figure 1. Chemo-mechanical valve.

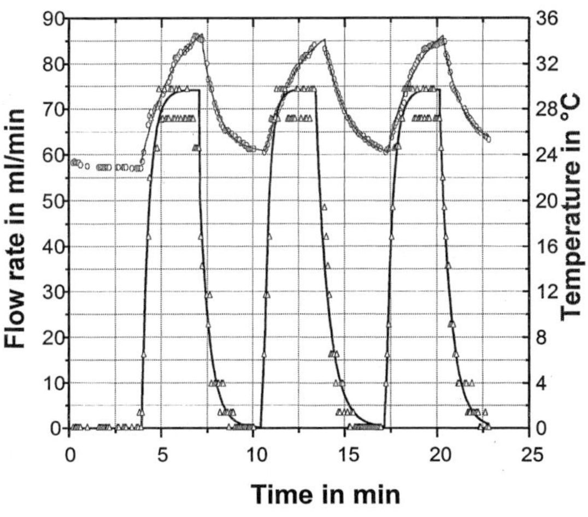

Figure 2. Flow-rate vs. time, upper curve: temperature gradient.

- Micro-valve, general remarks

The miniaturization of a chemical analysis system is supposed to yield a number of technological and economical benefits, which evolve from a drastic reduced sample and reagent consumption and reduction of analysis time. Conventional micro-valves mostly contains a flexible membrane, which can be deformed to block the flow in an adjacent channel. The actuators used to switch the membrane are based on, e.g., electromagnetic, electrostatic, piezoelectric, thermo-pneumatic, bimetallic effects, or on shape memory alloys. Disadvantages of membrane valves as well as of other conventional valves are an unavoidable leakage flow and an intolerance against small particles in the fluid. A possibility to overcome these disadvantages could be the design of gel-based micro-valves.

Some principles for regulating small flow rates based on environmental sensitive polymers are described in literature:

- A chemical valve effect in thin films with densely hexagonal packed nano-channels (diameter 17 nm) which were partially filled with poly(acrylic acid) (PAA) is reported in ref.[14]. The structural change of PAA chains with variations in pH, ionic strength, and ion type caused a change in the water permeability of the membrane.
- A flow sorter and a shut off valve (actuator size of approximately 1000 x 700 x 250 μm^3) with a response time of less than 10 seconds is described in ref.[15]. A layer of a pH-dependent hydrogel based on acrylic acid was synthesized via direct photo-patterning around prefabricated posts. The posts are inside of a micro-channel (500 to 2,000 μm wide, and 50 to 180 μm deep). Changing the pH of the liquid stimulates the swelling/deswelling of the gel layer.

- Design of micro-valve

In principal, the swelling of sensitive polymers can be stimulated by changing the properties of the fluid itself, or a separate source of energy. The change of temperature of the gel and therefore its swelling/shrinking is controllable by heat flow due to an applied electrical current.

The presented micro-valve consists of components with different functions (Figure 3):

- Flow channels and actuator chamber in different design were generated on a Silicon-wafer by a photo-lithographic process. The actuator chamber, or the actuator channel, contains the sensitive polymer.
- A heater and a temperature sensor are integrated in the cover. The cover close up the flow structure and protect them.

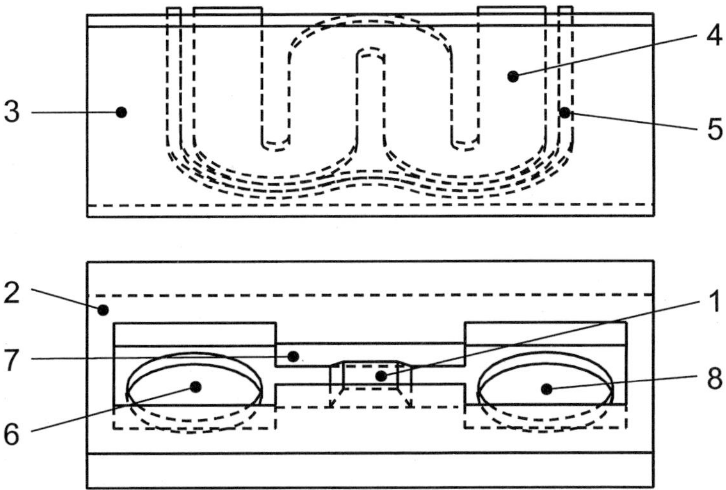

Figure 3. Design of micro-valves
1 – actuator chamber; 2 – flow structure; 3 – cover ; 4 – heat meander ; 5 – temperature sensor ; 6 – inlet ; 7 – flow channel; 8 – outlet.

Two different working principles were tested:

- particle-based actuator (Figure 4a),

  The valve chamber is filled with spherical particles of PNIPAAm-network with a mean value of diameter of (82 ± 8) μm in the dry state.

- dots of photo-patterned sensitive polymer (Figure 4b),

  The flow channel contains a certain number (mostly three) of sensitive dots.

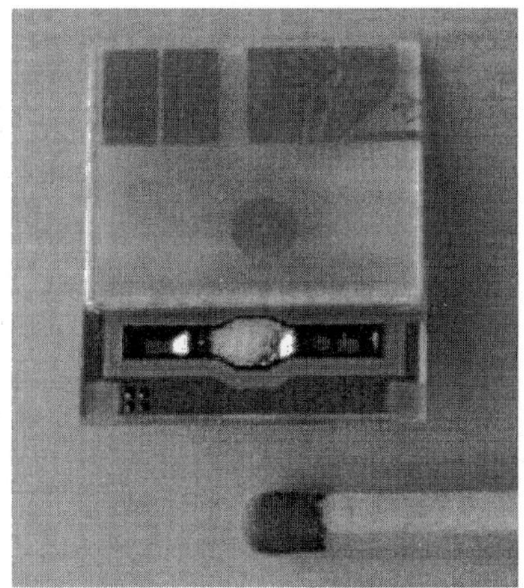

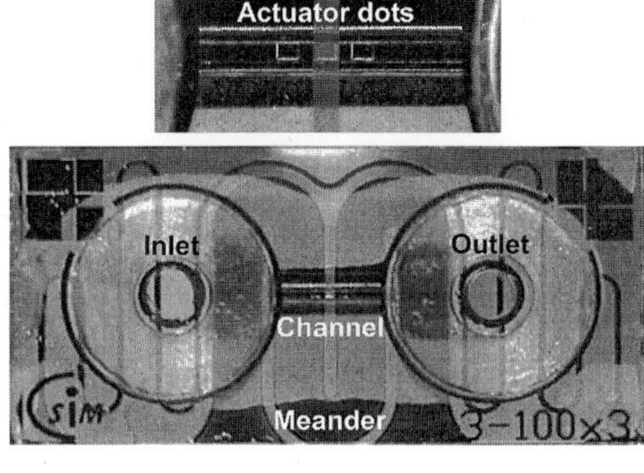

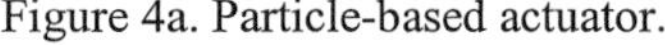

Figure 4a. Particle-based actuator.

Figure 4b. Actuator based on photo-patterned dots of smart gel.

Results and discussion

- Synthesis of gels with fast response

Figure 5 shows different pattern of thermo-sensitive polymeric gel. The thickness of a mechanical stable gel increases in 5 s from 10 μm (dry state) to 30 μm (swollen state).

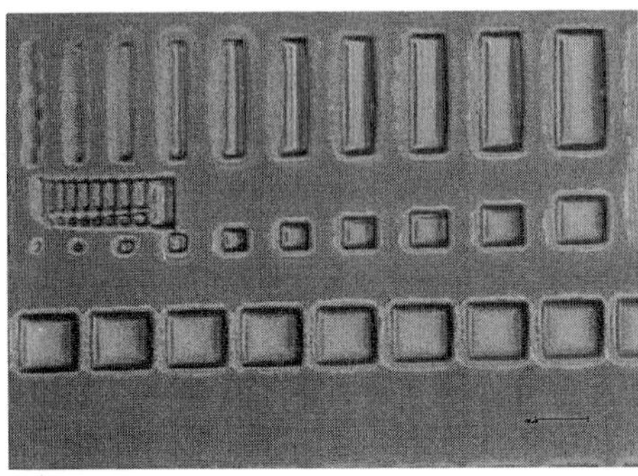

Figure 5. Patterned thin layer of PNIPAAm copolymers.

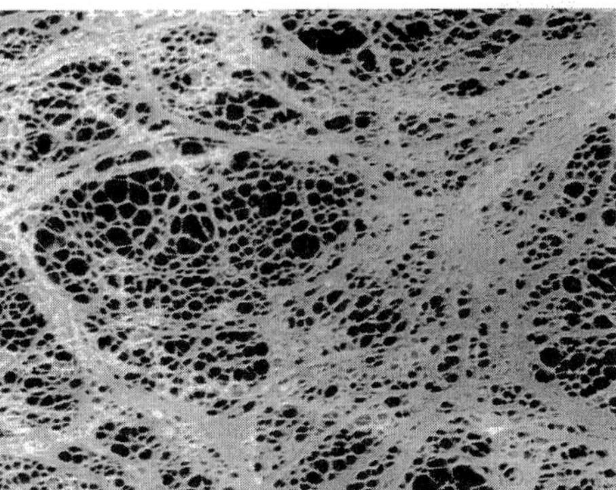

Figure 6. Bulky PVME in the high swollen state ($T = 23$ °C).

The experiments have shown, that e-beam irradiation is a suitable procedure for the synthesis of thermo-sensitive hydrogels. A porous structure, which contains a polymer-poor phase (cavities in μm-range) and a polymer-rich phase (full of holes in nm-range) is typical. Both, the swollen state of the hydrogel at 25 °C and at 40 °C are characterized by three-dimensional sponge-like patterns, which consist of many small cavities ($10^2...10^3$ nm) separated permeably from each other by a thin membrane-like layer (Figure 6). There structure can be observed by images at high magnification. The layer is porous with a typical pore-size of 10 nm. After heating to 40 °C, the gel collapses and shrinks. The micrographs show that the collapsed gel is porous (the cavities are smaller than at 25 °C), but the polymer layers which separate the cavities are more compact.

The gel particles synthesized by irradiation of a diluted PVME solution were almost globular with a sponge-like structure (Figure 7). They have a mean diameter in the range of 250 – 600 nm (swollen state).

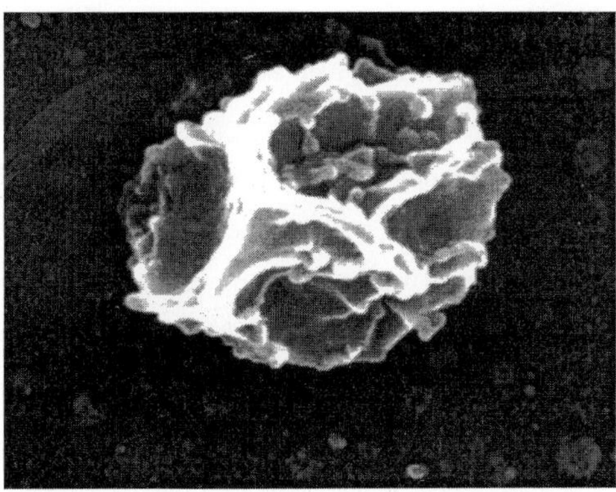

Figure 7. Thermo-sensitive PVME-particles in the swollen state.

The hydrodynamic radius depends on temperature (Figure 8). The volume phase transition is not as sharp as for the bulk gel and occurred at a higher temperature. The viscosity of solution of gel particles is temperature dependent too (two orders of magnitude for a semi-dilute solution).

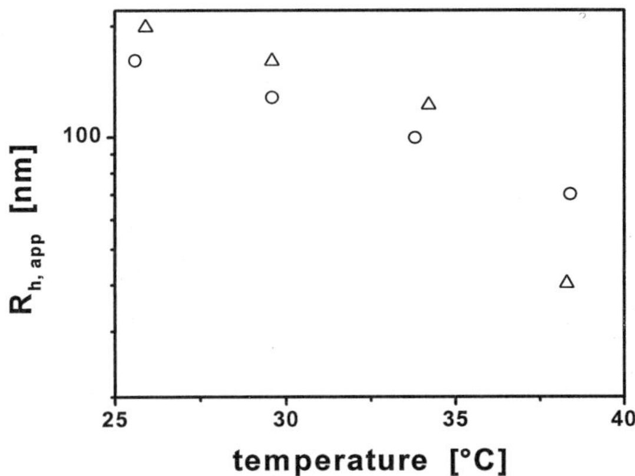

Figure 8. Temperature-dependence of apparent hydrodynamic radius.

- *Valve- characteristics*

We investigate the behavior of the micro-valve, especially the influence of heating power on the flow rate with a test equipment. Figure 9 shows the flow rate vs. time for different values of power and different current pulses at constant parameters of the pump. The upper part of both figures shows the temperature measured by the thermo-sensor. These curves reflect the profile of the power pulse. On Figure 9a the electrical current is given as a simple rectangular pulse (power amplitude: 120 mW).

The flow rate is reproducible only for a modified rectangular pulse and higher values of power, shown on Figure 9b. Heating the gel opens the valve in a short time (power amplitude 770 mW). After reaching the temperature for volume phase transition the power is adjusted to a lower level (180 mW). The heat losses are compensated and the gel remains in the shrunken state. To stop the flow, the gel must swell, or with other words, it must cool down. The current is regulated to zero.

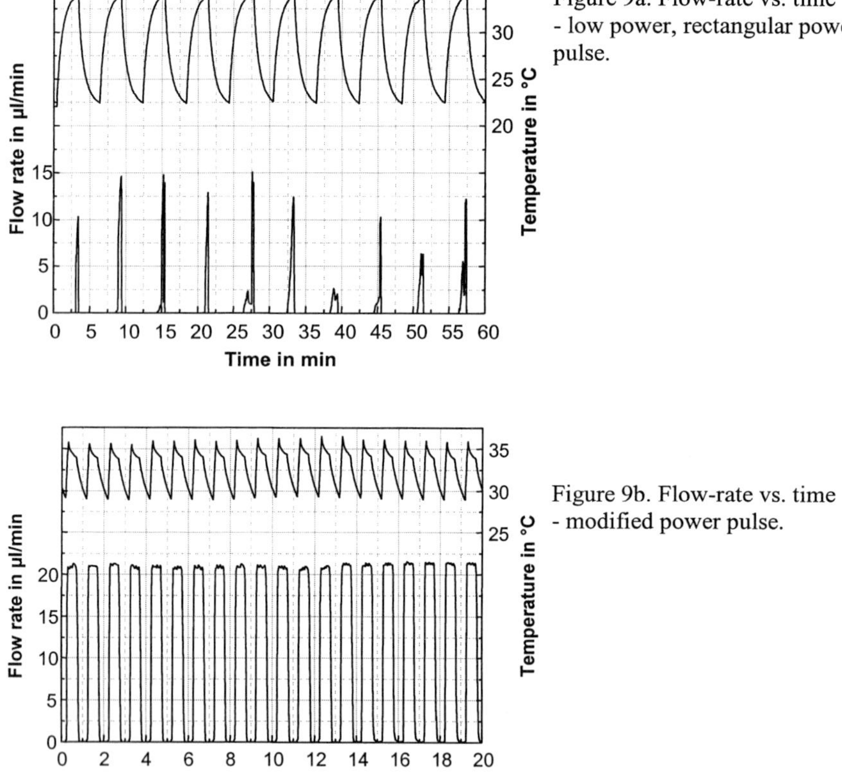

Figure 9a. Flow-rate vs. time - low power, rectangular power pulse.

Figure 9b. Flow-rate vs. time - modified power pulse.

With the modified power pulse we were able to realize a reproducible mode of function with a short cycle time (minimum opening time 300 ms; minimum closing time: 2 s).[16]

Furthermore we have designed a micro-valve based on dots of photo-patterned sensitive polymer. The minimum actuator size was 4 x 4 x 1 μm^3. The cycle time was higher than for the particle-based valve.[17]

The outcome of this part of our work is:

- it is possible to use sensitive polymer gels as materials for flow control;
- the valve behavior is reproducible, the repeat fault was less than 15 %;
- the valve was tested in a pressure range up to 5 bar.

Table 1 gives an overview on the advantages and disadvantages of conventional and gel-based valves.

Table 1. Comparison between conventional and gel-based micro-valves.

Conventional micro-valves mostly based on membrane principle	Gelbased micro-valves particles or dots of sensitive polymer
Advantages - universally applicable due to their electronic controllability - small cycle times, ms-range - manufacturing by established micro-technological processes	Advantages - simple fabrication and set-up, integration into known layouts - high degree of miniaturization - particle tolerant - zero leakage flow - pressure independent
Disadvantages - pressure sensitive - particle intolerant - possess a non-avoidable leakage	Disadvantages - higher cycle times - indirect mode of operation

Acknowledgement

Financial support of this work has been provided by the German National Science Foundation (DFG, Sonderforschungsbereich 287 "Reaktive Polymere"). S. Howitz (GeSiM mbh Großerkmannsdorf, Germany) is thanked for support design and testing of the micro-valve. We thank R. Reichelt (Westfälische Wilhelms-Universität Münster) for the excellent FESEM-micrographs and H. Dorschner (Institut für Polymerforschung Dresden e.V.) for the electron beam irradiation.

[1] M. Knörgen, K.-F. Arndt, S. Richter, D. Kuckling, H. Schneider, *J. Mol. Structure* **554**, 69 (2000).
[2] S. Hirotsu, *J. Chem. Phys*. **94,** 3949 (1991).
[3] M. Shibayama, M. Moritomo, S. Nomura, *Macromolecules* **27**, 5060 (1994).
[4] R. Yoshida, K. Uchida, Y. Kaneko, K. Sakai, A. Kikuchi, Y. Sakurai, T. Okano, *Nature* **374**, 240, (1995).
[5] M. Suzuki, O. Hirasa, *Advances in Polymer Sci.* **110**, 241 (1993).
[6] T. Gotoh, Y. Nakatani, S. Sakohara, *J. Appl. Polym. Sci.* **69**, 895 (1998).
[7] R. Pelton, *Adv. Colloid Interface Sci.* **85**, 1 (2000).
[8] A. E. English, E. R. Edelman, T. Tanaka, Polymer Hydrogel Phase Transition, ed. T. Tanaka, *Experimental Methods in Polymer Science*, Academic Press, 569, (2000).
[9] S. Hirotsu, Y. Hirokawa, T. Tanaka, *J. Chem. Phys*. **87**, 1392 (1987).
[10] D. Kuckling, H.-J. P. Adler, K. F. Arndt, J. Hoffmann, M. Plötner, Th. Wolff, *Polym. Adv. Techn.* 10, 345 (1999).
[11] K.-F. Arndt, T. Schmidt, H. Menge, *Macromol. Symp*. **64**, 313 (2001).
[12] K.-F. Arndt, T. Schmidt, R. Reichelt, *Polymer* **42**, 6785 (2001).
[13] K.-F. Arndt, D. Kuckling, A. Richter, *Pol. Adv. Techn*. **11**, 496 (2000).
[14] G. Liu, J. Ding, *Adv. Material Sci.* **10**, 69 (1998).
[15] D. J. Beebe, J. S. Moore, J. M. Bauer, Q. Yu, R. H. Lui, C. Devadoss, Byung-Ho Jo, *Nature* **404**, 588 (2000).
[16] A. Richter, K.-F. Arndt, S. Howitz, D. Kuckling, *Sensors and Actuators*, received.
[17] A. Richter, K.-F. Arndt, S. Howitz, D. Kuckling, in preparation.